Isotope Effects on Enzyme-Catalyzed Reactions

Proceedings of the Sixth Annual Harry Steenbock Symposium, held in Madison, Wisconsin, on June 4 and 5, 1976.

Other volumes in this series include:

DNA Synthesis In Vitro, edited by Robert D. Wells, Ph.D., and Ross B. Inman, Ph.D. (University Park Press, 1973)

Trace Element Metabolism in Animals-2, edited by W. G. Hoekstra, Ph.D., J. W. Suttie, Ph.D., H. E. Ganther, Ph.D., and Walter Mertz, M.D. (University Park Press, 1974)

Structure and Conformation of Nucleic Acids and Protein-Nucleic Acid Interactions, edited by M. Sundaralingam, Ph.D., and S. T. Rao, Ph.D. (University Park Press, 1975)

CO_2 Metabolism and Plant Productivity, edited by R. H. Burris, Ph.D., and C. C. Black, Ph.D. (University Park Press, 1976)

Isotope Effects on Enzyme-Catalyzed Reactions

Edited by

W. Wallace Cleland, Ph.D.
Professor of Biochemistry
University of Wisconsin
Madison

Marion H. O'Leary, Ph.D.
Associate Professor of Chemistry
University of Wisconsin
Madison

Dexter B. Northrop, Ph. D.
Associate Professor of Pharmaceutical Biochemistry
University of Wisconsin
Madison

HARRY STEENBOCK SYMPOSIUM

University Park Press
Baltimore • London • Tokyo

UNIVERSITY PARK PRESS
International Publishers in Science and Medicine
Chamber of Commerce Building
Baltimore, Maryland 21202

Typeset by Action Comp Co., Inc.
Manufactured in the United States of America by Universal
Lithographers, Inc., and the Optic Bindery Incorporated.

Library of Congress Cataloging in Publication Data
Harry Steenbock Symposium, 6th, Madison, Wis., 1976.
Isotope effects on enzyme-catalyzed reactions.

Bibliography: p.
Includes index.
1. Chemical reaction, Rate of—Congresses.
2. Catalysis—Congresses. 3. Enzymes—Congresses.
4. Isotopes—Congresses. I. Cleland, William
Wallace, 1930- II. O'Leary, Marion H.
III. Northrop, Dexter B. IV. Title.
QD502.H37 1976 547'.1'395 77-6650
ISBN 0-8391-0851-6

Contents

Contributors

Warren E. Buddenbaum
Department of Chemistry
Indiana University
Bloomington, Indiana 47401

W. Wallace Cleland
Department of Biochemistry
University of Wisconsin-Madison
Madison, Wisconsin 53706

Jack F. Kirsch
Department of Biochemistry
University of California
Berkeley, California 94720

Judith P. Klinman
Institute for Cancer Research
Philadelphia, Pennsylvania 19111

A. J. Kresge
University of Toronto
Scarborough College
West Hill, Ontario, Canada

Dexter B. Northrop
School of Pharmacy
University of Wisconsin-Madison
Madison, Wisconsin 53706

Marion H. O'Leary
Department of Chemistry
University of Wisconsin-Madison
Madison, Wisconsin 53706

Irwin A. Rose
Institute for Cancer Research
Philadelphia, Pennsylvania 19111

R. L. Schowen
Department of Chemistry
University of Kansas
Lawrence, Kansas 66045

V. J. Shiner, Jr.
Department of Chemistry
Indiana University
Bloomington, Indiana 47401

Preface

The first demonstrations of differences in reaction rate resulting from isotopic substitution came in the 1930s. A theoretical basis for understanding these rate differences was laid down by Bigeleisen and his associates in the 1940s and 1950s. Organic chemists soon became aware of the usefulness of isotope effects in studies of reaction mechanisms, and in the 1950s and 1960s they refined techniques for measuring and interpreting isotope effects and accumulated a large body of information on isotope effects in organic reactions. Today, these techniques are part of the standard practice of organic chemistry.

Perhaps because of the complexity of enzyme-catalyzed reactions, the use of isotope effects as tools for understanding mechanisms of enzymatic reactions has been slower to develop. Within the last decade an increasing number of enzymologists have become aware of the potential of isotope effects for probing the intricacies of enzyme-catalyzed reactions, and the number of publications in this area each year has substantially increased. Several developments have been responsible for this change. First, theoretical techniques for predicting isotope effects have been refined and simplified and have thus become more accessible to the laboratory chemist. Second, the large body of data on organic reactions provides an empirical foundation for understanding isotope effects in enzymatic reactions. Third, techniques for the measurement of isotope effects have been refined considerably. Particularly noteworthy are the development of the equilibrium perturbation method by Schimerlik and Cleland and the refinement of methods for the direct measurement of small isotope effects by Schowen and others. Fourth, enzyme kineticists have developed new approaches to enzyme kinetic data. Investigators have recognized the important and often complex role of substrate concentration in isotope effects. Also of note is Northrop's derivation of the relationship between deuterium and tritium isotope effects in enzymatic reactions.

Because of the rapid growth of this field and because of our own interest in isotope effects, we decided to convene a symposium on isotope effects and their application to enzyme-catalyzed reactions in order to bring together practitioners of isotope effects and interested enzymologists. Thanks to the generosity of the Harry Steenbock Foundation, the symposium was held June 4 and 5, 1976, in Madison, Wisconsin. We think the symposium was successful in its educational and communication functions, and we would like to acknowledge the help of all who made this symposium a success.

The present volume contains the papers given at the symposium, together with the discussion (somewhat edited) that followed the formal presentations, and a number of appendices relating to isotope effects.

Opening Remarks

On behalf of the Steenbock Symposium Organizing Committee, I would like to welcome you to the Madison campus. Many of you probably do not know a great deal about Harry Steenbock, who was a nutritional biochemist with a long and distinguished career at the University of Wisconsin. He was perhaps best known for his studies in the field of the fat-soluble vitamins, but he also contributed substantially to the field of trace mineral nutrition. He is particularly well known for establishing the relationship between carotene and vitamin A and for his demonstration that ultraviolet irradiation of food would develop vitamin D activity. Perhaps many of you are aware that it was this finding which was later patented and led to the start of the Wisconsin Alumni Research Foundation.

Harry Steenbock died in 1969. A year or so later there was a fat-soluble vitamin symposium held in his honor on the Madison campus. Following this successful meeting, his widow, Evelyn, very generously established a substantial trust fund that allows the Department of Biochemistry to fund a continuing series of symposia on this campus. We have had symposia dealing with trace minerals, DNA synthesis in vitro, structure-function relations of nucleic acids and proteins, and last year a symposium that dealt with various aspects of CO_2 fixation in plants. This year, of course, we are participating in a symposium that is probably somewhat more chemical than those held in the past. If the symposia in the past have been a forerunner to this, I trust this will also be an interesting and informative one, and we welcome you here.

John W. Suttie
University of Wisconsin

Isotope Effects on Enzyme-Catalyzed Reactions

Computation of Isotope Effects on Equilibria and Rates

W. E. Buddenbaum and V. J. Shiner, Jr.

The study of isotope effects on reaction rates provides one of the most powerful and subtle methods for the elucidation of reaction mechanisms (Collins and Bowman, 1971; Shiner, 1975; Wolfsberg, 1972). The method is applicable at several levels of sophistication from the simple qualitative determination of the presence or absence of a primary isotope effect on the rate to the detailed modeling of transition-state structure. This chapter examines two closely related methods of analyzing observed isotope rate effects. The first method uses calculated *fractionation factors* for stable molecules to estimate semiquantitatively secondary isotope effects attendant upon specified structural changes (Bigeleisen and Ishida, 1975; Hartshorn and Shiner, 1972). This method depends heavily on obvious structural analogies between simple molecules that are subject to complete vibrational analysis and more complex structures of interest to those studying reaction mechanisms. These fractionation factors have semiquantitative transferability between like structures and can be used by structure-wise chemists in a way closely analogous to the way bond dissociation energies are used. The last part of this chapter examines some recent results we have obtained from an effort that aims to establish a general, quantitative, programmable approach to the examination of transition state structure and the calculation of isotope rate effects.

ISOTOPE EFFECTS AS ISOTOPE EXCHANGE EQUILIBRIUM CONSTANTS

Isotope effects on the rate or equilibrium constants for chemical reactions can be expressed as equilibrium constants for hypothetical isotope exchange reactions. For example, if some chemical reactant R_1 is in equilibrium with some product P_1

$$R_1 \leftrightarrows P_1$$

the equilibrium constant, neglecting activity coefficients, can be expressed as

$$K_1 = [P_1]/[R_1]$$

If R_2 represents the molecule R containing a specific isotopic substituent, the equilibrium constant for its conversion to P_2 can be expressed as

$$K_2 = [P_2]/[R_2]$$

The isotope effect on the equilibrium is then formulated as

$$\frac{K_1}{K_2} = \frac{[P_1][R_2]}{[P_2][R_1]}$$

which is really the equilibrium constant for the isotope exchange reaction

$$R_1 + P_2 \leftrightarrows R_2 + P_1$$

Isotope effects on reaction rates can be treated in an analogous manner. In transition-state theory (Glasstone, Laidler, and Eyring, 1941; Laidler and Tweedale, 1971), reactants are considered to be in equilibrium with the transition state and all transition states decompose by the same universal rate constant, $\mathbf{k}T/h$. Thus, the observed rate constant, k, can be expressed as

$$k = \kappa \frac{\mathbf{k}T}{h} K^{\ddagger} \quad (1)$$

where $\mathbf{k}$ is Boltzmann's constant, T is the temperature (°K), h is Planck's constant, κ is the transmission coefficient for the transition state, and $K^{\ddagger}$ is a constant (see next section) representing the equilibrium between ground state (R) and transition state (P). If, as generally assumed, the transmission coefficient, κ, is independent of isotopic substitution, the isotope effect on a reaction rate reduces to the form of an isotope exchange equilibrium constant

$$\frac{k_1}{k_2} = \frac{\kappa_1 K_1^{\ddagger}}{\kappa_2 K_2^{\ddagger}} = \frac{K_1^{\ddagger}}{K_2^{\ddagger}} \quad (2)$$

where the light reactant (R_1) and the heavy transition state (P_2) are written on the left hand side of the equation if k_1/k_2 is to represent the isotope effect written in the usual way.

The next section considers how equilibrium constants for isotope exchanges which involve transition states differ from those for exchanges involving only molecules.

BIGELEISEN THEORY OF ISOTOPE EFFECTS

Both Bigeleisen and Mayer (1947) and Melander (1960) used the methods of statistical mechanics to develop equations for the evaluation of isotope rate and equilibrium effects. These equations are based on the following assumptions:

1. The Born-Oppenheimer approximation is valid: electronic potential energy is independent of isotopic substitution
2. Partition functions are factorable into independent contributions from translational, vibrational, and rotational motions
3. Translational and rotational motion is classical
4. Vibrational motion is harmonic
5. Quantum mechanical tunneling is absent
6. There is no isotope effect on the transmission coefficient κ.

Detailed theoretical analysis (Stern and Wolfsberg, 1966; Wolfsberg and Stern 1964a) as well as the comparison of theoretical and experimental results (Bell and Crooks, 1962; Kleinman and Wolfsberg, 1974a, b; Van Hook, 1971) indicate that none of these approximations leads to any significant error in the usual applications of the technique, with the exception of significant tunneling effects in some hydrogen transfer reactions (Bell, 1974).

The Bigeleisen equation for equilibrium isotope effects can be formulated in two general ways (Wolfsberg and Stern, 1964a)

$$K_1/K_2 = \text{MMI} \cdot \text{EXC} \cdot \text{ZPE} \tag{3}$$

or

$$K_1/K_2 = \text{VP} \cdot \text{EXC} \cdot \text{ZPE} \tag{4}$$

where ZPE is the contribution resulting from the effect of zero-point energy differences in initial and final state between the two isotopic molecules. EXC is the effect caused if the vibrational frequencies are low enough or the temperature is high enough to cause the vibrational energy levels above the zero level to be populated. MMI results from the ratio of ratios of translational and rotational partition functions between the two isotopic species and between initial and final states. By application of the Teller-Redlich product rule (Bigeleisen and Mayer, 1947) the MMI term, which involves molecular masses (M_i) and the three moments of inertia (I_A, I_B, and I_C) of R and P, can also be expressed solely in terms of the vibrational frequencies of R and P. This leads to the term VP, which is the ratio of ratios of all vibrational frequencies between the two isotopic molecules in the initial state and the final state. Thus, according to this theory the calculation of equilibrium isotope effects requires only the knowledge of all the normal-mode vibrational frequencies of both isotopic forms of both states.

For rate effects the form of the Bigeleisen equation is almost the same, but the normal-mode frequencies of transition states are treated in a slightly different way: transition state theory requires that one normal mode have a zero or imaginary frequency, ν_L. This mode is referred to as the *reaction coordinate* because motion along this coordinate (in phase space) corresponds to the transformation of the transition state back to reactants or forward into products. More importantly, the theory treats this aperiodic motion as a translation having no zero-point energy. Thus, the factors EXC‡, ZPE‡, and VP‡ in the Bigeleisen equations for rate effects are calculated using only the $3N-7$ (for a linear molecule, this is $3N-6$) real vibrational frequencies of the transition state, where N is the number of atoms

$$\frac{k_1}{k_2} = \mathrm{MMI}^\ddagger \cdot \mathrm{EXC}^\ddagger \cdot \mathrm{ZPE}^\ddagger \tag{5}$$

and MMI‡ is calculated using the assumed structure for the transition state. The dependence of k_1/k_2 on ν_L can be obtained by application of the Teller-Redlich product rule to MMI‡ to yield

$$\frac{k_1}{k_2} = \frac{\nu_{1L}}{\nu_{2L}} \cdot \mathrm{VP}^\ddagger \cdot \mathrm{EXC}^\ddagger \cdot \mathrm{ZPE}^\ddagger \tag{6}$$

where ν_{1L}/ν_{2L} is the effect of isotopic substitution on ν_L. Thus, the difference between isotope exchange equilibrium constants for exchanges between normal molecules and those involving transition states is the unique treatment given the reaction coordinate motion of the transition states.

The Bigeleisen equations for rates, expressed in full flower, can be written as

$$\frac{k_1}{k_2} = \frac{\left(\dfrac{M_2}{M_1}\right)^{3/2}\left[\dfrac{I_{A_2}I_{B_2}I_{C_2}}{I_{A_1}I_{B_1}I_{C_1}}\right]^{1/2}}{\left(\dfrac{M_2^\ddagger}{M_1^\ddagger}\right)^{3/2}\left[\dfrac{I_{A_2}^\ddagger I_{B_2}^\ddagger I_{C_2}^\ddagger}{I_{A_1}^\ddagger I_{B_1}^\ddagger I_{C_1}^\ddagger}\right]^{1/2}} \times$$

$$\frac{\prod_i^{3N-6} \frac{[1-\exp(-u_{1i})]}{[1-\exp(-u_{2i})]}}{\prod_i^{3N^{\ddagger}-7} \frac{[1-\exp(-u_{1i}^{\ddagger})]}{[1-\exp(-u_{2i}^{\ddagger})]}} \times \frac{\exp\left[\sum^{3N-6}(u_{1i}-u_{2i})/2\right]}{\exp\left[\sum^{3N^{\ddagger}-7}(u_{1i}^{\ddagger}-u_{2i}^{\ddagger})/2\right]}$$

and

$$\frac{k_1}{k_2} = \frac{\nu_{1L}^{\ddagger}}{\nu_{2L}^{\ddagger}} \times \frac{\prod^{3N-6} \frac{u_{2i}}{u_{1i}}}{\prod^{3N^{\ddagger}-7} \frac{u_{2i}^{\ddagger}}{u_{1i}^{\ddagger}}}$$

$$\times \frac{\prod^{3N-6} \frac{[1-\exp(-u_{1i})]}{[1-\exp(-u_{2i})]}}{\prod^{3N^{\ddagger}-7} \frac{[1-\exp(-u_{1i}^{\ddagger})]}{[1-\exp(-u_{2i}^{\ddagger})]}} \times \frac{\exp\left[\sum^{3N-6}(u_{1i}-u_{2i})/2\right]}{\exp\left[\sum^{3N^{\ddagger}-7}(u_{1i}^{\ddagger}-u_{2i}^{\ddagger})/2\right]}$$

where the last ratio is ZPE‡ and the next to the last large ratio is EXC‡, u_i equals $h\nu_i/\mathbf{k}T$, h is Planck's constant, and **k** is Boltzmann's constant. As indicated above, for equilibria the $3N^{\ddagger} - 7$ becomes $3N - 6$ and $\nu_{1L}^{\ddagger}/\nu_{2L}^{\ddagger}$ is not included; otherwise the equations have the same form.

ZERO-POINT ENERGY ANALYSIS OF ISOTOPE EFFECTS

Isotope effects are generally dominated by the ZPE term and this contribution is the focus of all qualitative discussions explaining isotope effects. Zero-point energy effects are generally represented in terms of potential energy diagrams as in Figure 1, with each light and heavy species indicated for convenience in the same well. The magnitudes of the zero-point energies in the figure are exaggerated for clarity. In this example the potential energy well for vibration of the isotopically substituted atom is assumed to be more shallow in the product than in the reactant and more shallow still in the transition state (although any other order of these relationships is a priori possible). The vibrations in the transition state must be orthogonal to the reaction coordinate mo-

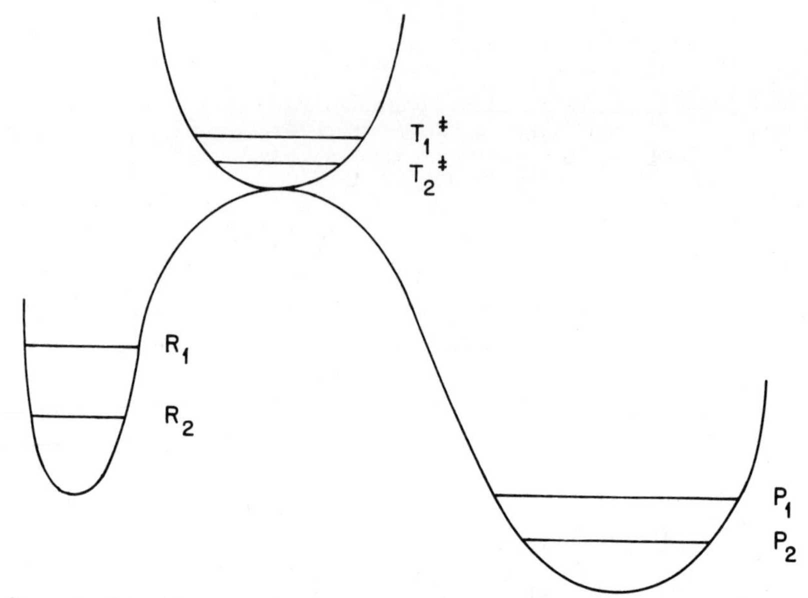

Figure 1. Potential energy diagram showing zero-point energy contributions to kinetic isotope effects. R_1 and R_2 are light and heavy reactants, respectively; $T_1^‡$ and $T_2^‡$ are corresponding transition states; P_1 and P_2 are corresponding products.

tion (see below). For this example, then, the isotopically sensitive vibration frequencies are lower and the zero-point energy difference between the isotopic product molecules is less than it is between the isotopic reactants. Thus, the light reactant loses more zero-point energy in conversion to product than does the heavy reactant. This illustrates the first law of isotope chemistry: the light isotopic molecule prefers the state in which the restrictions to vibration are lower, i.e., the state in which the bonding to the isotopic atom is "less stiff." The zero-point energy contribution to the decrease in enthalpy of reaction is simply

$$E_0 \text{ (light reactant)} - E_0 \text{ (heavy reactant)} - E_0 \text{ (light product)} + E_0 \text{ (heavy product)} = \Delta E_0 \text{ (reactants)} - \Delta E_0 \text{ (products)}$$

A similar expression with product quantities replaced by transition state quantities applies for the zero-point energy contribution to the decrease in energy of activation. The energy changes on reaction can be equivalently represented by an isotope exchange equilibrium process, with one potential energy well for each species, as in Figure 2. The isotope effect on the (initial) equilibrium is simply the equilibrium constant for the

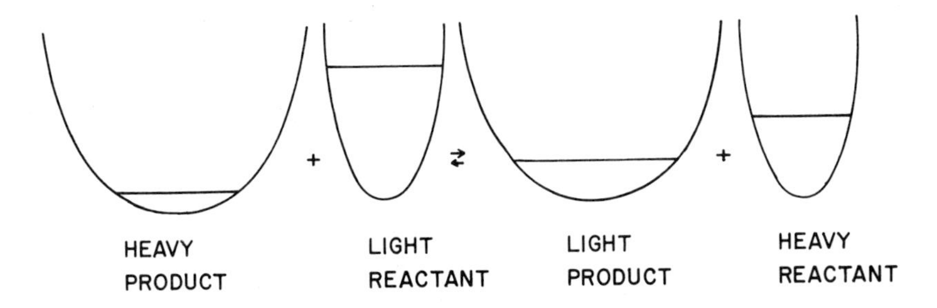

Figure 2. Potential energy curves for an isotope exchange equilibrium.

isotope exchange reaction. A similar isotope exchange equilibrium between reactants and transition state can be used to represent the zero-point energy contribution to the decrease in activation energy. Because these zero-point energy effects on rates and equilibria are determined by the vibrational motions of the molecules and do not directly depend on the total electronic energy, any molecule that adequately models the isotopically sensitive vibrational modes can serve as a model for the reactant, product, or transition state in estimating the zero-point energy contribution to the isotope effect.

The ZPE contribution to the isotope effect on an equilibrium (or rate) constant can be calculated by a very simple formula: the zero-point energy difference between reactants and products (or transition state) contributes directly to the enthalpy change in the reaction and has an exponential effect on the equilibrium (or rate) constant. Thus

$$\frac{K_1}{K_2} = \exp[h(\nu_{R_1} - \nu_{R_2} - \nu_{P_1} + \nu_{P_2})/2\mathbf{k}T] \tag{7}$$

where the ν's for each species, light (1) and heavy (2) reactants (R) and products (P), are the sums of all real normal mode frequencies; that is, $\nu_{R_1} = \Sigma\nu_{R_1}$, etc. If the frequencies are expressed in cm^{-1} and the temperature in °K the equation reduces to

$$\frac{K_1}{K_2} = [\exp[0.7193(\nu_{R_1} - \nu_{R_2} - \nu_{P_1} + \nu_{P_2})/T]$$

If all of the normal-mode frequencies are not available, then useful estimates for equilibrium effects for isotopes other than hydrogen are probably not possible. Because hydrogen is so much lighter than the other atoms that normally occur in organic molecules, its vibrations can be considered to some extent to be approximately independent of the rest of the molecule, and one can sometimes obtain a useful estimate for hydrogen isotope effects on equilibrium constants by using in this formula only the three principal kinds of hydrogen frequencies, i.e., the bond stretching and the two bond angle bending modes. Furthermore, corresponding H and D frequencies often occur in approximately the same ratio to each other that they would show in an analogous, hypothetical diatomic molecule. For a diatomic molecule in the harmonic oscillator approximation, the ratio of stretching frequencies of two isotopic molecules is simply the square root of the ratio of the reduced masses for each vibration. The reduced mass (m_r) of the AB diatomic is related to the masses of A and B by

$$1/m_r = 1/m_A + 1/m_B$$

thus

$$m_r = \frac{m_A m_B}{m_A + m_B}$$

For a hypothetical diatomic CH molecule

$$\nu_{CD} = \nu_{CH}\left[\frac{(m_C + m_D)/(m_C m_D)}{(m_C + m_H)/(m_C m_H)}\right]^{1/2}$$

$$= \nu_{CH}\left(\frac{14 \times 12 \times 1}{13 \times 12 \times 2}\right)^{1/2}$$

$$= \nu_{CH} \times 0.7338$$

or

$$\nu_{CH}/\nu_{CD} = 1.363$$

Observed CH and CD stretching frequencies in organic molecules generally give ratios not much different from this. CH and CD bending frequencies obey this relationship generally less closely because the triatomic angle bending motion does not correspond as closely to an isolated diatomic motion as does the CH stretch. Using the above ratio and substituting for ν_{R_2} and ν_{P_2} in the exponential, the equation reduces further to the approximation used by Streitwieser et al. (1958) for C—H versus C—D isotope effects

$$\frac{K_1}{K_2} = \exp[0.1331h(\nu_{R_1} - \nu_{P_1})/\mathbf{k}T]$$

or for ν in cm^{-1} and T in °K:

$$\frac{K_1}{K_2} = \exp[0.1915(\nu_{R_1} - \nu_{P_1})/T]$$

For the roughest approximation, the ν's for only the stretching mode could be used, or for a somewhat better estimate, the ν's for the stretching and the two bending motions could be used. Even though there are slight isotope effects in other vibrations of the molecule, they cannot be included in this formula because for these frequencies the isotope effect does not approach the factor of 1.363.

FRACTIONATION FACTOR ANALYSIS OF ISOTOPE EFFECTS

The accurate calculation of isotope effects on equilibria from the full Bigeleisen equation is relatively simple if all of the normal-mode vibrational frequencies are known for all four molecules. A number of such factors for small molecules have been calculated using observed frequencies for a variety of isotopes since the introduction of the equation in 1947. However, isotopic frequency shifts are generally small variations in much larger numbers (reminiscent of bond energies relative to total electronic energies) so that the normal-mode frequencies must be known with considerable accuracy ($1\ cm^{-1} = 2.84$ cal/mol) for all four molecules involved in the equilibrium if the calculation is to be very useful.

Thus, only a few organic molecules with well characterized spectra or large isotope effects could be treated with sufficient accuracy for the results to be of interest. Monse, Kauder, and Spindel (1964) pointed out that significant improvements in accuracy could be obtained if the observed frequencies for the molecules involved are first used to establish a force field from which calculated frequencies are obtained for use in the Bigeleisen equation. Errors in the observed frequencies then contribute roughly *proportional* errors in the differences between corresponding calculated isotopic molecular frequencies rather than contributing *directly* to those all-important, generally small, differences.

The calculation of molecular force fields and vibrational frequencies from observed vibrational spectra is far from a trivial matter (Duncan and Mallinson, 1973; Wolfsberg, 1969). Fortunately, computer programs for this have been developed by Schachtschneider and Snyder (1963) and others. Wolfsberg and Stern (1964a,b) have adapted the Schachtschneider and Snyder program so that isotope rate effects can be calculated via the Bigeleisen equation from specified structures and force fields for initial and final (or transition) states.

Hartshorn and Shiner (1972) used the Schachtschneider and Snyder program to develop force fields for a number of small organic molecules for which good spectral data could be found in the literature for two or more isotopic variants each (usually the hydrogen and perdeutero compounds). These force fields were then used with the Wolfsberg and Stern programs to calculate fractionation factors. The resulting H/D fractionation factors, with a few additions made subsequently, are shown in Table 1. The numbers are expressed as fractionations relative to acetylene and are equilibrium constants (neglecting symmetry contributions, i.e., they are effects per D per position) for reactions of the general class:

$$\text{H—C}\equiv\text{C—D} + \text{R—H} \leftrightarrows \text{H—C}\equiv\text{C—H} + \text{R—D}$$

Acetylene is used as the reference substance because its molecular structure and spectrum are simple and well established and because all of the exchange constants between it and nonacetylenic C—H bonds calculated so far are greater than unity. This result is general if the sum of the hydrogen sensitive vibrations in R—H is greater than in H—C≡C—H. Thus, under exchange conditions hydrogen tends to concentrate in acetylene, where its zero-point energy is lower, and deuterium tends to concentrate in R—H, where its zero-point energy is higher. We refer to this situation by the "rule of thumb" that hydrogen tends to concentrate in the "less stiff" bond and deuterium in the "stiffer" bond. The term stiffer is preferred to tighter because the latter connotes bond dissociation energies, which do not always parallel fractionation factors. The detailed calculations of Hartshorn show that although the simple approximation of Streitwieser is useful for general qualitative purposes, it can frequently be quite far off quantitatively because of the uncertainty of the isotope effect on the bending motions and because many other frequencies of the molecule may be affected by the substituent. It will also be in error in those cases where EXC and MMI make important contributions to the isotope effect.

Table 1 shows that secondary equilibrium isotope effects for hydrogen attached to carbon cover a greater than two-fold range. Examination of the results shows that among this group of compounds *the fractionation factors are determined in large measure by the number and kinds of atoms attached directly to the carbon carrying the exchanging hydrogen.* Examination of valence force fields and vibration frequencies generally shows these variations to be predominately the result of the valence bending motions of the hydrogen attached to carbon. A number of interesting trends are apparent in Table 1:

Table 1. H/D fractionation factors relative to acetylene, 25 °C[a]

$FC\equiv CD$	0.987	$CH_2=CDCl$	1.348[c,i]
$ClC\equiv CD$	0.994	$HC\equiv CCH_2D$	1.351
$BrC\equiv CD$	0.995	$BrCH_2D$	1.358
$HC\equiv CD$	1.000	CH_3CH_2D	1.361
cis-$DHC=CHBr$	1.183[c,h]	$CH_2=CHCH_2D$	1.362[b]
cis-$DHC=CHCH_3$	1.201[b]	$N\equiv CCH_2D$	1.373
cis-$DHC=CHCl$	1.211[c,i]	Cl_3CCH_2D	1.397
trans-$DHC=CHCH_3$	1.226[b]	$(CH_2)_2CHD$ (cyclic)	1.400
$Br_2C=CHD$	1.233	$(CH_2)_3CHD$ (cyclic)	1.400
H_3SiCH_2D	1.243	$ClCH_2D$	1.405
HCH_2D	1.246	CF_3CH_2D	1.427
trans-$DHC=CHBr$	1.250[c,h]	$H_3N^+-CH_2D$	1.439
$DCOO^-$	1.254[c,j]	FCH_2D	1.465
$H_2C=CHD$	1.257	O_2NCH_2D	1.471
$F_2C=CHD$	1.259	CH_3CHDCH_3	1.501
$DCHO$	1.273[c,e]	CH_3CHDCl	1.502
H_3GeCH_2D	1.275	CBr_3D	1.516
trans-$DHC=CHCl$	1.282[c,i]	CCl_3D	1.656
$CH_2=CDBr$	1.292[c,h]	CF_3D	1.993
ICH_2D	1.316		
$CH_3CH_2CH_2D$	1.324		
$DCFO$	1.33[c,d]		
$CH_2=CD-CH_3$	1.336[b]		
$ClCH_2CH_2D$	1.341		

[a] Hartshorn and Shiner, 1972.
[b] Hartshorn, unpublished results.
[c] This work.
[d] Force field of Venkateswarlu et al., 1963.
[e] Force field of Duncan and Mallinson, 1973.
[h] Frequencies of Scherer and Overend, 1960.
[i] Frequencies of El-Sabban and Zwolinski, 1968.
[j] Frequencies of Muller and Nagarajan, 1967.

1. Increasing the coordination number of the carbon atom bearing the hydrogen increases the H/D fractionation factor, e.g., in the series $H-C\equiv C-D$ (1.00), $H_2C=CHD$ (1.257), CH_3-CH_2D (1.361).
2. For XCH_2D compounds the stiffness of bonding increases for X groups in the order H_3Si-, $H-$, H_3Ge-, $I-$, CH_3-CH_2-, $ClCH_2-$, $H-C\equiv C-$, $Br-$, CH_3-, $N\equiv C-$, Cl_3C-, $Cl-$, CF_3-, H_3N-, $F-$, O_2N-. This order *roughly* parallels bond strength and electronegativity of X; H— is a notable exception.
3. For $C_\beta C_\alpha H_2D$ compounds the fractionation factor is not appreciably dependent on the coordination number of C_β or on the nature of groups

attached at C_{β}. This result confirms the effectiveness of the "cut-off procedure" of Stern and Wolfsberg (1966). Thus, it appears that one can estimate to a fair degree of accuracy the effects of structural changes on H/D fractionation factors in stable molecules (but not carbonium ions (Shiner, 1971)) by taking into account only the changes in groups attached to the α-carbon atom. The compounds CCl_3CH_2D and CF_3CH_2D do show slight increases in K_{HD} over CH_3CH_2D, which are in the direction and order expected for the operation of a small inductive effect from the substituents at the β-carbon atom.

4. C—CH_2—C groups in cyclopropane and cyclobutane rings show fractionation factors in between those for $H_2C=CHD$ and $CH_3CH_2CH_3$.

5. The effect of a given change for a group attached at the α-carbon atom is roughly independent of the other groups attached at that atom. For example, between CH_3D and CH_2DCl there is a fractionation factor of 1.405/1.246, or 1.128, while between CH_3CH_2D and CH_3CHDCl there is a factor of 1.502/1.361, or 1.104. The two estimations of the effect of the change from α-H to α-Cl on the CH/CD fractionation factor are probably the same within the combined limits of error of the calculations in the four molecules compared; however, the small difference might be real and indicative of the accuracy of this approximation. A corollary of this general relationship is that the effect of each successive replacement of α-H by a given atom should be the same. For example, between CH_3D and CH_3CH_2D the factor is 1.361/1.246, or 1.092, while between CH_3CH_2D and CH_3CHDCH_3 it is 1.501/1.361, or 1.102.

The trends noted in (5) are particularly important from a practical point of view because they greatly extend the utility of the tables of fractionation factors. However, some caution is necessary when using the tables in this way because the effects of α substituents may not always be cumulative. The fractionation factor of CH_2DF relative to CH_3D is 1.465/1.246, or 1.176. Assuming the effect of α-F to be cumulative, the expected fractionation factor of CDF_3 relative to CH_3D is $(1.176)^3$, or 1.626; the fractionation factor calculated directly is 1.992/1.246, or 1.600. On the other hand, K_{HD} for CH_2DCl relative to CH_3D is 1.128, which leads to an expected value of K_{HD} for $CDCl_3$ relative to CH_3D of 1.435; the value of K_{HD} calculated directly for $CDCl_3$ is 1.329. Similarly for $CHBr_3$, K_{HD} calculated from the value of K_{HD} for CH_3Br assuming cumulative behavior is 1.295; the direct calculation gives 1.217.

As implied in these comparisons, the equilibrium constants for isotope exchange between any two molecules in the table can be determined by dividing the factor for one molecule by the factor for the second; this gives the equilibrium constant of the isotope exchange of the light first molecule with the heavy second molecule, the reference molecules being removed from the equilibrium by the subtraction.

The new values in Table 1 for propene are from some calculations by Hartshorn (unpublished results), while we have done the ones for vinyl halides, formate, formyl fluoride, and formaldehyde as part of a continuing current effort to examine effects in carbonyl compounds systematically.

Figure 3 compares fractionation factors in similarly substituted molecular series $CH_2D—X$, $CH_3CHD—X$, $CH_2=CD—X$, and $O=CD—X$. The trends show significant parallels.

The fractionation factor is larger for ethane than methane because the C—C bond is stiffer than the C—H bond it replaces. The factors

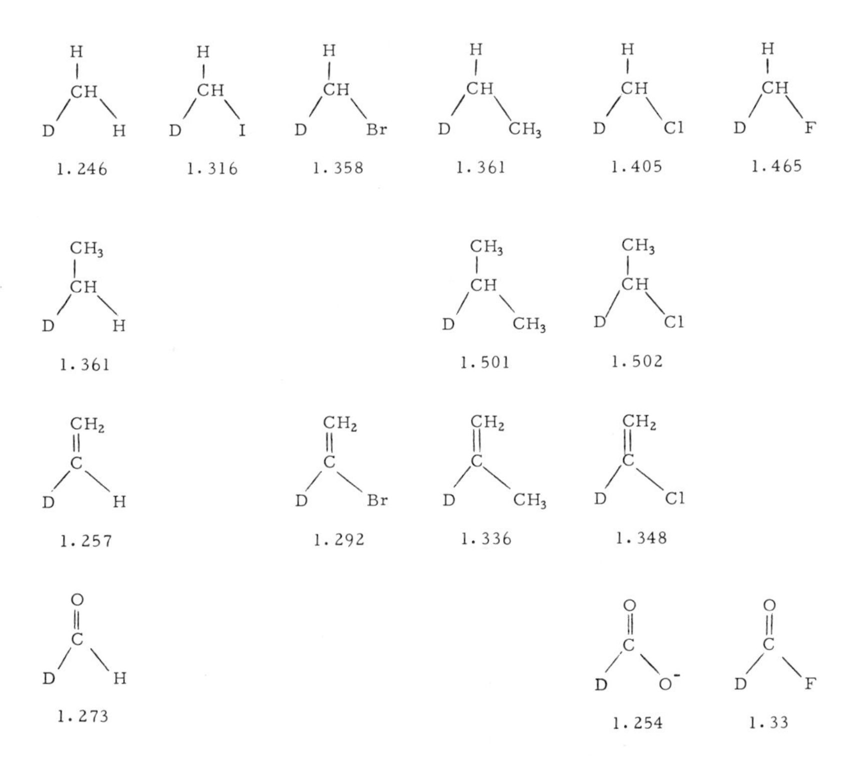

Figure 3. Fractionation factors relative to acetylene at 25°

for ethylene and formaldehyde are similar to that for methane because even though the double bonds increase the stiffness, there is one less α-bound atom in each of them than in methane. In all four series the factors generally increase from left to right as stronger bonding X groups are attached to the label-bearing carbon. An important exception seems to come in for the vinyl and formyl derivatives: if the X group can conjugate with the double bond the factors are lowered some 5–13%. Compare vinyl chloride (1.348) with methyl chloride (1.405) and formyl fluoride (1.33) with methyl fluoride (1.465). Also compare formate ion (1.254), which has a lot of conjugation, with formyl fluoride (1.33), which has less.

Table 2 gives calculated $^{12}C/^{13}C$ fractionation factors relative to acetylene. These cover a range of somewhat over 7% and depend primarily on the number and nature of the atoms attached directly to the isotopically substituted carbon. If other things are equal, more strongly bonding atoms cause larger fractionation factors, as expected, e.g., in the series CH_3I (0.9955), CH_3Br (1.0027), CH_3Cl (1.0058), and CH_3F (1.0259). Occasionally, remote substituents have a surprisingly large effect, apparently through their influence on the moments of inertia, for example, $^{13}CH_2=CBr_2$ (1.0107) and $^{13}CH_2=CH_2$ (1.0162).

Of increasing interest in recent years has been the application of *secondary* isotope rate effects in the analysis of mechanism. Secondary effects are defined as those caused by isotopic substitution for an atom which is not transferred during the course of the reaction, i.e., which remains bonded to the same atom throughout the course of the reaction.

Table 2. $^{12}C/^{13}C$ fractionation factors relative to acetylene, 25 °C

$$H{-}^{13}C{\equiv}C{-}H + {}^{12}X \overset{K}{\rightleftarrows} H{-}C{\equiv}C{-}H + {}^{13}X$$

^{13}X	K	^{13}X	K
$^{13}CH_3I$	0.9955	$^{13}CH_2=CF_2$	1.0144
$H{-}^{13}C{\equiv}C{-}H$	(1.0000)	$^{13}CH_3{-}CH_3$	1.0175
$^{13}CH_3Br$	1.0027	cyclopropane ($^{13}CH_2$ / \ $CH_2{-}CH_2$)	1.0163
$^{13}CH_4$	1.0050	$^{13}CH_3{-}CH_2{-}CH_3$	1.0219
$^{13}CH_3Cl$	1.0058	$^{13}CH_3F$	1.0259
$^{13}CH_2=CBr_2$	1.0107	$CH_3{-}^{13}CH_2{-}CH_3$	1.0286
$^{13}CH_2=CH_2$	1.0162	$^{13}CO_2$	1.0659
$^{13}CH_3CN$	1.0166	$^{13}CO_3^{=}$	1.0729

These effects are much smaller, of course, than the primary effects and their use has so far been restricted primarily to the isotopes of hydrogen. (See, however, Raaen et al. (1974) for some observed secondary ^{14}C effects.) In general, for most reactions secondary hydrogen effects are only greater than a few percent if the hydrogen is attached to an atom which undergoes changes in its bonding (α-effects). The only significant exceptions appear to be 1) where strong changes in hyperconjugative demand take place *beta* to the isotopically substituted bond (Shiner et al., 1968a) (in some cases this can be caused by conjugation of an unsaturated β-linkage with a more remote reaction center (Shiner, 1971)); or 2) the isotopic atom undergoes strong changes in steric compressions (Carter and Melander, 1973).

Many secondary effects can be modeled quite well by calculations based on observed spectra of small stable molecules. In these examples, the effect of reaction coordinate motion ν_{1_L}/ν_{2_L} must be small and the ideas above correlating structure with fractionation factors are qualitatively useful. However, for quantitative purposes one needs to use all of the frequencies and the full Bigeleisen-Mayer equation, particularly in the case of heavy-atom kinetic isotope effects, where the ν_{1_L}/ν_{2_L} factor is considerably different from unity and will be an important component of the isotope effect (O'Leary, this volume).

Thus, in summary, one can estimate isotope effects on equilibria or *secondary* effects on rates by choosing appropriate simplified model compounds for the two relevant states and using the fractionation factors for the model compounds from the table. To the extent that the trends shown by the molecules for which we have calculations are general, the basic requirement for a reasonable model is that it have the same α-bound groups as the compound it represents. A second generally valid but more risky modeling method can be used if appropriate model molecules with the same α-bound groups are not available in the tables; if an α-bound group does not change in the reaction, it can be ignored and the model compounds chosen so as to represent only the appropriate *change* in α-bound groups (see item 5 above where the exchange between methane and methyl chloride models the exchange between ethane and ethyl chloride reasonably well).

INFLUENCE OF REACTION COORDINATE MOTION ON ISOTOPE RATE EFFECTS

Next, a special problem is discussed that arises in the consideration of isotope effects on *reaction rates* but does not apply in equilibrium processes. That is the effect of reaction coordinate motion. As indicated above, the reaction coordinate motion is considered to be one of the normal coordinate motions of the transition state except that it has no restoring force (is *aperiodic*) and has no zero-point energy. It contributes directly to the isotope effect in the ν_{1L}/ν_{2L} term, which is simply the ratio of the two zero or imaginary frequencies of this motion. The lack of a restoring force and of vibrational energy terms associated with the reaction coordinate motion also influences the isotope effect by removing a real frequency from the EXC and ZPE terms.

If the dissociation of a diatomic molecule ^{1}A—B in comparison to ^{2}A—B is considered, the representation in terms of an isotopic exchange equilibrium would simply be

$$^1\text{A—B} + {}^2\text{A} \rightleftarrows {}^2\text{A—B} + {}^1\text{A}$$

there being no vibrational energy of monatomic B. The reaction coordinate motion is simply the translation of A and B away from each other. One can estimate the isotope effect for such a process using the zero-point energy approximation given above for equilibria

$$k_1/k_2 = \exp[h(\nu_{1AB} - \nu_{2AB})/2\mathbf{k}T]$$

For hypothetical diatomic "C—H" with a stretching frequency of 3,000 cm^{-1}, using the Streitwieser approximation one estimates the isotope effect on C—H dissociation at 25° as:

$$k_1/k_2 = \exp[(0.1915 \times 3{,}000)/298] = 6.87(298\,°\text{K})$$

Many observed H/D isotope effects for proton and hydrogen atom abstractions are found to be approximately seven, and this value is usually taken to be the "maximum" normal primary CH/CD isotope effect. However, these actual processes invariably involve the *abstraction* of H or D by an attacking reagent. Under these circumstances H or D is never really free to translate away by itself. If so, how is the zero-point energy lost? This can best be illustrated (Melander, 1960; Westheimer, 1961) with a triatomic transition-state model wherein an abstracting atom C is included. The reaction can be written as

$$\text{C} + {}^1\text{A—B} \rightleftarrows [\text{C---}{}^1\text{A---B}]^\ddagger \rightarrow \text{C—}{}^1\text{A} + \text{B}$$

and the corresponding isotope exchange equilibrium as

$$^1\text{A—B} + [\text{C---}{}^2\text{A---B}]^\ddagger \rightleftarrows {}^2\text{A—B} + [\text{C---}{}^1\text{A---B}]^\ddagger$$

One can see how the zero-point energy is lost if the motions of (C---A---B)‡ are examined. If this were a stable linear molecule, it would have $3N$-5, or 4 vibrations, where *N* is the number of atoms (3). Two of these four are bending modes, which for purposes of simplicity are neglected here; in a quantitative treatment they, of course, must be considered. Of the remaining two motions, one must be aperiodic, representing in one direction the conversion to reactants and in the reverse direction the conversion to products. The fourth motion is a stretching vibration. The stretching motions can be represented as follows, keeping in mind that these "internal" motions cannot result in the displacement of the center of mass or in any molecular rotation:

Symmetrical stretch: $\overleftarrow{C}$----A----$\overrightarrow{B}$

Because the three atoms cannot all separate without an increase in energy, this motion has a restoring force, is periodic, and has a zero-point energy. However, if the structure is symmetric with A in the center as indicated, A will not move in this vibration, the frequency will be the same for *both* isotopes of A, and there will be no zero-point energy difference between the isotopic transition states contributed by this motion.

Asymmetrical stretch: $\overrightarrow{C}$----$\overleftarrow{A}$----$\overrightarrow{B}$

This is the reaction coordinate motion; as it continues, A becomes bonded to C and B separates. There is no restoring force for this motion, no zero-point energy, and, of course, no zero-point energy differences contributed by it to the isotopic transition states. The reverse of this motion simply takes the transition state back to reactants.

The total absence of zero-point energy differences in the stretching modes between the isotopic molecules for the symmetrical transition state thus leads one to expect a maximum zero-point energy effect on the reaction rate. The reasons for the maximum loss of zero-point energy in the stretching mode are then two-fold: 1) the transition state is symmetric and 2) the reaction coordinate motion involves translation of the isotopic atom. Actually, these reasons are closely related; if the transition state is not symmetric there is less isotopic motion in the reaction coordinate and *more* isotopic motion in the other stretching mode, leading to isotope effects less than the nominal maximum. This fall-off from the expected maximum is greater the greater the degree of transition-state asymmetry. In proton transfers between bases of unequal basicity or in atom transfers between radicals of unequal bonding strength, the transition states should be asymmetric. Bell and Cox (1971) have correlated isotope effects in proton transfer with the pK_a difference between the two bases (ΔpK_a) and found a roughly bell-shaped curve with a maximum near $\Delta pK_a = 0$. Pryor and Kneipp (1971) have found a similar curve correlating the hydrogen-deuterium isotope effect in some hydrogen atom abstractions with the difference in bond energies between hydrogen and each of the two radicals involved in the transfer. One point that is hard for many people to grasp initially is the *direction* of the asymmetry; in the transition state the transferred atom is closest to the group to which it bonds *least* strongly in the corresponding stable molecule. In the extreme, one can visualize an attacking agent sufficiently powerful to react at every collision, in which case the reaction coordinate motion is simply the diffusion together of the reactants; the transition state is passed at a relatively large distance of separation of the reagent from the isotopic atom, there is no perturbation of the vibration frequencies and no isotopic effect.

This analysis of the isotope effects in an atom transfer reaction indicates that the diatomic model can be used for a simple calculation of the approximate maximum isotope rate effect to be expected in reactions (involving either dissociation or transfer) of various kinds of isotopically substituted bonds. One needs to know only the approximate stretching frequency of the bond for one isotopic species; the stretching frequency for the other can be calculated from the reduced mass relationship, and the effect can be estimated using the formula:

$$k_1/k_2 = \exp[0.7193(\nu_1 - \nu_2)/T]$$

Table 3 gives a list of such hypothetical diatomic molecules, the approximate frequency, the isotope effect on the frequency, and the calculated zero-point energy contribution to the isotope effect. Effects near these

Table 3. Calculated isotope effects for diatomic molecular dissociation model

Hypothetical isotopic molecules				
1	2	$\nu_1{}^a$	$\nu_1/\nu_2{}^b$	k_1/k_2 (298°K)
$C—{}^{1}H$	$C—{}^{2}H$	3000	1.363	6.870
$C—{}^{12}C$	$C—{}^{13}C$	1129	1.020	1.054
$C—{}^{14}N$	$C—{}^{15}N$	1134	1.016	1.044
$C—{}^{16}O$	$C—{}^{18}O$	1113	1.025	1.068
$C—{}^{32}S$	$C—{}^{34}S$	915	1.008	1.018
$C—{}^{35}Cl$	$C—{}^{37}Cl$	804	1.007	1.014

[a] Assumed frequency.
[b] Calculated via reduced mass equation (see text).

values have been observed in bond-breaking reactions and can be taken as reasonable estimates of typical, near-maximum, primary isotope effects for cleavage of the indicated bond types. If a reaction that apparently involves the breaking of an isotopically substituted bond does not show an effect near the size indicated from such a calculation, generally either one of two conditions applies: 1) some step other than the bond cleavage is rate-determining or partially rate-determining or 2) the cleavage occurs in the rate-determining step but the transition state is very unsymmetrical. Other mechanistic or general chemical information needs to be supplied to distinguish between these two possibilities. For example, it is very significant mechanistically if the isotope effect changes with a change in concentration of one reactant, because this indicates a change in the rate-determining step. Also of interest, but more difficult to interpret, is whether or not the isotope effect changes with a substituent change in one of the reactants.

TRANSITION-STATE MODELING AND ISOTOPE RATE EFFECTS IN THE S_N2 REACTION

Next, some of the problems associated with the quantitative modeling of transition states (Thornton and Thornton, 1971; Van Hook, 1971) are considered. Although our aim is to formulate a general computer-programmable approach, we will here use for specific illustration some of our results obtained for the S_N2 reaction (Ingold, 1953; Shiner, 1975)

$$I^- + CH_3Cl \rightleftarrows CH_3I + Cl^-$$

There are a number of reasons for selecting this example:

1. Experimental carbon, hydrogen, and chlorine isotope effects have been measured for both the rate and equilibrium constants (Maccoll et al., unpublished results; Shiner and Buddenbaum, 1975)
2. Accurate force fields for the organic reactants and products are available (Hartshorn and Shiner, 1972; Shiner et al., 1968b)
3. A number of calculations have already been published for closely related S_N2 reactions (Bron, 1974; Buddenbaum and Shiner, 1976; Sims et al., 1972; Won and Willi, 1972)
4. There are two reacting bonds and some α-bound atoms so that these features of the more general problem are included. Through application of the Wolfsberg-Stern cut-off procedure, many reactions could be simulated by models not much more complicated than these.

An important element of our approach which turns out to be operationally as well as conceptually useful is to separate out for each transition-state model the isotope effect it would show if it were a stable molecule and the isotope effect it shows on the introduction of the reaction coordinate motion, which is an otherwise normal mode that has no restoring force and no zero-point energy (Buddenbaum and Yankwich, 1967; Glasstone, Laidler, and Eyring, 1941). In modeling a transition state we are, of course, never satisfied with just one model but generally want to think about a series of models with a varying degree of progression toward product or a varying degree of reactant-like or product-like character. It is convenient to choose a particular reacting bond for which the bond order (or better, bond number) serves as the index of the fraction of reaction. Using Pauling's Rule (Pauling, 1947) and Badger's Rule (Badger, 1934) or some variation thereof, the bond length, adjacent bond angles, and bond force constants can be mathematically related to this bond number, assuming, of course, that the structures and force constants of reactants and products (or intermediates) are known or can be inferred. Pauling's rule relates the bond length at the transition state ($R^{\ddagger}$) to the length of the fully formed bond (R°) and the bond order of that bond at the transition state ($N^{\ddagger}$)

$$R^{\ddagger} = R^{\circ} - 0.26 \ln N^{\ddagger}$$

Badger's rule relates transition-state force constants ($F^{\ddagger}$) to ground-state force constants (F°) and bond order

$$F^{\ddagger} = F^{\circ} N^{\ddagger}$$

A rule analogous to Badger's rule can be derived to cover the change in geometry which occurs during the S_n2 reaction

$$A_{HCCl}^{\ddagger} = 70.5^{\circ} + 38.5^{\circ} N_{CCl}^{\ddagger}$$

where A is the angle.

If, as in this example, there is more than one reacting bond, one adds for each additional bond one more dimension to the reaction coordinate space to be searched; physical organic chemists have been enamored for some years now with the idea that reacting bond changes do not always occur in complete concert with one another. We have found it useful to think of reaction coordinate space for two reacting bond processes in terms of the O'Ferrall-Jencks diagram (Jencks, 1972; More O'Ferrall, 1970), which represents the bond numbers of the two reacting bonds on rectangular axes and represents the energy as contours. Figure 4 is an example, except that the energy contours are not included and the lines

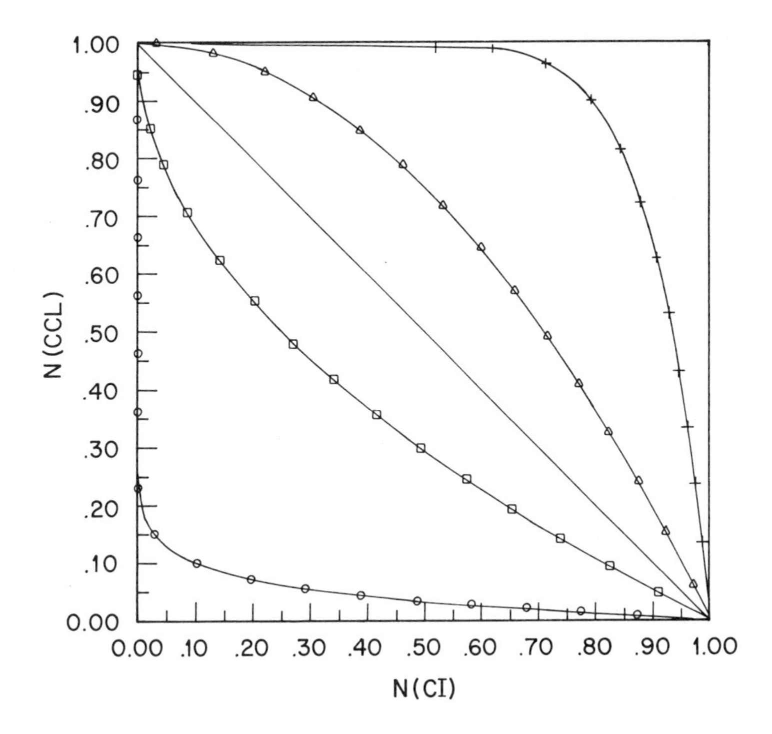

Figure 4. O'Ferrall-Jencks diagram for the reaction of iodide ion with methyl chloride.

Comments	*Legend*
Parabolic transit	$\alpha = 1.00$
Constant change in A_{HCCl}	$\alpha = 2.00$
X = degree reaction	$\alpha = 21.85$
$N(CCL) = 1.0 - X$	$\alpha = 0.44$
$N(CI) = X^{\alpha}$	$\alpha = 0.14$

represent several hypothetical transits (not necessarily reaction paths) from reactants to products (see below; see also Shiner and Seib, 1976; Winey and Thornton, 1975). For more than two reacting bonds one must think of more than three spatial dimensions or a series of O'Ferrall-Jencks diagrams.

One of the problems in systematically searching such reaction coordinate space in these calculations is to keep the models realistic and to formulate them mathematically so that they can be extrapolated with convenient variables all the way to reactants or products. However, in extrapolating such model structures back to reactants or forward to products, it is important to realize that fractionation factors are affected directly by the presence of a reagent molecule or by a by-product molecule especially through the MMI term, even though the reaction may not have *electronically* proceeded far enough to alter any force constants significantly. Thus, the isotope *rate* effect for a completely product-like *transition state* will, in general, not be the same as the isotope effect on the overall equilibrium constant, and the rate effect for a completely reactant-like transition state will not generally be unity except for unimolecular isomerizations. Although the sum of the partial bond orders for a two-bond process such as the one considered here does not have to be unity, it must extrapolate to unity in the limits. With this in mind, it occurred to us that such transits from reactants to products would be conveniently represented with parabolic functions, because parabolas can be formulated so that they always pass through two opposite corners of the unit square while all intermediate points remain within the square. Thus, we can represent the space in the O'Ferrall-Jencks diagram in terms of two parameters, N_{C-Cl} and α such that $N_{C-I} = (1 - N_{C-Cl})^{\alpha}$. The symbol α can be referred to as a "concertedness parameter." When equal to unity it represents a completely concerted reaction which conserves bond order. When different from unity, it represents a non-concerted reaction; for α values greater than unity C—Cl breaking "runs ahead" of C—I formation, while for values less than unity C—Cl bond breaking "runs behind" C—I formation. In the more general case, for each additional reacting bond another concertedness parameter can be introduced relating that bond number to the index bond number. In this way, bond numbers, structures, and force constants can be systematically varied between defined limits, and the entire reaction coordinate space can be covered without being exceeded. Concertedness parameters near 100 or 0.01 represent nearly non-concerted processes. If $\alpha = 0.01$, $N_{C-Cl} = 0.03$ when N_{C-I} is 0.03, i.e., the total partial bond order is only $\sim 6\%$.

Thus, from known initial and final state structures and force constants using Pauling's Rule and Badger's Rule, the reaction coordinate space of an O'Ferrall-Jencks diagram can be searched completely with two parameters, one representing the degree of reaction of one bond and the other representing the degree of concert between the two reacting bonds. For each additional reacting bond one more concertedness parameter is needed. *The structures so derived, however, represent stable molecules rather than transition states because the reaction coordinate motion has not been introduced and all 3N — 6 normal modes behave as vibrations.* Calculations equivalent to those for equilibria could, however, be done on these structures. Reaction coordinate motion can be introduced through assignment of off-diagonal or interaction force constants. For example, a simple triatomic model

$$A\overset{F_1}{---}B\overset{F_2}{---}C$$

held together by force constants F_1 and F_2 would show two periodic stretching vibrations for all real positive values of F_1 and F_2. It can be made to undergo aperiodic stretching motion by assignment of an interaction force constant F_{12} which makes the stretching of one bond much easier if it occurs simultaneously with the compression of the other bond. In the general case, the interaction constants must be assigned in such a way that the resulting atomic motions represent a realistic reaction coordinate motion because the isotope effects are very much dependent on the relative motions of the atoms. Thus, there are really two problems to be solved: first, what atomic motions represent a realistic reaction coordinate, and second, if one knows what a realistic reaction coordinate is, how can the off-diagonal force constants be assigned so that the model follows this motion?

In the simple three-atom case, the interaction constant is usually assigned according to the formula suggested by Johnston, Bonner, and Wilson (1957); see Buddenbaum and Yankwich (1967)

$$F_{12} = \sqrt{(F_1 - F_0)(F_2 - F_0)} \tag{8}$$

where F_0 is a measure of the curvature of the reaction barrier. If F_0 is chosen as zero, then ν_L, the resulting reaction coordinate mode frequency, is zero; if $F_0 < 0$, then ν_L becomes imaginary. Of course ν_L cannot have a real positive value because it represents an aperiodic motion. For the ν_L equals zero case, this assignment of F_{12} fixes the relative changes in bond lengths R_{AB} and R_{BC} that constitute the reaction coordinate motion according to the equation (Buddenbaum and Yankwich, 1967)

$$\frac{\Delta R_{AB}}{\Delta R_{BC}} = -\sqrt{\frac{F_2}{F_1}} \tag{9}$$

In the more general multiatomic S_n2 transition state represented by

```
        H
        |
Cl --- C --- I
       / \
      H   H
```

it is possible through assignment of properly determined off-diagonal force constants to select freely any desired reaction coordinate motion. This bit of mathematical magic was discovered by Buddenbaum and Yankwich (1967), and for the present case the formulas necessary for assignment of the interaction constants take the following form:

$$F_{12} = \frac{[(\Delta R_{CCl})^2(F_0 - F_{CCl}) + (\Delta R_{Cl})^2(F_0 - F_{Cl}) - (\Delta X_{HCCl})^2(F_0 - F_{HCCl}{}^X)]}{2\Delta R_{CCl}\Delta R_{Cl}}$$

where ΔX_{HCCl} is the A_1 HCCl, HCl, HCH—symmetry coordinate of C_{3V} symmetry (Holmes, 1967), and $F_{HCCl}{}^X$ is its force constant. Similar equations are necessary for F_{13} and F_{23}, the interaction constants between each carbon-halogen stretching motion and the X_{HCCl} bending motion

$$F_{13} = \frac{[(\Delta R_{CCl})^2(F_0 - F_{CCl}) - (\Delta R_{Cl})^2(F_0 - F_{Cl}) + (\Delta X_{HCCl})^2(F_0 - F_{HCCl}{}^X)]}{2(\Delta R_{CCl})(\Delta X_{HCCl})}$$

$$F_{23} = \frac{[-(\Delta R_{CCl})^2(F_0 - F_{CCl}) + (\Delta R_{Cl})^2(F_0 - F_{Cl}) + (\Delta X_{HCCl})^2(F_0 - F_{HCCl}{}^X)]}{2(\Delta R_{Cl})(\Delta X_{HCCl})}$$

To complete these transition-state modeling formulas there remains, then, only the chemically very interesting problem of determining the individual atomic movements that constitute the reaction coordinate motion. If the points in the O'Ferrall-Jencks diagrams are referred to as possible transition-state structures, the reaction coordinate motion corresponds to the trajectories of these points. There are obviously a lot of possibilities and these must be examined much more extensively than the present authors have been able to do so far. One simple and appealing approach is to use as the reaction coordinate the parabolic functions that we have used to search the space of the diagram. This has been incorporated into the program and makes a neatly defined, consistent package. It has the drawback of not being symmetrical about both diagonals. Mainly, it is important to avoid totally unrealistic motions because these can have strong effects on the results. It would seem unreasonable, for example, to have a reaction coordinate motion that does not take the molecule in the *general direction* of the products. Even a 90° "window" pointing toward the product state contains a wide range of results. One might accept some sort of "least motion" principle (Ehrenson, 1974) and have the trajectory be simply the line that connects the point to the product state. Of course, other functions than these parabolic ones can be devised to search the reaction coordinate space systematically and to describe reaction coordinate motion. We are in the process of investigating some of these.

At this point it will be useful to review what we have outlined so far by describing the calculation procedure before presenting some of the results. Briefly, the procedure is as follows:

1. Set up structures for reactants and products. Include bond lengths, bond angles, and valence force constants.
2. Set up a generalized transition state structure so that by variation of appropriate bond lengths and angles it can be converted to reactant (plus reagent) and product (plus by-product).
3. Select one reacting bond as the "index bond."
4. Relate the bond numbers of the other reacting bonds to the bond number of the index bond using an appropriate analytical function such as the parabolic one described above.
5. Relate bond lengths, angles, and force constants (including off-diagonal force constants) to the appropriate bond numbers using Pauling's Rule, Badger's Rule, etc.
6. For selected values of the index bond number covering an appropriate extent of reaction, calculate structures and force fields for hypothetical transition states.
7. For each hypothetical transition state, identify the atomic motions *implicitly* defined by the above assumptions.
8. Calculate additional off-diagonal force constants necessary to make the determined atomic motions constitute the reaction coordinate for each hypothetical transition state.
9. Calculate isotope effects via the Bigeleisen equation using the program of Wolfsberg and Stern.
10. Print out isotope effects, vibrational eigenvalues (frequencies), and eigenvectors (internal and cartesian coordinates).
11. Make a three-dimensional plot of structure and reaction motions (Hilderbrandt, 1969).
12. Plot isotope effects as functions of bond numbers and concertedness parameters.

In Table 4 are given the $^{12}C/^{13}C$ isotope effects calculated for the $I^- + CH_3Cl$ reaction as a function of the carbon-chlorine bond number for the completely concerted reaction ($N_{C-Cl} + N_{C-I} = 1$) with $\nu_L = 0$; various force field and reaction coordinate choices are included. In the first line the complete valence force field for reactants and products (Hartshorn and Shiner, 1972) is used but no reaction coordinate motion

Table 4. Calculated k_{12}/k_{13} effects[a] for concerted[b] $I^- + CH_3Cl$ reaction, 25°C

Force field[c]	reaction coord.[d]	N_{C-Cl} 1.0	0.85	0.60	0.50	0.40	0.15	0.0	Equil. value
T	None	1.001	1.004	1.009	1.010	1.011	1.011	1.011	1.010
D	None	0.997	1.000	1.003	1.005	1.006	1.008	1.009	
D	2	1.004	1.029	1.051	1.058	1.060	1.038	1.010	
D	C	1.004	1.033	1.066	1.075	1.064	1.024	1.010	
T	C	1.007	1.028	1.062	1.072	1.059	1.018	1.012	

[a] Experimental values: rate, 1.072; equil. 1.011.
[b] $N_{C-Cl} + N_{C-I} = 1$; T.S. structures and force constants interpolated by reacting bond number (see text); $\nu_L = 0$.
[c] T, total harmonic force field; D, only diagonal harmonic force field.
[d] 2, only the two reacting bond lengths change; C, includes all changes required by structure rules.

is included; as the C—Cl bond breaks and the C—I bond forms, the isotope effect goes steadily up to 1.011. The equilibrium value calculated is 1.010, in good agreement with the experimental value of 1.011 (Shiner and Buddenbaum, 1975). The second line shows that in this case all of the values are reduced only a little if the off-diagonal force constants of the transition-state model are set to zero. The third line shows that the introduction of the Johnston two-coordinate reaction motion (Johnston, Bonner, and Wilson, 1957) causes a significant primary carbon isotope effect that reaches a maximum value near the symmetrical state that is less than the experimental value of 1.072. The fourth line shows that with the diagonal force field and the reaction coordinate motion now including the angle bends, the isotope effects are similar to those in the third line but reach a higher maximum. The last line shows that for the full force field and full reaction coordinate motion the numbers are generally slightly smaller. This maximum is in good agreement with the experimental value. Thus, as expected, the biggest ^{13}C effect is caused by the carbon reaction coordinate motion with smaller influences as a result of the "inversion" motion and the changes in off-diagonal force constants.

Figure 5 shows a graph of the values from the last line of Table 4 and, in addition, corresponding values obtained for the parabolic reaction coordinate motion when α is 2.0 and 0.44. The maximum may not have been closely approached in the latter calculation, for which only three values have been obtained. When $\alpha = 2.0$, C—Cl breaking runs ahead of C—I formation, so that the symmetrical transition state occurs

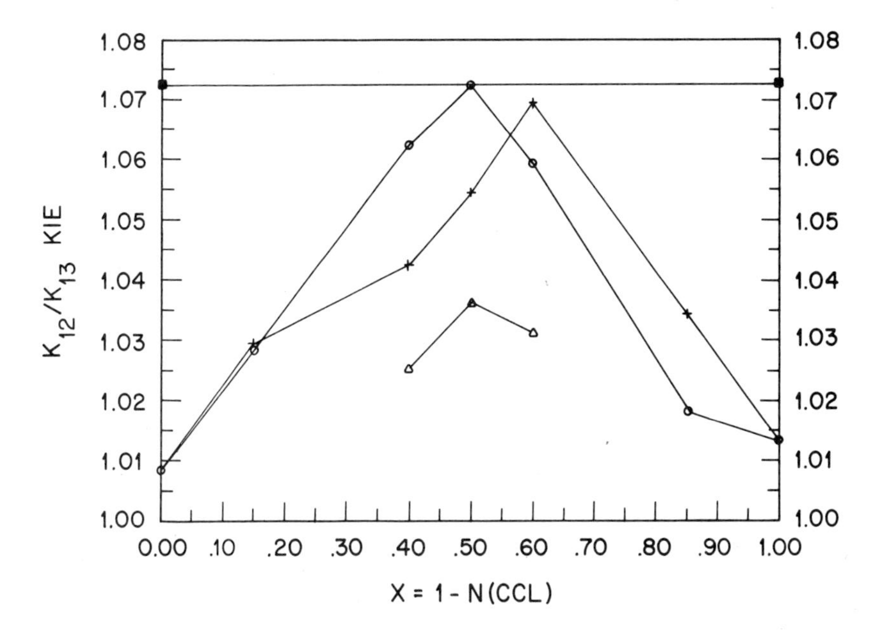

Figure 5. Calculated carbon-13 kinetic isotope effects at 25° for the reaction of iodide ion with methyl chloride.

Comments	*Legend*
Parabolic transits	■ k_{12}/k_{13} exp. KIE
$N(\text{Cl}) = X^{\alpha}$	○ KIE, α = 1.00
Constant change in A_{HCCl}	△ KIE, α = 0.44
k_{12}/k_{13} exp. KIE = 1.072	+ KIE, α = 2.00
K_{12}/K_{13} exp. EIE = 1.010	
K_{12}/K_{13} calc. EIE = 1.011	

with a smaller C—Cl bond order than in the concerted case and the maximum carbon isotope effect occurs at a smaller value of $N_{\text{C—Cl}}$. The situation is quite analogous to that which is well known for proton transfers (see above) and the explanation (Fry, 1964) is the same as that given by Westheimer (1961) for those examples. The carbon isotope effect reaches its maximum when the reaction coordinate motion is most dominated by carbon and the other atoms move as little as possible with respect to the center of mass. This requires that the heavy Cl and I atoms be disposed symmetrically around the carbon with respect to both distance and force constant, especially the latter, and that the hydrogen atoms are undergoing the "inversion" motion rather than being carried along by the carbon as a rigid methyl group (Buddenbaum and Shiner, 1976).

Table 5 gives the calculated H/D isotope effects analogous to the $^{12}C/^{13}C$ effects of Table 4. The first line shows that the hydrogen effect is significantly *secondary* and that the calculated equilibrium effect agrees with the experimental value. Line two shows that the hydrogen effects are influenced significantly by setting the off-diagonal force constants to zero, because these numbers are 6–10% larger than the corresponding values of line one. The third line shows that the introduction of the two-coordinate reaction motion causes a small increase in the hydrogen isotope effect. The last two lines (in comparison to the first two) show that the full reaction coordinate motion does *not* introduce much hydrogen isotope effect; e.g., the maximum value of 1.091 is not much bigger than the values of 1.080 obtained with the full force field but *no* reaction coordinate motion. The calculated value of 1.091 is somewhat larger than the observed value of 1.042. This may mean that the C—H stretching force constants are higher in the transition state than the weighted average of the reactant and product states (Kresge and Preto, 1967); this is not unexpected because they might be considered to be formed from sp^2 hybrid orbitals on carbon. Alternatively, Badger's rule may not adequately describe the alterations in bending force constants, possibly leading to transition state values that are too small (Wolfsberg and Stern, 1964b). Our calculations at present do not allow for this and this problem needs to be examined further.

Table 6 shows the calculated $^{35}Cl/^{37}Cl$ effects corresponding to the ^{13}C and D values shown in Tables 4 and 5. The first line shows that there is a significant non-reaction coordinate effect as the C—Cl bond is weakened. The calculated equilibrium value (1.0085) is larger than the experimentally observed 1.007, probably indicating a solvation effect on

Table 5. Calculated (k_K/k_D)/D effects[a] for concerted[b] $I^- + CH_3Cl$ reaction, 25 °C

Force field[c]	reaction coord.[d]	N_{C-Cl} 1.0	0.85	0.60	0.50	0.40	0.15	0.0	Equil. value
T	None	1.000	1.024	1.068	1.080	1.084	1.074	1.068	1.067
D	None	1.106	1.115	1.137	1.145	1.151	1.163	1.172	
D	2	1.113	1.131	1.164	1.176	1.184	1.179	1.172	
D	C	1.113	1.112	1.138	1.154	1.160	1.164	1.172	
T	C	1.007	1.024	1.072	1.091	1.095	1.080	1.068	

[a] Experimental values: rate, 1.042; equil., 1.067.
[b, c, d] See Table 4.

Table 6. Calculated k_{35}/k_{37} effects[a] for concerted[b] $I^- + CH_3Cl^*$ reaction, 25 °C

Force field[c]	reaction coord.[d]	N_{C-Cl} 1.0	0.85	0.60	0.50	0.40	0.15	0.0	Equil. value
T	None	1.000	1.001	1.003	1.004	1.005	1.007	1.008	1.0085
D	None	1.000	1.001	1.003	1.004	1.005	—	1.008	
D	2	1.014	1.010	1.005	1.004	1.010	1.023	1.031	
D	C	1.014	1.013	1.006	1.0045	1.015	1.029	1.031	
T	C	1.014	1.013	1.006	1.004	1.014	1.029	1.031	

[a] Experimental values; rate, 1.0087; equil., 1.007.
[b, c, d] See Table 4.

Cl^- not modeled in our calculations (Howald, 1960). The second line indicates no significant influence on the results as a result of the off-diagonal force constants. The third, fourth, and fifth lines indicate that the reaction coordinate motion has important influence in the asymmetric transition states, where Cl is moving as part of CH_3Cl toward I^- or as Cl^- away from CH_3I, but that the effect in the near-symmetric situations is small. The calculated values in line five for N_{C-Cl} around 0.40 to 0.60 are near the experimental value and would probably fit quite well if some fraction of the solvation characteristic of the product state were introduced (Cremaschi, Gamba, and Simonetta, 1972).

Figure 6 is a plot of the isotope effects from the last lines from Tables 4, 5, and 6 against N_{C-Cl}. It shows reasonable, but not perfect, agreement with the observed values near 50% reaction. The aim of this chapter has not been to discover a transition state that gives a precise fit to the experimental values but rather to look at the effects of larger variations in transition state structure carried out according to our generally formulated rules. We hope to develop our programs further so that they become generally applicable to a variety of structures in the way that the Schachtschneider and Snyder and the Wolfsberg and Stern programs on which they are built are applicable.

To a large extent we have put together several of what in hindsight seem like straightforward ideas that have been separately used by others in the field into a generalized approach that is extensively computer assisted, if not yet completely automated. We have tried to steer away from empirical adjustment of force constants and discussions of how they relate to the results but instead have concentrated our approach on parameters of more general chemical meaning.

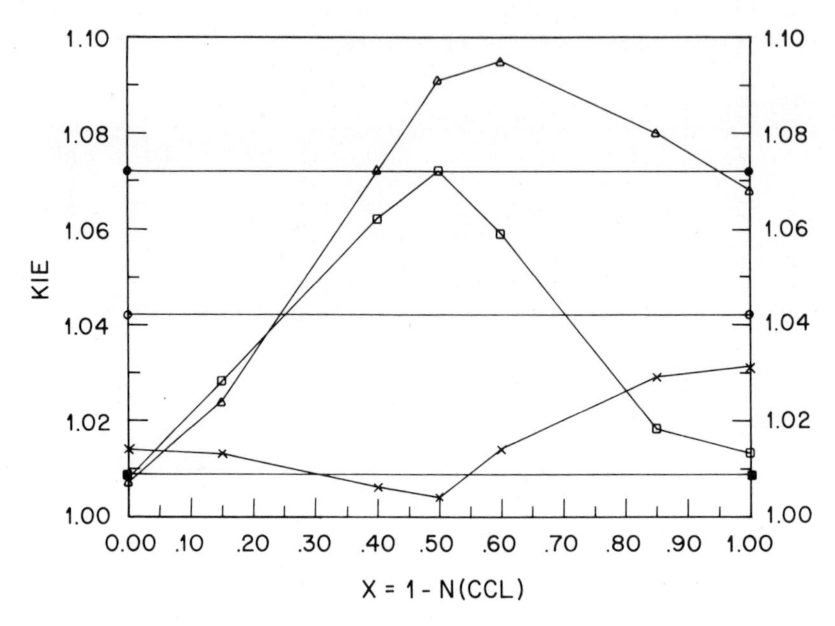

Figure 6. Carbon, hydrogen, and chlorine kinetic isotope effects at 25° for the reaction of iodide ion with methyl chloride.

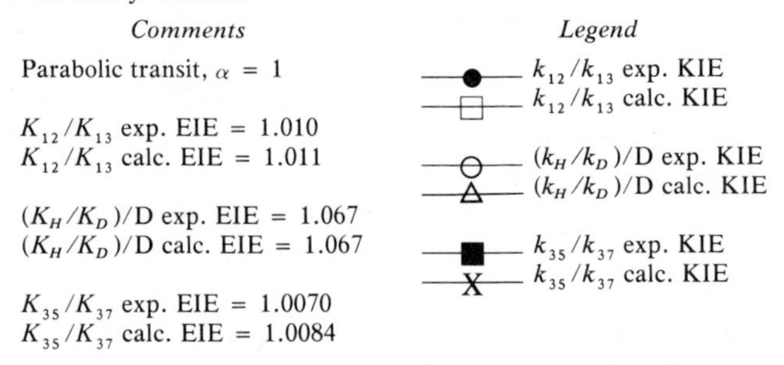

Johnston (1960; 1966), in developing the BEBO method, applied Pauling's and Badger's rules to transition states and reaction coordinates and examined isotope effects as a function of degree of reaction. Bell (1974), Bigeleisen (1964), Chiang, Kresge, and Wiseman (1976), More O'Ferrall and Kouba (1967), Willi and Wolfsberg (1964), and Wolfsberg and Stern (1964a,b) have examined proton transfer isotope effects as a function of degree of reaction. Won and Willi (1972), Sims et al. (1972), Bron (1974), and Buddenbaum and Shiner (1976) have calculated isotope effects for S_n2 reactions as a function of degree of reaction. The last two groups also examined cases where the total partial bond order is not conserved. Yankwich and coworkers (Huang et al., 1968; Kass and Yankwich, 1969; Wei and Yankwich, 1974) have systematically examined the influence of assigned reaction coordinate motion on carbon isotope effects in unimolecular decarboxylation reactions. Although Katz and Saunders (1969) studied the effects of systematic variation of off-diagonal force constants on nitrogen and hydrogen isotope effects in elimination reactions, they did not separately examine the influences of the resulting reaction coordinate motions on the isotope effects.

To our knowledge no one has used functions analogous to our parabolic ones to search systematically reaction coordinate space or to generate reaction coordinate motions. Buddenbaum and Yankwich in an early important paper (1967) realized the significance of reaction coordinate motion and devised the mathematical method we have used to impose it on transition-state models. This method, however, has not yet been adopted by other workers in the field. An important feature of our approach is that it avoids extrapolation to very unrealistic structures and motions. As long as a straightforward approach gives a result that fits experimental values, extreme values of force constants and geometries, it would seem, should not be considered. More fundamental examination of transition-state structures selected by the current method could be carried out quantum mechanically (Dedieu and Veillard, 1972; Raff, 1974; Schowen, this volume) to determine if the behavior predicted by our rules is fundamentally sound. Such theoretical calculations might also be used to formulate improved versions of Pauling's and Badger's Rules for partial bonds (Burgi et al., 1974; Wiberg and Ellison, 1974).

DISCUSSION

M. H. O'Leary. How large is the chlorine isotope effect on the solvation of chloride ion?

V. J. Shiner. It is not known for the DMF solvent in which we did our studies. Howald (1960) measured the fractionation between gaseous hydrogen chloride in equilibrium with chloride ion in acetic acid and estimated that $^{35}Cl^-$ was concentrated over $^{37}Cl^-$ in acetic acid relative to the *ions* in the gas phase by a factor of 1.004. However, our observed equilibrium isotope effect for $^{35}Cl/^{37}Cl$ exchange between methyl chloride and chloride ion in DMF agrees with the calculated value, assuming no solvation effects on either chloride ion or methyl chloride, within a factor of 1.001.

M. H. O'Leary. For enzyme-catalyzed reactions, is it possible that there is an isotope effect on the binding of the substrate to the enzyme?

V. J. Shiner. There should be an effect on binding, although I think it would be small. I don't know what will happen to the rotational moments of inertia when you bind an isotopic substrate to something as massive as an enzyme.

On the other hand, the isotope effects on the solubilities of most compounds are very small, consistent with the observation that the infrared frequency shifts observed on going from the gas phase to solution are generally small, except in cases where there is hydrogen bonding. There hydrogens may shift several hundred cm^{-1} in a 3,600 cm^{-1} frequency. Otherwise, the frequency shifts on dissolution of covalent compounds are generally only a few cm^{-1} and not strongly solvent dependent. Thus, if binding is analogous to phase transfer, such as hydrophilic to hydrophobic, this analogy suggests that the isotope effects would be small unless there are specific interactions at the isotopic site.

I. Rose. Do you think the equilibrium isotope effect in going from sp^3 to sp^2 is a valid estimate of the upper limit on the kinetic isotope effect?

V. J. Shiner. Of course, as we have shown, the equilibrium value of the α-d effect will depend on the nature of the leaving group in going from sp^3 carbon to sp^2. For a given leaving group the equilibrium α-d effect should be close to the upper limit for the kinetic isotope effect but it will depend on how the hydrogen moves; this influence should be small because, by analogy with the S_N2 reaction, we expect that the carbon and leaving group atoms would mostly dominate the reaction coordinate motion; accordingly we expect and find that the *carbon* isotope *rate* effects are much larger than the effects expected for any related equilibrium process.

I. Rose. What are the limitations of that comparison?

V. J. Shiner. The main limitation in estimating α-d effects is finding appropriate models for the sp^3 and sp^2 states, i.e., those having the same α-bound groups as occur in the reaction. Another limitation is estimating the effect of reaction coordinate motion. We should be able to do this satisfactorily using the programs we have described but so far we only have results for the S_N2 reaction. I should also point out that the relationship between the degree of reaction and the α-d isotope effect is probably not linear; the involvement of the isotopic atom in reaction coordinate motion always makes the isotope effect more normal (less inverse) so that a linear interpolation in the present example would indicate too much sp^3 character for the transition state.

LITERATURE CITED

Badger, R. M. 1934. A relation between internuclear distances and bond force constants. J. Chem. Phys. 2:128-131.

Bell, R. P. 1974. Recent advances in the study of kinetic hydrogen isotope effects. Chem. Soc. Revs. 3:513-544.

Bell, R. P., and Cox, B. G. 1971. Primary hydrogen isotope effects on the rate of ionization of nitroethane in mixtures of water and dimethyl sulphoxide. J. Chem. Soc. B:783-785.

Bell, R. P., and Crooks, J. E. 1962. Secondary hydrogen isotope effect on the dissociation constant of formic acid. Trans. Faraday Soc. 58:1409-1411.

Bigeleisen, J. 1964. Correlation of kinetic isotope effects with chemical bonding in three-center reactions. Pure Appl. Chem. 8:217-223.

Bigeleisen, J., and Ishida, T. 1975. Isotope chemistry and molecular structure. Total deuterium isotope effects. J. Chem. Phys. 62:80-88.

Bigeleisen, J., and Mayer, M. G. 1947. Calculations of equilibrium constants for isotope exchange reactions. J. Chem. Phys. 15:261-267.

Bron, J. 1974. Model calculations of kinetic isotope effects in nucleophilic substitution reactions. Can. J. Chem. 52:903-909.

Buddenbaum, W. E., and Shiner, V. J., Jr. 1976. ^{13}C kinetic isotope effects and reaction coordinate motions in transition states for S_N2 displacement reactions. Can. J. Chem. 54:1146-1161.

Buddenbaum, W. E., and Yankwich, P. E. 1967. Theoretical calculation of kinetic isotope effects. Effect of reaction coordinate variations on the temperature-independent factor, TIF. J. Phys. Chem. 71:3136-3143.

Burgi, H. B., Dunitz, J. D., Lehn, J. M., and Wipff, G. 1974. Stereochemistry of reaction paths at carboxyl centers. Tetrahedron 30:1563-1572.

Carter, R. E., and Melander, L. 1973. Experiments on the nature of steric isotope effects. Adv. Phys. Organic Chem. 10:1-27.

Chiang, Y., Kresge, A. J., and Wiseman, J. R. 1976. Kinetics and mechanism of hydration of strained bridgehead bicyclic olefins. J. Am. Chem. Soc. 98: 1564-1566.

Collins, C. J., and Bowman, N. S. 1971. Isotope Effects in Chemical Reactions. ACS Monogr. 167, Van Nostrand Reinhold Co., New York.

Cremaschi, P., Gamba, A., and Simonetta, M. 1972. The influence of solvation on the calculated activation energy for the reaction $CH_3F + F^-$. Theor. Chim. Acta (Berl.). 25:237-247.

Dedieu, A., and Veillard, A. 1972. A comparative study of some S_N2 reactions through ab initio calculations. J. Am. Chem. Soc. 94:6730-6738.

Duncan, J. L., and Mallinson, P. D. 1973. The general harmonic force field of formaldehyde. Chem. Phys. Lett. 23:597-599.

Ehrenson, S. 1974. Application of analytical least motion forms to organic reactivities. J. Am. Chem. Soc. 96:3784-3793.

El-Sabban, M. Z., and Zwolinski, B. J. 1968. Normal coordinate treatment and assignment of fundamental vibrations of vinyl chloride and the seven deuterovinyl chlorides. J. Mol. Spectr. 27:1-16.

Fry, A. 1964. Application of the successive labelling technique to some carbon, nitrogen, and chlorine isotope effect studies of organic reaction mechanisms. Pure Appl. Chem. 8:409-419.

Glasstone, S., Laidler, K., and Eyring, H. 1941. The Theory of Rate Processes. McGraw-Hill, New York.

Hartshorn, S. R., and Shiner, Jr., V. J. 1972. Calculation of H/D, $^{12}C/^{13}C$, and $^{12}C/^{14}C$ fractionation factors from valence force fields derived for a series of simple organic molecules. J. Am. Chem. Soc. 94:9002-9012.

Hilderbrandt, R. L. 1969. Cartesian coordinates of molecular models. J. Chem. Phys. 51:1654–1659.

Holmes, R. R. 1967. Potential field and molecular vibrations of the trigonal bipyramidal model AX_3YZ: Pentacoordinated molecules. X. J. Chem. Phys. 46:3724-3729.

Howald, R. A. 1960. Ion pairs. I. Isotope effects shown by chloride solutions in glacial acetic acid. J. Am. Chem. Soc. 82:20-24.

Huang, T. T. S., Kass, W. J., Buddenbaum, W. E., and Yankwich, P. E. 1968. Anomalous temperature dependence of kinetic carbon isotope effects and the phenomenon of crossover. J. Phys. Chem. 72:4431-4446.

Ingold, C. K. 1953. Structure and Mechanism in Organic Chemistry. 1st Ed. Cornell University Press, Ithaca, N.Y.

Jencks, W. P. 1972. General acid-base catalysis of complex reactions in water. Chem. Rev. 72:705-718.

Johnston, H. S. 1960. Large tunnelling corrections in chemical reaction rates. Adv. Chem. Phys. 3:131-170.

Johnston, H. S. 1966. Gas Phase Reaction Rate Theory. Ronald Press, New York.

Johnston, H. S., Bonner, W. A., and Wilson, D. J. 1957. Carbon isotope effect during oxidation of carbon monoxide with nitrogen dioxide. J. Chem. Phys. 26:1002-1006.

Kass, W. J., and Yankwich, P. E. 1969. Three-element reaction coordinates and intramolecular kinetic isotope effects. J. Phys. Chem. 73:3722-3735.

Katz, A. M., and Saunders, Jr., W. H. 1969. Mechanisms of elimination reactions. XI. Theoretical calculations of isotope effects in elimination reactions. J. Am. Chem. Soc. 91:4469-4472.

Kleinman, L. I., and Wolfsberg, M. 1974a. Corrections to the Born-Oppenheimer approximation and electronic effects on isotopic exchange equilibria. II. J. Chem. Phys. 60:4740-4748.

Kleinman, L. I., and Wolfsberg, M. 1974b. Shifts in vibrational constants from corrections to the Born-Oppenheimer approximation: Effects on isotope exchange equilibria. J. Chem. Phys. 60:4749-4754.

Kresge, A. J., and Preto, R. J. 1967. The relative importance of inductive and steric effects in producing secondary hydrogen isotope effects on triphenyl cation formation. J. Am. Chem. Soc. 89:5510-5511.

Laidler, K., and Tweedale, A. 1971. The current status of Eyring's rate theory. Adv. Chem. Phys. 21:113-125.

Melander, L. 1960. Isotope Effects on Reactions Rates. Ronald Press, New York.

Monse, E. U., Kauder, L. N., and Spindel, W. 1964. Analysis of isotope exchange reactions among nitrogen oxides involving N_2O_3. J. Chem. Phys. 41:3898-3905.

Muller, A., and Nagarajan, G. 1967. Mean amplitudes of vibration in some ZXY_2 molecules and ions of C_{2v} symmetry. Z. Phys. Chim., L. 235:113-126.

More O'Ferrall, R. A. 1970. Relationships between E2 and ElcB mechanisms of β-elimination. J. Chem. Soc. B:274-277.

More O'Ferrall, R. A., and Kouba, J. 1967. Model calculations of primary hydrogen isotope effects. J. Chem. Soc. B:985-990.

Pauling, L. 1947. Atomic radii and interatomic distances in metals. J. Am. Chem. Soc. 69:542-553.

Pryor, W. A., and Kneipp, K. G. 1971. Primary kinetic isotope effects and the nature of hydrogen-transfer transition states. The reaction of a series of free radicals with thiols. J. Am. Chem. Soc. 93:5584-5586.

Raaen, V. F., Juhlke, T., Brown, F. J., and Collins, C. J. 1974. Do S_N2 reactions go through ion pairs? The isotope effect criterion. J. Am. Chem. Soc. 96:5928-5930.

Raff, L. M. 1974. Theoretical investigations of the reaction dynamics of polyatomic systems: Chemistry of the hot atom ($T^* + CH_4$) and ($T^* + CD_4$) systems. J. Chem. Phys. 60:2220-2244.

Schachtschneider, J. H., and Snyder, R. G. 1963. Vibration analysis of the *n*-paraffins-II. Normal coordinate calculations. Spectrochim. Acta 19:117-168.

Scherer, J. R., and Overend, J. 1960. Transferability of Urey-Bradley force constants. III. The vinylidene halides. J. Chem. Phys. 32:1720-1733.

Shiner, V. J., Jr. 1971. Deuterium isotope effects in solvolytic substitution at saturated carbon. In: C. J. Collins and N. S. Bowman (eds.), Isotope Effects in Chemical Reactions, pp. 90-159. Van Nostrand Reinhold Co., New York.

Shiner, V. J., Jr. 1975. Isotope effects and reaction mechanisms. In: P. A. Rock (ed.), Isotopes and Chemical Principles, pp. 163-183. ACS Symposium Series No. 11.

Shiner, V. J., Jr., and Buddenbaum, W. E. 1975. The role of mass spectroscopy in the study of heavy atom kinetic isotope effects. In: A. C. Maccoll (ed.), Mass Spectrometry. International Review of Science, Physical Chemistry, Series Two, Vol. 5, Ch. 4.

Shiner, V. J. Jr., and Seib, R. C. 1976. Influence of neighboring phenyl participation on the α-deuterium isotope effect on solvolysis rates. Neophyl esters. J. Am. Chem. Soc. 98:862-864.

Shiner, V. J., Jr., Buddenbaum, W. E., Murr, B. L., and Lamaty, G. 1968a. Effects of deuterium substitution on the rates of organic reactions. XI. α- and β-deuterium effects on the solvolysis rates of a series of substituted 1-phenylethyl halides. J. Am. Chem. Soc. 90:418-426.

Shiner, V. J., Jr., Rapp, M. W., Halevi, E. A., and Wolfsberg, M. 1968b. Solvolytic α-deuterium effects for different leaving groups. J. Am. Chem. Soc. 90:7171-7172.

Sims, L. B., Fry, A., Netherton, L. T., Wilson, J. C., Reppond, K. D., and Crook, S. W. 1972. Variations of heavy-atom kinetic isotope effects in S_N2 displacement reactions. J. Am. Chem. Soc. 94:1364-1365.

Stern, M. J., and Wolfsberg, M. 1966. Simplified procedure for the theoretical calculation of isotope effects involving large molecules. J. Chem. Phys. 45: 4105-4124.

Streitwieser, A., Jr., Jagow, R. H., Fahey, R. C., and Sukuki, S. 1958. Kinetic isotope effects in the acetolyses of deuterated cyclopentyl tosylates. J. Am. Chem. Soc. 80:2326-2332.

Thornton, E. K., and Thornton, E. R. 1971. Origin and interpretation of isotope effects. In: C. J. Collins and N. S. Bowman (eds.), Isotope Effects in Chemical Reactions, pp. 213-285. Van Nostrand-Reinhold Co., New York.

Van Hook, W. A. 1971. Kinetic isotope effects: Introduction and discussion of the theory. In: C. J. Collins and N. S. Bowman (eds.), Isotope Effects in

Chemical Reactions, pp. 1-89. Van Nostrand-Reinhold Co., New York.
Venkateswarlu, K., Jagatheesan, S., and Rajalakshmi, K. V. 1963. Molecular vibration of HCOF and DCOF. Proc. Indian Acad. Sci. A58:373-380.
Wei, G. J., and Yankwich, P. E. 1974. Calculation of heavy-atom kinetic isotope effects: Barrier curvature and internal coordinate effects with three-element reaction coordinates. J. Chem. Phys. 60:3619-3633.
Westheimer, F. H. 1961. The magnitude of the primary kinetic isotope effect for compounds of hydrogen and deuterium. Chem. Rev. 61:265-273.
Wiberg, K. B., and Ellison, G. B. 1974. Distorted geometries at carbon. Tetrahedron 30:1573-1578.
Willi, A. V., and Wolfsberg, M. 1964. The influence of "Bond Making and Bond Breaking" in the transition state on hydrogen isotope effects in linear three center reactions. Chem. and Ind. 2097-2098.
Winey, D. A., and Thornton, E. R. 1975. Elimination reactions. Deuteroxide/hydroxide isotope effects as a measure of proton transfer in the transition states for E2 elimination of 2-(*p*-trimethylammoniophenyl) ethyl onium ions and halides. Mapping of the reaction-coordinate motion. J. Am. Chem. Soc. 97:3102-3108.
Wolfsberg, M. 1969. Correction to the effect of anharmonicity on isotope exchange equilibrium. Adv. Chem. Ser. 89:185-191.
Wolfsberg, M. 1972. Theoretical evaluation of experimentally observed isotope effects. Acct. Chem. Res. 5:225-233.
Wolfsberg, M., and Stern, M. J. 1964a. Validity of some approximation procedures used in the theoretical calculation of isotope effects. Pure Appl. Chem. 8:225-242.
Wolfsberg, M., and Stern, M. J. 1964b. Secondary isotope effects as probes for force constant changes. Pure Appl. Chem. 8:325-338.
Won, C. M., and Willi, A. V. 1972. Kinetic deuterium isotope effects in the reactions of methyl iodide with azide and acetate ions in aqueous solution. J. Phys. Chem. 76:427-432.

Magnitude of Primary Hydrogen Isotope Effects

A. J. Kresge

It is now well understood that kinetic isotope effects originate in isotopically sensitive zero-point energy differences between initial states and transition states. In the case of primary isotope effects, it seems natural to identify this difference with the loss of zero-point energy which accompanies the conversion of vibrational into translational motion as the reacting bond breaks. This picture of primary isotope effects leads to the well known estimate of $k_H/k_D = 6$ for the primary kinetic isotope effect on C—H bond breaking: the zero-point energy of a typical C—H stretching vibration with $\nu = 3{,}000\ \mathrm{cm}^{-1}$ is 4.3 kcal/mol; deuterium substitution reduces the frequency to $2{,}200\ \mathrm{cm}^{-1}$ and the zero-point energy to 3.2 kcal/mol; and the difference, $\Delta E_0 = 1.1$ kcal/mol, assuming no compensating effect in the transition state, gives $k_H/k_D = \exp(\Delta E_0/RT) = 6.4$.

This model of primary hydrogen isotope effects is attractive in its simplicity, but unfortunately it does not account for the well known wide variation in the magnitude of these effects. It predicts, for example, that reactions of the same C—H bond in a given substrate with different reagents will always give the same isotope effect, but, as the pairs of entries in Table 1 show, this is not the case. The model also restricts to a relatively narrow range the variation among isotope effects for different C—H bonds in different substrates, inasmuch as the range of variation in C—H stretching frequencies of ordinary stable molecules is itself small. With $2{,}800\ \mathrm{cm}^{-1}$ and $3{,}300\ \mathrm{cm}^{-1}$ as reasonable limits for this vibration, the model gives $k_H/k_D = 5\text{-}8$; but, as the examples of

Table 1. Examples of primary hydrogen isotope effects of variable magnitude

Reaction	T(°C)	k_H/k_D	Reference
$C_6H_5CH_3 + Br \rightarrow C_6H_5CH_2 + HBr$	77°	4.6	a
$C_6H_5CH_3 + Cl \rightarrow C_6H_5CH_2 + HCl$	77°	1.3	a
$(CH_3)_2CHNO_2 \xrightarrow{HO^-} (CH_3)_2CNO_2^-$	25°	7.4	b
$(CH_3)_2CHNO_2 \xrightarrow{\text{2,6-lutidine}} (CH_3)_2CNO_2^-$	25°	19.5	b
$p\text{-}NO_2C_6H_4CH_2NO_2 \xrightarrow{Et_3N} p\text{-}NO_2C_6H_4CHNO_2^-$	25°	11.0	c
$p\text{-}NO_2C_6H_4CH_2NO_2 \xrightarrow{TMG^a} p\text{-}NO_2C_6H_4CHNO_2^-$	25°	45.0	c

[a] TMG, tetramethyl guanidine.
a) Wiberg and Slaugh, 1958; b) Bell and Goodall, 1966; c) Caldin and Mateo, 1973.

Table 1 show, primary hydrogen isotope effects on C—H bond breaking can range down to essentially unity, and examples as great as k_H/k_D = 45 are also known.

The problem of accounting for this large variation in the magnitude of experimentally observed hydrogen isotope effects may conveniently be divided into two parts 1) large isotope effects and 2) small isotope effects. It was recognized quite early in the history of the subject that proton tunneling could magnify the size of hydrogen isotope effects and lead to anomalously large values. R. P. Bell (1933) was among the first to point this out, and he has continued to be a leader in ongoing research into the scope and importance of this effect. A reasonable explanation for small isotope effects did not come until considerably later when Melander (1960) and Westheimer (1961) drew attention to the fact that an unsymmetrical three-center transition state for hydrogen transfer will have additional isotopically sensitive zero-point energy which will serve to offset that of the initial state and thus reduce the isotope effect. It was quickly realized that this explanation requires isotope effects to change systematically with transition state structure and with the free energy of reaction, and to pass through a maximum near $\Delta G° = 0$. No clear evidence for an isotope effect maximum was available at the time, but the first example was quickly provided (Kresge, 1965; Longridge and Long, 1967) and the phenomenon is now well documented (More O'Ferrall, 1975). Other treatments of the isotope effect maximum have also been proposed, a particularly useful one being the phenomenological aproach of Marcus rate theory (Marcus, 1968) which provides an analytical expression relating k_H/k_D to $\Delta G°$.

In this chapter, the tunnel effect and the Melander-Westheimer treatment are discussed. Marcus rate theory is then described and its application to isotope effects is illustrated with several sets of data for hydrogen transfer involving carbon. Finally, the theory is applied to some recent work on proton transfer from nitrogen to oxygen, and the information thus obtained is shown to give new insight into the general paucity of large isotope effects on proton transfer between electronegative atoms.

LARGE ISOTOPE EFFECTS

Tunnel Effect

According to classical mechanics, the reactants in a chemical process must pass over the top of an energy barrier in order to be converted into products. Only those systems whose energy is equal at least to the top of the barrier can react; others will be reflected back. Quantum mechanics, on the other hand, admits the possibility that systems with insufficient energy to surmount the barrier may leak or "tunnel" through.

The tunnel effect is significant only for particles whose deBroglie wavelength $\lambda = h/\text{mv}$ is comparable to the thickness of the barrier that must be penetrated. For most atoms at usual velocities, therefore, it is of no consequence. Protons at ordinary temperatures, however, have wavelengths of the order of 1-2Å, which is similar to the width expected for the energy barrier to a proton transfer reaction. The tunnel effect should therefore have some influence on the rates of proton transfer reactions.

Although this has been known for some time and a good deal of both experimental and theoretical work has been done on the problem, it is still not certain just how widespread proton tunneling really is. A good part of the difficulty is knowing whether or not a given reaction is faster than it would be classically, and much effort has gone into developing other, less direct, methods for detecting tunneling.

These need not concern us here; the interested reader is referred to several excellent reviews that are now available (Bell, 1973, pp. 270–289, 1974; Caldin, 1969; Lewis, 1975). It is sufficient for our purposes here to note that the deBroglie wavelength of deuterium, because of its greater mass, will be shorter than that of protium, and as a consequence there will be less tunneling in reactions of the heavier isotope. This will augment the isotope effect, making k_H/k_D greater that it would otherwise be. The mass dependence, moreover, enters into the expression for the tunnel correction to the rate constant in a complicated way which amplifies the effect of the mass difference, especially at low temperatures.

In addition to low temperatures, tunneling is favored by high, thin barriers, and steric hindrance appears to promote tunneling as well. It was once thought that steric hindrance operated by increasing the barrier height, and especially its steepness, but there is now evidence that solvent exclusion is responsible for at least a part of the effect. Tunneling is also most likely to occur in reactions with symmetrical transition states, i.e., near $\Delta G^\circ = 0$. That is where protonic motion makes its greatest contribution to the reaction coordinate and it is also where that part of the barrier which lies above initial and final states—and is therefore available for tunneling—is greatest.

Bending Vibrations

It was implied above that the maximum value of k_H/k_D for breaking a C—H bond with $\nu = 3{,}000\ \text{cm}^{-1}$ in the absence of tunneling is 6.4. This estimate was based upon the stretching vibration of this bond alone, but C—H bonds have bending as well as stretching modes and the bending vibrations are also isotopically sensitive. Loss of bending zero-point energy will raise the isotope effect appreciably; for example, the isotope effect corresponding to complete loss of the zero-point energy of one stretching vibration with $\nu = 3{,}000\ \text{cm}^{-1}$ and two bending vibrations at $\nu = 1{,}400\ \text{cm}^{-1}$ is $k_H/k_D = 18$.

This isotope effect could only be realized, however, if there were no compensating isotopically sensitive bending vibrations in the transition state, and that could come about only if there were essentially no restoring force to the lateral as well as to the forward motion of the particle being transferred. Such a loosely bound transition state seems intuitively unlikely, and there is also some experimental evidence to support this. It has been found that the hydrolysis of several vinyl ethers, which is known to occur by rate-determining proton transfer from the catalyzing acid to the substrate, shows a much smaller isotope effect when the catalyst is HF than when it is a carboxylic acid of similar strength (Kresge and Chiang, 1969; Kresge, Chen, and Chiang, 1977). Because HF has no bending vibrations but carboxylic acids do, this difference implies that the transition states for both acids have strong bending vibrations themselves. In the case of the carboxylic acid, these are offset by initial-state bending vibrations, and a normal isotope effect results. In the case of HF, however, no compensation is possible, and the isotope effect is reduced. A quantitative treatment gives two transition-state bending frequencies of 1,200–1,300 cm^{-1}, in good agreement with the bending vibrational frequency of the FHF^- ion, $\nu = 1{,}206\ cm^{-1}$ (Jones and Penneman, 1954; Newman and Badger, 1951). Moreover, potential energy surfaces calculated for a number of simple hydrogen atom transfer reactions as well as several electrostatic models for proton transfer processes all give transition-state bending frequencies in the range 750–1,500 cm^{-1} (Bell, 1974).

If, following Bell (1974), 750 cm^{-1} is taken as the maximum total decrease in isotopically sensitive bending frequencies that can occur upon formation of the transition state, then the maximum possible isotope effect at 25° with no tunneling becomes $k_H/k_D = 10$. Using this criterion, Bell lists some 30 hydrogen transfer reactions in solution whose isotope effects are too big not to have significant contributions from tunneling. The values of k_H/k_D for these reactions go as high as 45 at 25°, and there is one value of 250 at $-52°$.

Solvent Effects on Tunneling

One group of isotope effects in Bell's list is especially interesting, for it shows proton tunneling to be quite sensitive to changes in the reaction medium. The data, provided by Caldin and Mateo (1975), are for proton transfer from *p*-nitrophenylnitromethane to tetramethylguanidine to form an ion pair (equation 1) in a series of aprotic solvents.

$$NO_2C_6H_4CH_2NO_2 + NH{=}C(NMe_2)_2 \rightarrow NO_2C_6H_4CHNO_2^- \cdot NH_2C(NMe_2)_2^+ \quad (1)$$

Some of the values are reproduced in Table 2. It may be seen that as polar solvents such as acetonitrile are replaced by the less polar hydrocarbons mesitylene and toluene, the isotope effect rises from ca. 10 to 45. It should be noted that $\Delta G°$ is close to zero for all of these reactions, which is a condition conducive to tunneling; the near constancy of $\Delta G°$

Table 2. Isotope effects on the reaction of 4-nitrophenylnitromethane with tetramethylguanidine at 25° in various solvents[a]

Solvent	k_H/k_D	$\Delta G_H°$	D[b]
Acetonitrile	12	4.8	38.0
Dichloromethane	11	5.2	9.1
Tetrahydrofuran	13	5.0	7.4
Mesitylene	31	2.8	2.3
Toluene	45	3.0	2.4

[a] Caldin and Mateo, 1975.
[b] Dielectric constant.

also indicates that the change in isotope effect is unrelated to changes in transition-state symmetry. By analyzing these isotope effects and their temperature dependence, Caldin and Mateo were able to calculate tunnel corrections, Q_H, to the rate constant k_H, which rose from 2.6 for the reaction in acetonitrile solution to 28 for that in toluene (the corresponding change in Q_D was 1.6–4.3).

This striking change in tunnel effect is thought to be caused by a difference in the extent of solvent reorganization that accompanies the reaction in the two different, i.e., polar and nonpolar, media. As the proton is transferred, charges are developed on the ion-pair product. Polar solvent molecules will reorient themselves to stabilize these developing charges, and the protonic motion, therefore, will be coupled with motion of the solvent. This will increase the effective mass of the system for passage across the energy barrier, and that will decrease its deBroglie wavelength and reduce the amount of tunneling. In nonpolar solvents, on the other hand, there will be little tendency for the solvent molecules to follow charge generation; protonic motion, therefore, will not be coupled with solvent motion, and the tunnel effect will not be reduced. This explanation is supported by the fact that the reaction is more rapid in polar than in nonpolar solvents. The effective masses of the particle being transferred, moreover, which Caldin and Mateo were able to calculate for the H transfer reaction, were uniformly unity for the nonpolar solvents and significantly greater than unity for the nonpolar media.

These results suggest a reason for the increased incidence of tunneling in sterically hindered systems, such as proton transfer from 2-nitropropane to 2,6-lutidine (equation 2).

$$(CH_3)_2CHNO_2 + \text{2,6-}(CH_3)_2C_5H_3N \rightarrow (CH_3)_2CNO_2^- + \text{2,6-}(CH_3)_2C_5H_3NH^+ \quad (2)$$

The isotope effect here, $k_H/k_D = 20$ in water (Bell and Goodall, 1966) and $k_H/k_D = 24$ in a water-*t*-butanol mixture (Funderburk and Lewis, 1964; Lewis and Funderburk, 1967), is twice that for reaction of the same proton donor with pyridine itself. The role played by the methyl groups that flank the proton accepting site in 2,6-lutidine may be to exclude polar solvent molecules from the reaction vicinity. This creates an immobile nonpolar microenvironment for the reaction, which promotes tunneling by reducing the coupling between protonic and solvent motions.

SMALL ISOTOPE EFFECTS

Melander-Westheimer Effect

The explanation of small isotope effects offered by Melander (1960) and Westheimer (1961) is based upon a three-center model for the transition state, i.e., it assumes the presence of an acceptor, B, for the hydrogen being transferred (equation 3).

$$\text{A—H} + \text{B} \rightarrow [\text{A---H---B}]^{\ddagger} \rightarrow \text{A} + \text{HB} \quad (3)$$

This is certainly a valid assumption for proton transfer reactions in solution, inasmuch as bare protons cannot exist in ordinary solvents and the proton, therefore, must be transferred over to some acceptor molecule. Acceptors probably also take part in most, if not all, hydrogen atom and hydride ion transfers.

When such a three-center transition state is formed, new vibrational modes are created. Consider, for example, the case in which A and B consist of single atoms. The diatomic molecule A—H will then have only one normal vibration and B will have none, but the triatomic transition state will have three if it is bent and four if it is linear. One of these transition-state modes will, of course, be the "imaginary vibration," or translation, that corresponds to the reaction coordinate, but in each case some new vibrations will have been created. In the general situation where both A and B are polyatomic, five new vibrations (or internal rotations) plus the reaction coordinate will be formed.

Some of these new vibrations will be isotopically sensitive. They, therefore, will supply isotope-dependent zero-point energy to the transition state, which will offset that of the initial state and reduce the isotope effect. Consider again the triatomic case in which the arrangement of atoms is linear. The four normal vibrations will consist of an asymmetric stretch, ν_a, a symmetric stretch, ν_s, and two orthogonal bends, ν_b.

$$\nu_a: \overset{\leftarrow}{\text{A}}\text{—}\overset{\rightarrow}{\text{H}}\text{—}\overset{\leftarrow}{\text{B}} \qquad \nu_b: \overset{\uparrow}{\text{A}}\text{—}\overset{\downarrow}{\text{H}}\text{—}\overset{\uparrow}{\text{B}}$$

$$\nu_s: \overset{\leftarrow}{\text{A}}\text{—}\overset{?}{\text{H}}\text{—}\overset{\rightarrow}{\text{B}} \qquad \nu_b: \overset{+}{\text{A}}\text{—}\overset{-}{\text{H}}\text{—}\overset{+}{\text{B}}$$

The asymmetric stretch can be identified with the reaction coordinate, which does not contribute zero-point energy. The isotopically sensitive zero-point energy of the bends, as shown above, will be offset largely or completely by that of initial-state bending vibrations in all but diatomic reactants. The symmetrical stretching vibration, however, has no initial state counterpart, and any isotopically sensitive zero-point energy that it can generate will diminish the isotope effect.

The isotopic sensitivity of this symmetrical stretching vibration will depend upon the relative values of the force constants of the two partial bonds, f_1 and f_2. When these are equal, the end groups will move in and out without affecting the position of the central atom; replacing H by D, therefore, will not change the moving mass, and ν_H will be the same as ν_D. When the force constants are not equal, on the other hand, the central atom will be pulled along by the motion of the end group to which it is bound by the stronger bond; replacing H by D will now raise the moving mass, and ν_H consequently will be greater than ν_D.

To summarize:

$$f_1 = f_2: \quad \overleftarrow{\mathrm{A}} \underset{f_1}{-} \mathrm{H} \underset{f_2}{-} \overrightarrow{\mathrm{B}} \qquad \nu_H = \nu_D$$

$$f_1 > f_2: \quad \overleftarrow{\mathrm{A}} \underset{f_1}{-} \overleftarrow{\mathrm{H}} \underset{f_2}{-} \overrightarrow{\mathrm{B}} \qquad \nu_H > \nu_D$$

$$f_1 < f_2: \quad \overleftarrow{\mathrm{A}} \underset{f_1}{-} \overrightarrow{\mathrm{H}} \underset{f_2}{-} \overrightarrow{\mathrm{B}} \qquad \nu_H > \nu_D$$

This description thus predicts that hydrogen isotope effects will be greatest when the partial bonds holding the hydrogen being transferred to the donor and to the acceptor are of equal strength, i.e., when the hydrogen is half transferred. Any deviation from this symmetrical situation in either direction will make the isotope effect smaller. Because the strengths of the two partial bonds in a three-center transition state can be expected to vary in a regular way as the energetics of the reaction change, with the symmetrical situation occurring near $\Delta G^\circ = 0$, this implies that isotope effects will be small when ΔG° is appreciably negative, will pass through a maximum as ΔG° goes through zero, and will diminish again as ΔG° becomes appreciably positive.

There are now a number of examples of such behavior (More O'Ferrall, 1975), the most extensively documented being isotope effects on the ionization of some 100 carbonyl and nitro compounds.[1] The data are displayed in Figure 1. The line drawn there is that provided by least

[1] The substrates include ethyl malonate, acetylacetone, ethyl acetoacetate, and their derivatives (Bell and Crooks, 1965), menthone (Bell and Cox, 1970), tricarbomethoxymethane, acetone-1-sulfonate ion, 2-acetylcyclohexanone, and ethyl nitroacetate (Barnes and Bell, 1970), nitromethane (Reitz, 1936), nitroethane (Bell and Goodall, 1966; Bell, Sachs, and Tranter, 1971; Bell and Cox, 1971; Dixon and Bruice, 1967), 2-nitropropane and nitrocyclohexane (Bell and Goodall, 1966), 1-phenylnitroethane and its ring-substituted derivatives (Bordwell and Boyle, 1975) and phenylnitromethane (Keefe and Munderloh, 1974).

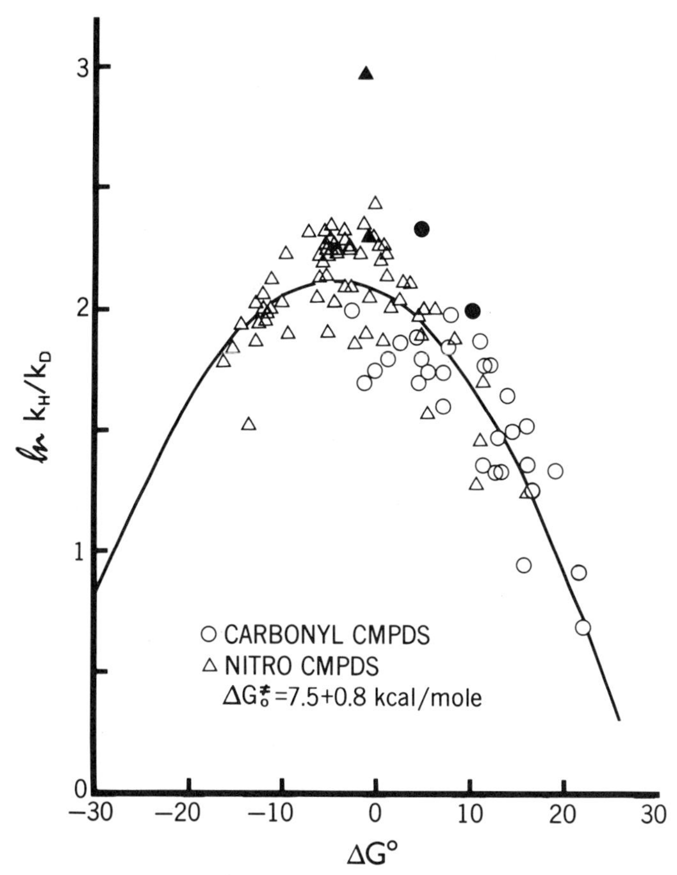

Figure 1. Isotope effect maximum for the ionization of carbonyl and nitro compounds; filled points are for 2,6-lutidine as the proton acceptor.

squares fit to an expression given by Marcus theory, which, together with other examples of isotope effect maxima, will be taken up in the following section.

Before moving on to that, however, it should be mentioned that serious objections have been lodged against this explanation of isotope effect maxima, principally by Bell. Bell (1965) has noted that the symmetric stretching vibration in the three-center transition-state model shows only a weak isotopic dependence for even quite unsymmetric transition states when realistic values of a crucial interaction force constant are used in the calculation; this leads to a very broad, flat-topped isotope effect maximum, quite unlike those actually observed. More detailed calculations by Willi and Wolfsberg (1964) lead to the same conclusion, as does also a recently proposed electrostatic charge-cloud model for the proton transfer process (Bell, Sachs, and Tranter, 1971). In fact, the inability of this charge-cloud model to account for the variation in k_H/k_D provided by the data on carbonyl and nitro compound ionization available at the time (roughly half of the examples shown in Figure 1) led Bell to suggest that the variation is entirely the result of a changing tunnel effect. Inclusion of a tunnel correction in the calculation of k_H/k_D based upon this charge-cloud model accounted for the data very nicely.

This charge cloud model, on the other hand, uses internuclear distances that appear to be improbably large. For example, the distances specified for the shorter of the two symmetrical transition states for which data are given lead, via the Pauling (1947) relationship, to bond orders of 0.05 for both the breaking and the forming bonds, whereas the values expected for this situation are 0.50 for each bond. The tunnel correction that Bell uses in these calculations, moreover, when applied to the isotope effect on proton transfer from HF to vinyl ethers, which was discussed above, gives the unrealistically large value of $\nu_H = 1{,}935$ cm^{-1} for the two transition-state bending vibrations (Kresge, unpublished calculations).

Marcus Theory

Marcus rate theory provides a useful formalism for treating proton and hydrogen atom transfer reactions (Marcus, 1968; Cohen and Marcus, 1968). It was first developed for outer-sphere electron transfers in solution (Marcus, 1956, 1964), where it seems to have some foundation in theory. Its basic relationship can be derived from the Leffler principle and the Hammond postulate (Murdoch, 1972) as well as from simple solvent-polarization (German et al., 1971) and intersecting-parabolae (Koeppl and Kresge, 1973) models for the proton transfer process. It is, moreover, the simplest relationship between rate and equilibrium constants which can give a curved Brønsted relation. More sophisticated models such as the BEBO method (Kresge and Koeppl, unpublished work) and intersecting parabolae with variable curvature and changing distance of separation (Koeppl and Kresge, 1973) give important quantitative differences from simple Marcus theory. Nevertheless, the simple Marcus expression (equation 5 below) does provide a valuable framework for interpreting and understanding experimental results.

Marcus separates the hydrogen transfer process into three discrete steps: encounter of the reactants, hydrogen transfer, and separation of the products (equation 4).

$$\mathrm{AH} + \mathrm{B} \xrightarrow[\text{encounter}]{w^r} \mathrm{AH}\cdot\mathrm{B} \xrightarrow[\text{transfer}]{\Delta G_R{}^\circ} \mathrm{A}\cdot\mathrm{HB} \xrightarrow[\text{separation}]{-w^p} \mathrm{A} + \mathrm{HB} \quad (4)$$

The energy required to bring the reactants together is called the work term w^r, with w^p being the corresponding quantity for the products, and the standard free energy of reaction within the encounter complex is $\Delta G_R{}^\circ$. The barrier to hydrogen transfer within this complex, $\Delta G_R{}^\ddagger$, is then a simple quadratic function of $\Delta G_R{}^\circ$ and one other variable, $\Delta G_0{}^\ddagger$ (equation 5).

$$\Delta G_R{}^\ddagger = \Delta G_0{}^\ddagger (1 + \Delta G_R{}^\circ/4\Delta G_0{}^\ddagger)^2 \quad (5)$$

The latter may be seen to be the barrier to transfer within the complex when $\Delta G_R{}^\circ = 0$; as such, it is an "intrinsic" or purely kinetic barrier free of any thermodynamic drive or impediment, which makes it an important kinetic parameter especially useful for characterizing chemical reactions (Kresge, 1975).

It is possible, by fitting experimentally determined values of $\Delta G^\ddagger$ and ΔG° to a quadratic expression, to obtain values of $\Delta G_0{}^\ddagger$, w^r, and w^p for the system, and a number of proton transfer reactions have now been treated in this way (Kresge, 1973). Some of the results obtained are summarized in Table 3. It may be seen that systems in which the proton transfer is accompanied by extensive structural reorganization and charge delocalization, such as aromatic protonation or carbonyl compound ionization, have large values of $\Delta G_0{}^\ddagger$; these proton transfers

Table 3. Marcus theory parameters for some proton transfer reactions

Reaction	$\Delta G_0{}^{\ddagger a}$	$w^{r\,a}$	Reference
1. Aromatic protonation	10	10	a
2. Ionization of carbonyl compounds	8	6	b
3. Dehydration of acetaldehyde hydrate	5	13	c
4. Protonation of diazo compounds	1-5	8-14	d
5. Proton transfer from acetic acid or phenol to "normal" oxygen and nitrogen bases	2	3	e

[a] kcal/mol.

a) Kresge et al., 1971; b) Bell, 1973, p. 203; c) Bell and Higginson, 1949; d) Kreevoy and Konasewich, 1971; Kreevoy and Oh, 1973; Albery, Campbell-Crawford, and Curran, 1972; e) Eigen, 1963, 1964; Ahrens and Maass, 1968.

are thus intrinsically slow, as expected. Proton transfer from normal (as opposed to pseudo) acids to normal bases, on the other hand, has a low value of $\Delta G_0{}^{\ddagger}$, again in keeping with the very fast, essentially diffusion-controlled nature of these processes. The work terms of Table 3 are also interesting. Those for which the proton transfer is to or from carbon are all quite large, of the order of 10 kcal/mol. This has been interpreted as meaning that w^r includes not only the energy of localizing the reactant molecules together, but also the work required to desolvate them and form them into a reaction complex. In these carbon systems, which include a normal acid or base in addition to the carbon base or acid partner, this desolvation will consist principally of sacrificing one hydrogen bond between the normal reactant and the solvent (Kreevoy and Konasewich, 1971; Kreevoy and Oh, 1973). Proton transfer between normal acids and bases, on the other hand, is known to occur by proton jumps down a solvent bridge (Eigen, 1964); desolvation, therefore, is not necessary and w^r is thus much smaller.

Isotope Effect Correlations Using Energies of Reaction

Marcus theory predicts that the barrier to an exothermic reaction will increase as the exothermicity decreases; it predicts also that the barrier to an endothermic reaction with the endothermicity subtracted off, i.e., $\Delta G^{\ddagger} - \Delta G°$, will decrease as the endothermicity increases. This is illustrated by the figures presented in Table 4, which were obtained with the aid of equation 5 using $\Delta G_0{}^{\ddagger} = 5$ kcal/mol. Differences in reaction rate because of differences in intrinsic reactivity, i.e., in $\Delta G_0{}^{\ddagger}$ as opposed to $\Delta G_R°$, therefore, will be greatest when $\Delta G_R° = 0$, and it follows from

Table 4. Solutions of the Marcus equation with $\Delta G_0{}^{\ddagger} =$ 5 kcal/mol

$\Delta G_R°$	$\Delta G_R{}^{\ddagger}$	$\Delta G_R{}^{\ddagger} - \Delta G_R°$
−20	0.00	—
−15	0.31	—
−10	1.25	—
−5	2.81	—
0	5.00	5.00
+5	7.81	2.81
+10	11.25	1.25
+15	15.31	0.31
+20	20.00	0.00

this that changes made to a system which affect the rate without having much influence upon the equilibrium constant will be felt most strongly near $\Delta G^\circ = 0$. Isotopic substitution, of course, generally has a much greater effect on rate than on equilibrium constants: equilibrium isotope effects are usually quite small (see, e.g., More O'Ferrall, 1975), and Marcus theory therefore predicts that isotope effects will vary systematically with changes in $\Delta G_R{}^\circ$, passing through a maximum at $\Delta G_R{}^\circ = 0$.

A quantitative relationship between the isotope effect and $\Delta G_R{}^\circ$ may be obtained by writing two Marcus theory expressions, one for *H* transfer (equation 6) and one for *D* transfer (equation 7)

$$\Delta G_H{}^\ddagger = w_H{}^r + \Delta G_{0,H}{}^\ddagger (1 + \Delta G_{R,H}{}^\circ / 4\Delta G_{0,H}{}^\ddagger)^2 \tag{6}$$

$$\Delta G_D{}^\ddagger = w_D{}^r + \Delta G_{0,D}{}^\ddagger (1 + \Delta G_{R,D}{}^\circ / 4\Delta G_{0,D}{}^\ddagger)^2 \tag{7}$$

and then subtracting one from the other (Marcus, 1968). The result reduces to a particularly simple expression (equation 8)

$$\Delta G_H{}^\ddagger - \Delta G_D{}^\ddagger = (\Delta G_{0,H}{}^\ddagger - \Delta G_{0,D}{}^\ddagger) \left[1 - \frac{(\Delta G_R{}^\circ/4)^2}{\Delta G_{0,H}{}^\ddagger \Delta G_{0,D}{}^\ddagger} \right] \tag{8}$$

on two assumptions: 1) that $w_H{}^r = w_D{}^r$ and 2) that $\Delta G_{R,H}{}^\circ = \Delta G_{R,D}{}^\circ = \Delta G_R{}^\circ$. This is equivalent to assuming that there will be no isotope effect on formation of the reaction complex nor on the free energy change within this complex. This will certainly not be generally correct, but the isotope effects on the equilibria represented by w^r and $\Delta G_R{}^\circ$ usually will be much smaller than the kinetic isotope effects corresponding to ($\Delta G_H{}^\ddagger - \Delta G_D{}^\ddagger$) and ($\Delta G_{0,H}{}^\ddagger - \Delta G_{0,D}{}^\ddagger$), and at the present level of approximation, these would seem to be tolerable assumptions. Expressing the free energy differences of equation 8 as isotope effects and recognizing that $\Delta G_{0,H}{}^\ddagger - \Delta G_{0,D}{}^\ddagger$ corresponds to the maximum isotope effect, then gives equation 9, in which $\Delta G_0{}^\ddagger = \sqrt{\Delta G_{0,H}{}^\ddagger \Delta G_{0,D}{}^\ddagger}$.

$$\ln(k_H/k_D) = \ln(k_H/k_D)_{\max} [1 - (\Delta G_R{}^\circ / 4\Delta G_0{}^\ddagger)^2] \tag{9}$$

This equation predicts that $\ln(k_H/k_D)$ will have a parabolic dependence upon $\Delta G_R{}^\circ$ and that the isotope effect will attain its maximum value at $\Delta G_R{}^\circ = 0$. It states also that the curvature of the parabola, or the rapidity with which the isotope effect passes through its maximum, will depend upon the value of $\Delta G_0{}^\ddagger$, i.e., upon the geometric mean of the intrinsic barriers for H and D transfer. Systems with small intrinsic barriers will show relatively sharp isotope effect maxima, while those with large values of $\Delta G_0{}^\ddagger$ will have broader maxima and consequently will require greater changes in $\Delta G_R°$ to show significant isotope effect variation.

Figure 1 shows that this model fits the data for the ionization of some 100 carbonyl and nitro compounds reasonably well. There are some deviations, such as the striking one provided by the very large isotope effect, $k_H/k_D = 20$, on the ionization of 2-nitropropane when 2,6-lutidine is the proton acceptor, but in this case there is clear evidence of proton tunneling. Least squares analysis of the data, leaving out this and three other cases in which 2,6-lutidine is the catalyst, gives $\Delta G_0^‡ = 8.1 \pm 0.4$ kcal/mol, which, through $(k_H/k_D)_{max}$, may be converted into an intrinsic barrier for H transfer of $\Delta G_{0,H}^‡ = 7.5 \pm 0.8$ kcal/mol. This is in excellent agreement with the value of 8 kcal/mol listed in Table 3, which was obtained from the curvature of a Brønsted plot using carbonyl compounds only. The value of the maximum isotope effect obtained by this analysis is $k_H/k_D = 8.4 \pm 0.2$, which again is a reasonable value for this system; it may be taken to indicate that, in addition to loss of the zero-point energy of an initial-state stretching vibration with $\nu = 3{,}000$ cm^{-1}, there is some small reduction, ca. 400 cm^{-1}, in total bending frequencies between initial and transition states.

It is interesting that the isotope effect maximum in this system occurs slightly but significantly below $\Delta G° = 0$ at $\Delta G° = -4.4 \pm 0.7$ kcal/mol. This is because the analysis was necessarily carried out using measured or overall free energies of reaction, $\Delta G° = w^r + \Delta G_R° - w^p$, rather than (unaccessible) values of $\Delta G_R°$. Using the condition that the maximum occurs at $\Delta G_R° = 0$, it follows that $w^p - w^r = 4.7 \pm 0.7$ kcal/mol, which means that the work performed in assembling a reaction complex for the reverse reaction using the reaction products as reactants exceeds that for the forward process by about 5 kcal/mol. This is understandable in terms of the fact that all of the carbonyl and nitro compound proton donors and most of their reaction partner proton acceptors were neutral substances. The reactants in the backward process, therefore, were charged ions whose desolvation attending formation of a reaction complex would likely require additional energy.

Another example of a correlation using equation 9 is shown in Figure 2. The system this time is a free radical reaction (equation 10) transfer of a hydrogen atom from one of several thiols to a series of free radicals in solution (Pryor and Kneipp, 1971).

$$RSH + R'\cdot \rightarrow RS\cdot + HR' \qquad (10)$$

The analysis this time uses heats of reaction, ΔH, estimated from bond dissociation energies of the reactants and products. The data are not as

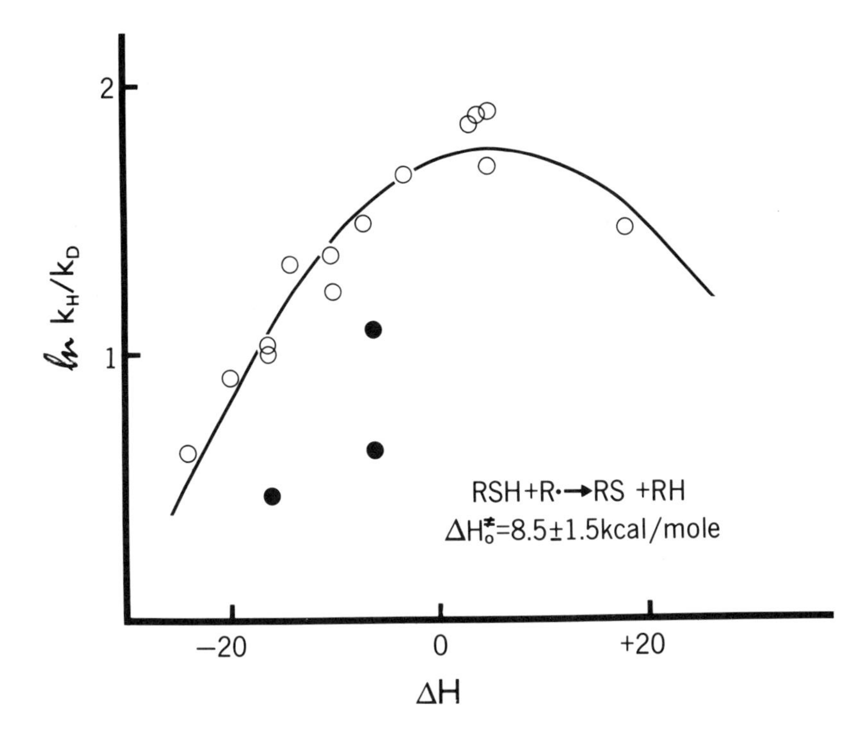

Figure 2. Isotope effect maximum for hydrogen atom transfer from thiols to a series of free radicals.

extensive as those of Figure 1, but they are not inconsistent with the model: least squares analysis excluding the three points shown as filled circles in the figure gives $\Delta H_{0,H}{}^{\ddagger}$ = 8.5 ± 1.5 kcal/mol and k_H/k_D = 5.8 ± 0.3. Two of the points left out are for reactions in which cyclohexyl and 1-adamantyl radicals are the hydrogen atom acceptors, where, as the authors suggest, unusual steric factors may be operative. In the third deviant reaction, a hydrogen atom is the acceptor, and that probably could not be expected to fit in with a series in which all of the other acceptors are large organic radicals. The low value of $(k_H/k_D)_{max}$ obtained in this correlation is very likely a result of the low frequency of the S—H stretching vibration in thiols.

These results, as well as additional values for several other systems, are summarized in Table 5. In some of these additional examples the limited data available indicated only an upward or downward trend in isotope effect rather than a clear maximum; nevertheless, the model

Table 5. Results of some isotope effect correlations using Marcus theory

Reaction	$\Delta G_{0,H}^{\ddagger}$ (kcal/mol)	$(k_H/k_D)_{max}$	w^r (kcal/mol)	Reference
Equilibrium correlations				
1. Ionization of carbonyl and nitro compounds	8	8	—	a
2. Hydrogen atom transfer from thiols to organic radicals	8*	6	—	b
3. Protonation of diazoacetate ion by trialkylammonium salts	2	9	—	c
4. Diazo-coupling of *p*-chlorophenyl diazonium ion with 2-naphthol-6,8-disulfonic acid	4	8	—	d
5. Diazo-coupling of *m*-chlorophenyl diazonium ion with 2-naphthol-6,8-disulfonic acid	6	11	—	d
6. Diazo-coupling of *p*-nitrophenyl diazonium ion with 2-naphthol-6,8-disulfonic acid	4	5	—	d
7. Diazo-coupling of 2-OH-4-$SO_3H—C_6H_3N_2^+$ with 1-naphthol-2-sulfonic acid	2	5	—	d
Rate correlations				
8. Ionization of carbonyl and nitro compounds	7	8	10	a
9. Protonation of vinyl ethers by hydronium ion	4	4	10	e

*$\Delta H_{0,H}^{\ddagger}$

a) Barnes and Bell, 1970; Bell and Cox, 1970, 1971; Bell and Crooks, 1965; Bell and Goodall, 1966; Bell, Sachs, and Tranter, 1971; Bordwell and Boyle, 1975; Dixon and Bruice, 1967; Keefe and Munderloh, 1974; Reitz, 1936; b) Pryor and Kneipp, 1971; c) Kreevoy and Konasewich, 1971; d) Hanna et al., 1974; e) Kresge and Sagatys, in preparation.

yielded reasonable values of intrinsic barriers and maximum isotope effects.

Isotope Effect Correlations Using Energies of Activation

Equilibrium constants for a series of reactions on which isotope effects have been measured are not always available, and therefore correlations involving $\Delta G°$ cannot always be made. In some such cases equilibrium constants for related reactions are known, such as acidity or basicity constants for acidic or basic catalysts in a proton transfer process, and a relative scale of $\Delta G°$ values can be set up. These serve just as well as absolute values for the purpose of obtaining $\Delta G_0^‡$ and $(k_H/k_D)_{max}$, and, in fact, in examples 3 through 7 of Table 5, the analysis was carried out in this way.

In some other cases, however, not even relative values of $\Delta G°$ can be obtained, and in these situations a relationship using free energies of activation rather than free energies of reaction is useful. By eliminating $\Delta G_R°$ between equations 6 and 9, equation 11 may be obtained.

$$\ln(k_H/k_D) = \ln(k_H/k_D)_{max} \left[2 \left(\frac{\Delta G^‡ - w^r}{\Delta G_0^‡} \right)^{1/2} - \frac{\Delta G^‡ - w^r}{\Delta G_0^‡} \right] \quad (11)$$

In this expression, $\Delta G^‡$ is the geometric mean of $\Delta G_H^‡$ and $\Delta G_D^‡$, in keeping with the definition of $\Delta G_0^‡$ as $(\Delta G_{0,H}^‡ \, \Delta G_{0,D}^‡)^{1/2}$, but w^r pertains to either H or D transfer because the assumption was made that $w_H^r = w_D^r$.

Correlations using this equation are not as good as those based upon equation 9, perhaps because transition states are subject to extraneous interactions that affect $\Delta G^‡$ but do not appear in $\Delta G°$ (Kresge, 1970) and cancel out in k_H/k_D. The increased scatter is illustrated in Figure 3, which uses the same ionization reactions of carbonyl and nitro compounds as are shown in Figure 1. Nevertheless, the analysis gives values of $\Delta G_{0,H}^‡ = 6.5 \pm 1.1$ kcal/mol and $(k_H/k_D)_{max} = 8.1 \pm 0.3$, which are nicely consistent with the results, $\Delta G_{0,H}^‡ = 7.5 \pm 0.8$ kcal/mol and $(k_H/k_D)_{max} = 8.4 \pm 0.2$, obtained using equation 9.

A system for which equation 11 must be used is the hydrolysis of vinyl ethers catalyzed by H_3O^+ (equation 12).

$$R_2C{=}CROR \xrightarrow[\text{slow}]{H_3O^+} R_2CHCROR^+ \xrightarrow[\text{fast}]{} R_2CHCOR + ROH \quad (12)$$

Isotope effects for many examples of this process using ethers of different structure have been measured (Kresge and Sagatys, in preparation). These refer to the first step of the reaction, protonation of a carbon atom of the C—C double bond, which is nonreversible and for which equilibrium constants therefore cannot be measured. The product of this step, moreover, is a highly reactive species whose stability in

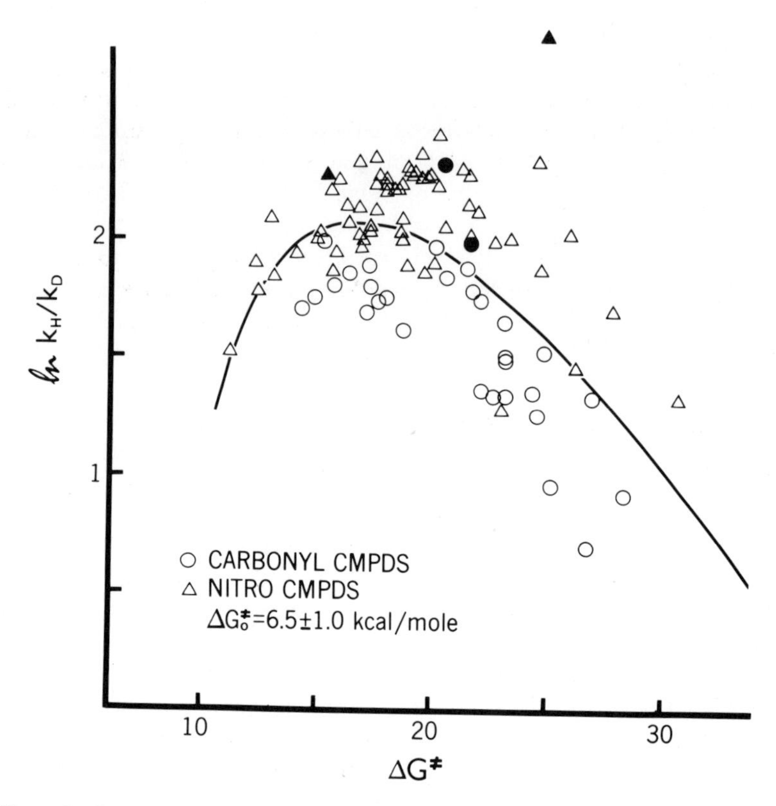

Figure 3. Correlation of k_H/k_D with $\Delta G^‡$ for the ionization of carbonyl and nitro compounds; filled points are for catalysis by 2,6-lutidine.

aqueous solution has so far proved to be unavailable even by indirect methods. As Figure 4 shows, however, the data are consistent with the model provided by equation 11. Least squares analysis of the approximately 30 points gives $\Delta G_{0,H}{}^‡ = 4.3 \pm 2.9$ kcal/mol and $(k_H/k_D)_{\max} = 3.7 \pm 0.4$.

The maximum isotope effect in this case is unusually small inasmuch as it includes an inverse secondary effect in addition to the normal primary effect. This is because the comparison necessarily involved H_3O^+ and D_3O^+, and two nonreacting bonds as well as the one that is broken were therefore isotopically substituted (equations 13 and 14).

$$H_2O{-}H^+ + S \rightarrow [\overset{\delta+}{H_2O}\cdots H \cdots \overset{\delta+}{S}]^‡ \rightarrow H_2O + HS^+ \quad (13)$$

$$D_2O{-}D^+ + S \rightarrow [\overset{\delta+}{D_2O}\cdots D \cdots \overset{\delta+}{S}]^‡ \rightarrow D_2O + DS^+ \quad (14)$$

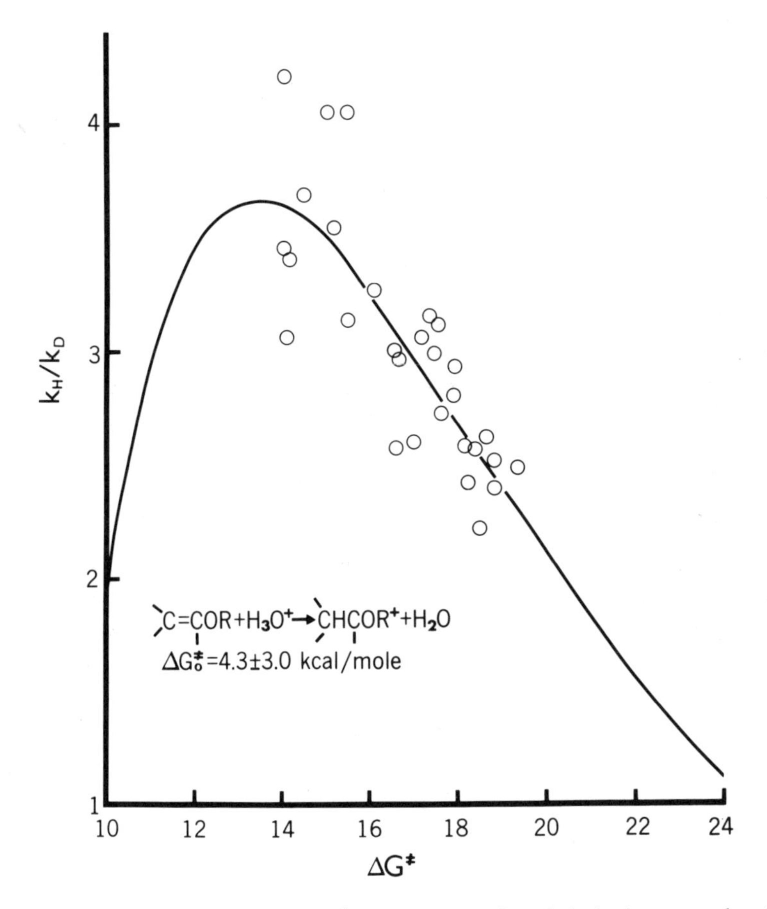

Figure 4. Correlation of k_H/k_D with $\Delta G^\ddagger$ for the protonation of vinyl ethers on carbon by H_3O^+.

In the transition state of this reaction, these nonreacting bonds are on the way to becoming O—H bonds of a water molecule, which are known to be stronger than the O—H bonds of H_3O^+ (More O'Ferrall, Koeppl, and Kresge, 1971a), and an inverse contribution to the overall isotope effect therefore is to be expected. This inverse secondary effect has in fact been measured directly by comparing rates of incorporation of tritium into aromatic substrates from acidified H_2O and D_2O where the tritiating reagents are H_2TO^+ and D_2TO^+ respectively (equations 15 and 16) (Kresge and Onwood, 1964; Kresge, Onwood, and Slae, 1968).

$$H_2O{-}T^+ + ArH \rightarrow H_2O + TArH^+ \quad (15)$$

$$D_2O{-}T^+ + ArH \rightarrow D_2O + TArH^+ \quad (16)$$

The value obtained using both dimethoxy- and trimethyoxybenzene as the aromatic substrate was $k_H/k_D = 0.59$. This effect in general, however, is expected to be variable, increasing regularly with the extent of proton transfer in the transition state up to a limit of $k_H/k_D = 0.48$ (More O'Ferrall, Koeppl, and Kresge, 1971b), and such variation was in fact allowed for in analyzing the data of Figure 4.

The results permit calculation of a pure primary effect as well, and the maximum value obtained for this was $k_H/k_D = 5.2 \pm 0.5$. This is not inconsistent with the rather imprecisely known O—H stretching vibration of the aqueous hydronium ion, $\nu = 2{,}600\text{-}2{,}900\ cm^{-1}$ (More O'Ferrall, Koeppl, and Kresge, 1971a).

Isotope Effects on Proton Transfer between Oxygen and Nitrogen

An interesting application of isotope effect correlation using Marcus theory is provided by some recent measurements on the base-catalyzed decomposition of nitramide. This is a much investigated reaction, and it is the system, in fact, that Brønsted and Pedersen (1924) used in their classic study which produced the Brønsted relation. The generally accepted mechanism for this reaction, first proposed by Pedersen (1934), involves rapid isomerization of nitramide to the *aci*-nitro form, followed by rate-determining elimination of H^+ and HO^- to give the reaction products (equation 17).

$$H_2NNO_2 \rightleftarrows HNNO_2H \xrightarrow{B} BH^+ + N_2O + HO^- \quad (17)$$

Some isotope effects on this reaction were measured a number of years ago (LaMer and Greenspan, 1937; Liotta and LaMer, 1938), but a systematic study changing $\Delta G°$ in a regular way was not performed until the recent work of Jones and Rumney (1975). Their results, plus some additional data supplied by Dr. Jones, are presented in Table 6.

These isotope effects are remarkable in that they provide some unusually large values for proton transfer from nitrogen to oxygen. Although large isotope effects are commonly found for proton transfer between oxygen or nitrogen and carbon, those involving oxygen and/or nitrogen alone are generally small enough to be regarded as purely secondary effects. This phenomenon has been widely remarked (see, e.g., Schowen, 1972), and it has led Swain, Kuhn, and Schowen (1965) to suggest that

Table 6. Isotope effects on the base-catalyzed decomposition of nitramide

Catalyst	pK_{BH^+}	k_H/k_D	Reference
2,4-Dinitrophenoxide ion	4.09	3.5	a
Pentachlorophenoxide ion	5.25	10.2	a
2,4,6-Trichlorophenoxide ion	6.00	9.0	a
2-Nitrophenoxide ion	7.17	6.3	a
2,3-Dichlorophenoxide ion	7.44	4.0	b
2,4-Dichlorophenoxide ion	7.74	4.8	b
Phenoxide ion	9.97	1.7	b

a) Jones and Rumney, 1975; b) Jones, personal communication.

proton transfer between electronegative atoms (such as oxygen and nitrogen) will not in general be concerted with heavy-atom rearrangement in reactions where both protonic and heavy-atom motion must occur. In other words, in such systems proton transfer will take place in a rapid step either coming before or following after the slow heavy-atom reorganization, and therefore there will be no primary isotope effect.

Schowen (1972; see also Thornton, 1967) later elaborated upon this idea by pointing out that in an E-2 olefin-forming elimination (equation 18),

$$B + HCCX \rightarrow BH^+ + C{=}C + X \qquad (18)$$

a good part of the driving force for reaction comes from delocalization of the electrons of the breaking C—H bond through the incipient double bond into the orbital being vacated by the leaving group. This forces a coplanar arrangement on the system in which mechanical coupling of hydrogenic and heavy-atom motion is possible, and the reaction coordinate can in fact be approximated as a combination of C—X and C—H stretching motions, **1**; large hydrogen isotope effects therefore can be realized.

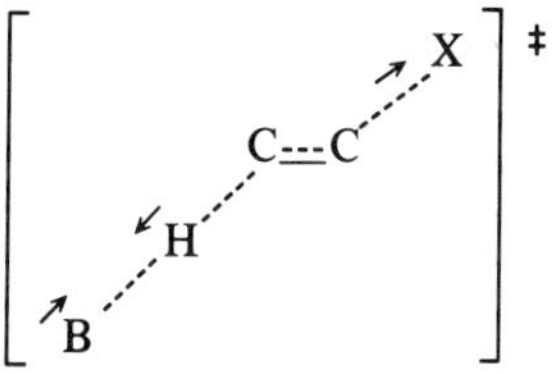

$$\left[B \cdots H \cdots C \overset{\cdots}{=} C \cdots X \right]^{\ddagger}$$

1

On the other hand, in an analogous elimination involving proton transfer from oxygen, such as the base-catalyzed conversion of a tetrahedral intermediate into a carbonyl group (equation 19), the oxygen atom bearing the proton to be transferred also has several nonbonded electron pairs.

$$\mathrm{B + HO{-}C{-}X \rightarrow BH^+ + O{=}C + X} \quad (19)$$

Greater stabilization will result from delocalization of one of these nonbonded pairs into the incipient π-system than from delocalization of a pair still partially bound to hydrogen, and that will force the system into a geometry in which this nonbonded pair and not the O—H bond is antiperiplanar to the leaving group. In such a configuration, 2, the

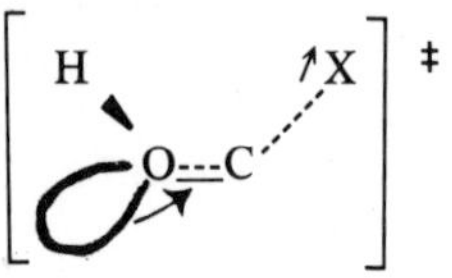

$$\left[\begin{array}{c} \mathrm{H} \quad\quad \nearrow\mathrm{X} \\ \mathrm{O{---}C} \end{array}\right]^{\ddagger}$$

2

O—H bond is approximately perpendicular to the O—C—X plane, and little coupling of O—H and C—X stretching motions can occur. The proton will consequently not move (translate) in the reaction coordinate motion, and large isotope effects will not be realized.

The structure of *aci*-nitramide is not known, but it is probably a planar molecule with a nonbonded electron pair occupying an sp^2 orbital on the nitrogen bearing the proton. A mechanism, therefore, can be written in which this electron pair serves to drive out the leaving group, equation 20;

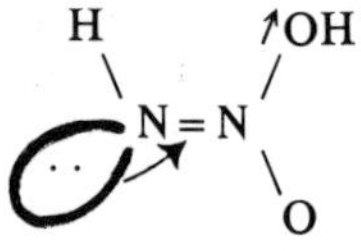

$$\mathrm{H(N{=}N(\nearrow OH)O)} \rightarrow \mathrm{HN^+{\equiv}NO + HO^-} \searrow \mathrm{BH^+ + N_2O} \quad (20)$$

but that would leave protonated nitrous oxide, which is likely to be a rather unstable species, as the immediate product. A more satisfactory scheme would seem to be one in which the electrons of the N—H bond are delocalized into the forming π-system at the same time as the proton is transferred to a base, equation 21.

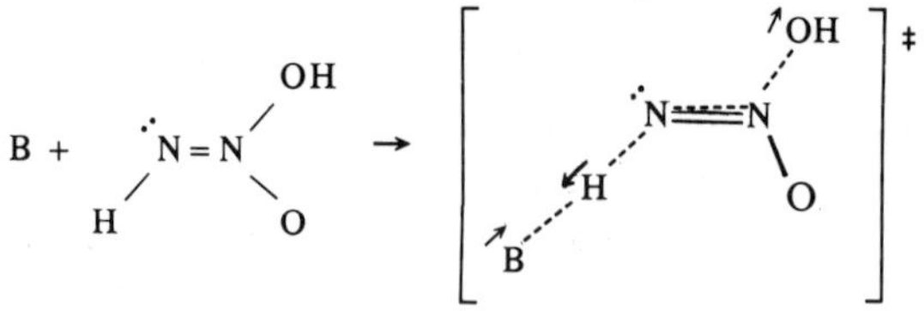

$$\mathrm{B + H{-}\ddot{N}{=}N(OH)O} \rightarrow \left[\mathrm{B{\cdots}H{\cdots}\ddot{N}{=\!=}N(\nearrow OH)O}\right]^{\ddagger} \rightarrow \mathrm{N{\equiv}NO + BH^+ + HO^-} \quad (21)$$

This gives a transition state which is the exact analog of **1**, that for E-2 olefin-forming elimination with proton transfer from carbon, and the concerted N—H and N—O stretching motions possible in such a species provide an attractive rationale for the large isotope effects actually observed in nitramide decomposition.

These isotope effects, however, despite their similarity to effects on proton transfer from carbon, differ from the latter in an important respect. As Figure 5 shows, the isotope effect maximum which they give is much sharper than that for any C—H system. Analysis of the data using equation 9 yields a very small intrinsic barrier, $\Delta G_{0,H}^{\ddagger} = 1.0 \pm 2.6$ kcal/mol.

It is interesting that this is of the same order of magnitude as the intrinsic barrier obtained for proton transfer between simple oxygen and nitrogen acids and bases without any bond-making or bond-breaking to heavy atoms (Table 3). This suggests that even when proton transfer

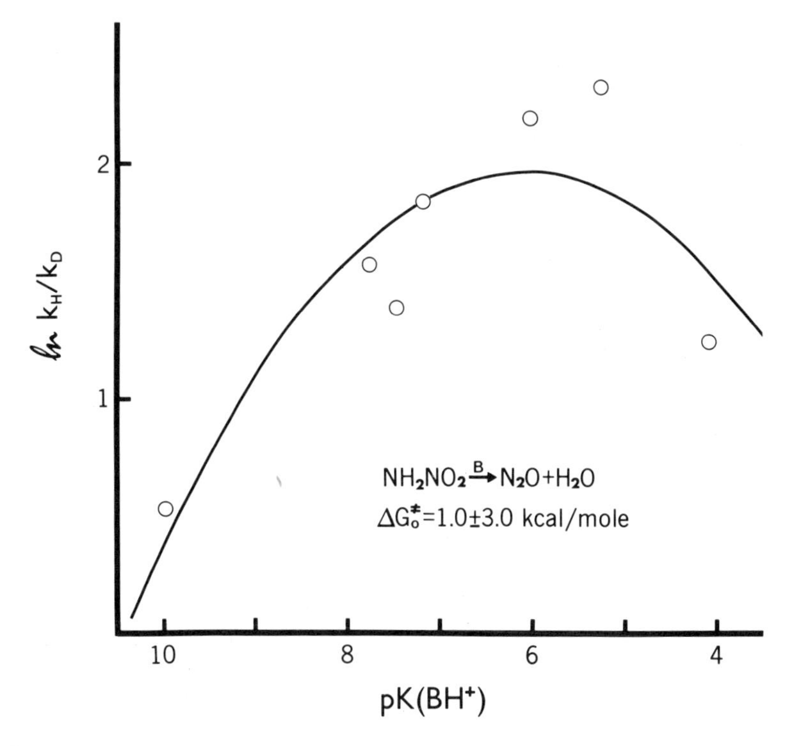

Figure 5. Correlation of k_H/k_D with pK_a of the catalyzing base in the decomposition of nitramide.

from nitrogen occurs by a mechanism in which protonic motion can be a part of the reaction coordinate, relatively small changes in ΔG° will serve to introduce enough asymmetry into the transition state to reduce the protonic contribution to the reaction coordinate to a relatively low level.

LITERATURE CITED

Ahrens, M. L., and Maass, G. 1968. Elementary steps in acid-base catalysis. Proton transfer reactions in aqueous solutions. Angew. Chem. Intl. Edn. 7:818–819.

Albery, W. J., Campbell-Crawford, A. N., and Curran, J. S. 1972. Kinetic isotope effects and aliphatic diazo-compounds. Part VI. The values of α and the Marcus theory. J. Chem. Soc. Perkin II:2206–2214.

Barnes, D. J., and Bell, R. P. 1970. Kinetic hydrogen isotope effects in the ionization of some carbon acids. Proc. Roy. Soc. A. 318:421–440.

Bell, R. P. 1933. Application of quantum mechanics to chemical kinetics. Proc. Roy. Soc. A. 139:466–474.

Bell, R. P. 1965. Isotope effects and the nature of proton-transfer transition states. Discussions Faraday Soc. 39:16–24.

Bell, R. P. 1973. The Proton in Chemistry, p. 203; pp. 270–289. 2nd Ed. Cornell University Press, Ithaca, N.Y.

Bell, R. P. 1974. Recent advances in the study of kinetic hydrogen isotope effects. Chem. Soc. Rev. 3:513–544.

Bell, R. P., and Cox, B. G. 1970. Hydrogen isotope effects on the inversion of (–)-menthone in mixtures of water and dimethylsulfoxide. J. Chem. Soc. B: 194–196.

Bell, R. P., and Cox, B. G. 1971. Primary hydrogen isotope effects on the rate of ionization of nitroethane in mixtures of water and dimethylsulfoxide. J. Chem. Soc. B.:783–785.

Bell, R. P., and Crooks, J. E. 1965. Kinetic hydrogen isotope effects in the ionization of some ketonic substances. Proc. Roy. Soc. A. 286:285–299.

Bell, R. P., and Goodall, D. M. 1966. Kinetic hydrogen isotope effects in the ionization of some nitroparaffins. Proc. Roy. Soc. A. 294:273–297.

Bell, R. P., and Higginson, E. 1949. The catalyzed dehydration of acetaldehyde hydrate and the effect of structure on the velocity of protolytic reactions. Proc. Roy. Soc. A 197:141–159.

Bell, R. P., Sachs, W. H., and Tranter, R. L. 1971. Model calculations of isotope effects in proton transfer reaction. Trans. Faraday Soc. 67:1995–2003.

Bordwell, F. G., and Boyle, Jr., W. J. 1975. Kinetic isotope effects for nitroalkanes on their relationship to transition-state structure in proton-transfer reactions. J. Am. Chem. Soc. 97:3447–3452.

Brønsted, J. N., and Pedersen, K. 1924. Die katalytische Zersetzung des Nitramids und ihre physikalisch-chemische Bedeutung. Z. physik. Chem. 108:185–235.

Caldin, E. F. 1969. Tunnelling in proton transfer reactions. Chem. Rev. 69: 135–156.

Caldin, E. F., and Mateo, S. 1973. Kinetic isotope effects in various solvents for the proton-transfer reactions of 4-nitrophenylnitromethane with bases. J. Chem. Soc. Chem. Comm. 854:855.

Caldin, E. F., and Mateo, S. 1975. Kinetic isotope effects and tunneling in the proton-transfer reaction between 4-nitrophenylnitromethane and tetramethylguanidine in various aprotic solvents. J. Chem. Soc. Faraday Trans. I:71: 1876–1904.

Cohen, A. O., and Marcus, R. A. 1968. On the slope of free energy plots in chemical kinetics. J. Phys. Chem. 72:4249–4256.

Dixon, J. E., and Bruice, T. C. 1967. Dependence of the primary isotope effect (k^H/k^D) on base strength for the primary amine catalyzed ionization of nitroethane. J. Am. Chem. Soc. 92:905–909.

Eigen, M. 1963. Fast elementary steps in chemical reaction mechanisms. Pure Appl. Chem. 6:97–115.

Eigen, M. 1964. Proton transfer, acid-base catalysis, and enzymatic hydrolysis. Angew. Chem. Intl. Edn. 3:1–19.

Funderburk, L. H., and Lewis, E. S. 1964. Tunnelling in a proton transfer. A large isotope effect. J. Am. Chem. Soc. 86:2531–2532.

German, E. D., Dogonadze, R. R., Kuznetsov, A. M., Levich, V. G., and Kharkats, Y. I. 1971. Kinetics of chemical reactions in polar liquids. I: Theory. J. Res. Inst. Catalysis, Hokkaido Univ. 19:99–114.

Hanna, S. B., Jermini, C., Loewenschuss, H., and Zollinger, H. 1974. Indices of transition state symmetry in proton-transfer reactions. Kinetic isotope effects and Brønsted's β in base-catalyzed diazo-coupling reactions. J. Am. Chem. Soc. 96:7222–7228.

Jones, J. R., and Rumney, T. G. 1975. Hydrogen isotope effects for nitramide, a nitrogen acid. J. Chem. Soc. Chem. Comm. 995–996.

Jones, L. H., and Penneman, R. A. 1954. Infrared absorption spectra of aqueous HF_2^-, DF_2^-, and HF. J. Chem. Phys. 22:781–782.

Keefe, J. R., and Munderloh, N. H. 1974. Rates and kinetic hydrogen isotope effects for the ionization of phenylnitromethane. J. Chem. Soc. Chem. Comm. 17–18.

Koeppl, G. W., and Kresge, A. J. 1973. Marcus theory and the relationship between Brønsted exponents and energy of reaction. J. Chem. Soc. Chem. Comm. 371–373.

Kreevoy, M. M., and Konasewich, D. E. 1971. The Brønsted α and the primary hydrogen isotope effect. A test of the Marcus theory. Adv. Chem. Phys. 21: 243–252.

Kreevoy, M. M., and Oh, S. W. 1973. Relation between rate and equilibrium constants for proton-transfer reactions. J. Am. Chem. Soc. 95:4805–4810.

Kresge, A. J. 1965. Discussion. Trans. Faraday Soc. 39:48–49.

Kresge, A. J. 1970. Deviant Brønsted relations. J. Am. Chem. Soc. 92:3210–3211.

Kresge, A. J. 1973. The Brønsted relation—recent developments. Chem. Soc. Rev. 4:475–503.

Kresge, A. J. 1975. What makes proton transfer fast? Accts. Chem. Res. 8: 354–360.

Kresge, A. J., Chen, H. J., and Chiang, Y. 1977. Vinyl Ether Hydrolysis VII. Isotope effects on catalysis by hydrofluoric acid. J. Am. Chem. Soc. 99: 802–805.

Kresge, A. J., and Chiang, Y. 1969. The effect of bending vibrations on the magnitude of hydrogen isotope effects. J. Am. Chem. Soc. 91:1025.

Kresge, A. J., Mylonakis, S. G., Sato, Y., and Vitullo, V. P. 1971. The kinetic protonation of hydroxy- and alkoxybenzenes. J. Am. Chem. Soc. 93:6181-6188.

Kresge, A. J., and Onwood, D. P. 1964. The secondary isotope effect on proton transfer from the hydronium ion in aqueous solution. J. Am. Chem. Soc. 86: 5014-5016.

Kresge, A. J., Onwood, D. P., and Slae, S. 1968. Aromatic protonation V. Secondary hydrogen isotope effects on hydrogen ion transfer from the hydronium ion. J. Am. Chem. Soc. 90:6982-6988.

Kresge, A. J., and Sagatys, D. S. Vinyl ether hydrolysis IX. Isotope effects on proton transfer from the hydronium ion. In preparation.

LaMer, V. K., and Greenspan, J. 1937. Kinetics of the solvent decomposition of nitramid in H_2O—D_2O mixtures. Trans. Faraday Soc. 33:1266-1272.

Lewis, E. S. 1975. Tunnelling in hydrogen-transfer reactions. In: E. F. Caldin and V. Gold (eds.), Proton Transfer Reactions, pp. 317-338. Chapman and Hall, London.

Lewis, E. S., and Funderburk, L. H. 1967. Rates and isotope effects in the proton transfers from 2-nitropropane to pyridine bases. J. Am. Chem. Soc. 89: 2322-2327.

Liotta, S., and LaMer, V. K. 1938. The temperature coefficients of the base catalyzed decomposition of nitramide in deuterium oxide. J. Am. Chem. Soc. 60:1967-1974.

Longridge, J. L., and Long, F. A. 1967. Hydrogen exchange of azulenes. VII. Magnitude of kinetic isotope effects. J. Am. Chem. Soc. 89:1292-1297.

Marcus, R. A. 1956. On the theory of oxidation-reduction reactions involving electron transfer. I. J. Chem. Phys. 24:966-978.

Marcus, R. A. 1964. Chemical and electrochemical electron-transfer theory. Annu. Rev. Phys. Chem. 15:155-196.

Marcus, R. A. 1968. Theoretical relations among rate constants, barriers, and Brønsted slopes of chemical reactions. J. Phys. Chem. 72:891-899.

Melander, L. 1960. Isotope Effects on Reaction Rates, pp. 24-32. Ronald Press, New York.

More O'Ferrall, R. A. 1975. Substrate Isotope Effects. In: E. F. Caldin and V. Gold (eds.), Proton Transfer Reactions, pp. 201-261. Chapman and Hall, London.

More O'Ferrall, R. A., Koeppl, G. W., and Kresge, A. J. 1971a. Vibrational analysis of liquid water and the hydronium ion in aqueous solution. J. Am. Chem. Soc. 93:1-9.

More O'Ferrall, R. A., Koeppl, G. W., and Kresge, A. J. 1971b. Solvent isotope effects upon proton transfer from the hydronium ion. J. Am. Chem. Soc. 93: 9-20.

Murdoch, J. R. 1972. Rate-equilibria relationships and proton transfer reactions. J. Am. Chem. Soc. 94:4410-4418.

Newman, R., and Badger, R. M. 1951. The polarized infrared spectrum of potassium bifluoride at $-185\,^{\circ}C$. J. Chem. Phys. 19:1207-1208.

Pauling, L. 1947. Atomic radii and interatomic distances in metals. J. Am. Chem. Soc. 69:542-553.

Pedersen, K. J. 1934. The theory of protolytic reactions and prototropic isomerization. J. Phys. Chem. 38:581-600.
Pryor, W. A., and Kneipp, K. G. 1971. Primary kinetic isotope effects and the nature of hydrogen transfer transition states. The reaction of a series of free radicals with thiols. J. Am. Chem. Soc. 93:5584-5586.
Reitz, O. 1936. Über die loslösung von Protonen und Deuteronen aus organischen Molekulen bei allgemeiner Basenkatalyse, untersucht an Hand der Bromierung des Nitromethaus. Z. Phys. Chem. A. 176:363-387.
Schowen, R. L. 1972. Mechanistic deductions from solvent isotope effects. Prog. Phys. Org. Chem. 9:275-332.
Swain, C. G., Kuhn, D. A., and Schowen, R. L. 1965. Effect of structural changes in reactants on the position of hydrogen-bonding hydrogens and solvating molecules in transition states. The mechanism of tetrahydrofuran formation from 4-chlorobutanol. J. Am. Chem. Soc. 87:1553-1561.
Thornton, E. R. 1967. A simple theory for predicting the effects of substituent changes on transition-state geometry. J. Am. Chem. Soc. 89:2915-2927.
Westheimer, F. H. 1961. The magnitude of the primary kinetic isotope effect for compounds of hydrogen and deuterium. Chem. Rev. 61:265-273.
Wiberg, K. B., and Slaugh, L. H. 1958. The deuterium isotope effect in the side chain halogenation of toluene. J. Am. Chem. Soc. 80:3033-3039.
Willi, A. V., and Wolfsberg, M. 1964. Influence of "bond making and bond-breaking" in the transition state on hydrogen isotope effects in linear three centre reactions. Chem. Ind. 2997-2998.

Solvent Isotope Effects on Enzymic Reactions

R. L. Schowen

Many enzymic properties, including catalytic rates, differ when the enzyme is dissolved in-deuterium oxide from their values in protium oxide (Bender and Kezdy, 1975; Katz and Crespi, 1970). These differences are systematic in character, suggesting that they contain valuable information about the enzyme and its mechanism of action. This chapter gives a brief overview of what is known from studies of simpler systems about the effect of heavy water on rates and equilibria in order to show how this information can be used to interpret enzymic solvent isotope effects in terms of mechanistic models and to illustrate this application by some examples.

Table 1 lists a few solvent isotope effects observed with enzymes. The selection shows that catalytic rates may be depressed by deuterium oxide, as is seen in the action of trypsin on a peptide substrate (and frequently in hydrolytic reactions), but also that this is not universally true: methyl transfer catalyzed by COMT is unaffected by deuterium oxide. Similarly, protein structural equilibria may be unaffected, as is true of a conformational isomerization of ribonuclease-A, but they also may be drastically altered. Thus, the cation-induced formation of the enzymically active tetramer of formyltetrahydrofolate synthetase is favored in deuterium oxide by more than 50-fold. Rates of reaction may be increased or decreased: the ribonuclease isomerization is reduced in rate by a factor of 4.8, while subunit association of the synthetase is accelerated by a factor of about 3.

To discern how to extract mechanistic information from these observations, the effects of deuterium oxide in simple chemical systems

Table 1. Selected examples of solvent isotope effects on enzyme reactions

Process	Solvent isotope effect	Reference
Hydrolysis of *N*-benzoyl-L-phenylalanyl-L-valyl-L-arginyl *p*-nitroanilide by trypsin	V_{H_2O}/V_{D_2O} (sat) $= 4$	a
Conformation change of ribonuclease	$k_{H_2O}/k_{D_2O} = 4.8$ $K_{H_2O}/K_{D_2O} \sim 1$	b
Subunit association of formyltetrahydrofolate synthetase	$k_{D_2O}/k_{H_2O} \sim 3$ $K_{D_2O}/K_{H_2O} > 50$	c
Methyl transfer to 3,4-dihydroxyacetophenone from *S*-adenosylmethionine, catalyzed by catechol-*O*-methyltransferase (COMT)	V_{H_2O}/V_{D_2O} (sat) ~ 1	d

a) Quinn et al., in preparation; b) Wang et al. (1975); c) Harmony, Himes, and Schowen (1975); d) M. F. Hegazi, R. T. Borchardt, and R. L. Schowen, unpublished data.

and the simplest of the physical models currently used to interpret them are examined first.

ORIGINS OF SOLVENT ISOTOPE EFFECTS

Properties of Light and Heavy Water

A good, simple rule of thumb that nearly universally serves for the interpretation of all kinds of isotope effects is, "the lighter isotope accumulates where the binding is weaker." This makes it easy to understand and to remember properties of deuterium and protium oxides (data from Katz and Crespi, 1970). Vaporization represents transfer from the more tightly bound liquid state to the more weakly bound vapor state: thus, it is easier with protium oxide, which boils at 100.00°, than with deuterium oxide, which boils at 101.42°. Transfer to the more tightly bound solid state occurs more readily with deuterium oxide (freezing point 3.81°) than with protium oxide (freezing point 0.00°). The viscosity of D_2O is higher than that of H_2O (1.25 centipoise at 20° vs. 1.005 centipoise), showing that binding to the isotopic centers is weakened in the transition state for viscous flow (the fluidity is greater for the lighter isotope). This may be because an interstitial water molecule hydrogen-bonds to two adjacent structural aggregates of the solvent, facilitating momentum transfer between them. Another useful general principle is that the geometrical and electronic structures of isotopic molecules are much the same because isotopic nuclei differ only in the number of neutrons, while it is the number of nuclear protons which determines the nuclear charge and thus electron distribution and molecular structure. Accordingly, deuterium oxide is more dense than protium oxide (1.1056 g/cc at 20° vs. 0.9982 g/cc) simply because each D_2O molecule has a higher molecular weight; the molar volumes of the two solvents differ by less than 0.5% (18.047 cc/mol for H_2O, 18.117 cc/mol for D_2O). The dielectric constants of protium oxide (78.54) and deuterium oxide (78.93) differ only slightly at 25°.

With these characteristics of light and heavy water in mind, some possible contributions to solvent isotope effects on rates and equilibria may be enumerated (Albery, 1975; Arnett and McKelvey, 1969; Laughton and Robertson, 1969; Schowen, 1972):

1. Water may be a reactant. If one of the hydrogens of the water is being transferred in the rate-determining step, it may have substantial amplitude in the reaction coordinate and thus generate a *primary isotope effect* on the rate. Otherwise, bonding changes may still alter the force constants at a hydrogenic site of a reacting water without any large participation of the isotopic center in the reaction coordinate, producing a *secondary isotope effect.*
2. Reaction may occur at a center labeled by exchange with water. This would include many examples from enzyme reactions in which active-site functional groups become labeled by exchange. Depending on the reaction-coordinate amplitude of the isotopic center, effects of this kind may also be primary or secondary in character.
3. Changes in reactant structure may be induced by effects of the deuterated solvent. This may be of particular concern with enzymes, where hydrogen bonds at exchangeable positions are important in the maintenance of secondary, tertiary, and quaternary structure. The point is discussed in detail below.

These three points cover the major sources that have, to date, been required to interpret and explain most solvent isotope effects in both enzymic and nonenzymic systems. Several other factors are logical to consider but emerge in the end as of smaller importance:

4. Bonding changes may occur in the primary solvation shell. Whenever strongly basic species such as methoxide ion and hydroxide ion are involved, the hydrogen bonds to them from solvent molecules surely con-

tribute to solvent isotope effects (Gold and Grist, 1972). These interactions may also be found at some point in enzymic reactions but have not yet had to be invoked.

5. Bonding changes may occur at more remote sites. The large sizes of hydrogen-bonded domains in aqueous solutions suggest naturally that isotope effects might arise from an accumulation of small changes in regions rather distant from the reaction site. In enzymes, the long chains of hydrogen bonds stabilizing the three-dimensional polypeptide array give rise to a similar concern. Models with such features can be made consistent with some experimental results (Kresge, 1973) but no compelling evidence favors the importance of such models over simpler ones.

6. Differences in viscosity, dielectric constant, etc., may produce rate changes. If diffusion processes are involved, rate differences may derive from the relative ease of diffusion in the isotopic solvents. These are ordinary kinetic isotope effects and can be interpreted in terms of primary and secondary contributions, as in any other case. Effects of dielectric constant are small enough to be negligible in general.

Distribution of Isotopic Species: Fractionation Factors

Any compound with exchangeable hydrogenic sites dissolved in a mixture of light and heavy water will, at equilibrium, contain in each site a mixture of protium and deuterium. The ratio of deuterium to protium in a particular site will be equal to the ratio present in the bulk solvent only if the binding in the site equals the binding in an average water site. If the solute site binds its hydrogen more weakly, protium will accumulate there, while if the binding is tighter, deuterium will be preferred. A quantitative measure of the deuterium preference (and thus of the tightness of binding) at any hydrogenic site is given by the *isotopic fractionation factor*. If the *atom fraction of deuterium* in a mixed isotopic solvent is called n, then the isotopic fractionation factor ϕ_i is given by equation 1; it is simply the deuterium-to-protium ratio at the ith hydrogenic site, relative to the ratio for water. It is, therefore, the equilibrium constant of the exchange reaction of equation 2, where L means "either H or D."

$$\phi_i = [[D]_i/[H]_i]/[n/(1 - n)] \tag{1}$$

$$\mathrm{H}_i + \mathrm{LOD} \rightleftarrows \mathrm{D}_i + \mathrm{LOH} \tag{2}$$

It is a generally good approximation that, if a molecule has two exchangeable sites i and j, the fractionation factor ϕ_i may be taken to be independent of which isotope occupies site j and vice versa. For example, the deuterium preference of one hydroxyl group of a sugar molecule should be the same when the other hydroxyl groups in the molecule contain deuterium as when they contain protium. This approximation is known as the rule of the geometric mean (RGM). This rule is extremely useful in making mechanistic deductions from solvent isotope effects, and it is thus fortunate that even when the RGM is not exactly true, this rarely causes much trouble (Albery and Davies, 1969; Gold, 1968). The rule is not obeyed precisely by the water molecule itself, which can exist in three isotopic modifications, viz., HOH, HOD, and DOD. The RGM holds that introduction of the first deuterium to form HOD should result in the same free-energy change as introduction of the second deuterium to convert HOD to DOD. If this were so, the equilibrium constant of the exchange reaction of equation 3 should be determined wholly by statistical considerations and would therefore be equal to 4.00.

$$\mathrm{HOH} + \mathrm{DOD} \rightleftarrows 2\mathrm{HOD} \qquad (3)$$

The value has been measured and is found to be around 3.78 (Friedman and Shiner, 1966; Van Hook, 1972). This discrepancy turns out to have only the slightest effect on the distribution of the three modifications of water. Figure 1 shows a plot of the mole fractions of HOH, DOD, and DOH in mixtures of light and heavy water. To within the precision of the figure, the same lines are obtained whether a value of 4.00 or a value of 3.78 is used for the equilibrium constant of equation 3! It is, therefore, a reasonable conclusion and has been demonstrated algebraically (Albery and Davies, 1969) that violations of the RGM need arouse no concern in mechanistic applications of solvent isotope effects at ordinary levels of precision.

Isotopic fractionation factors can be determined by several means, with a nuclear magnetic resonance (NMR) method of Allred, Gold, and Kresge (Gold, 1963; Kresge and Allred, 1963) being frequently very convenient. A considerable number of determinations have now been made and they support a further simplifying generalization which might be called the *functional-group rule.* By this is meant that values of ϕ_i depend only on the functional group in which the i*th* site is located and not on more remote features of the molecular environment. Thus, all alcohols can be expected to show the same fractionation factor, all thiols the same one, etc.

Table 2 gives a list of fractionation factors that are of use in mechanistic interpretation. The approximate values of ϕ^{-1} will be used later.

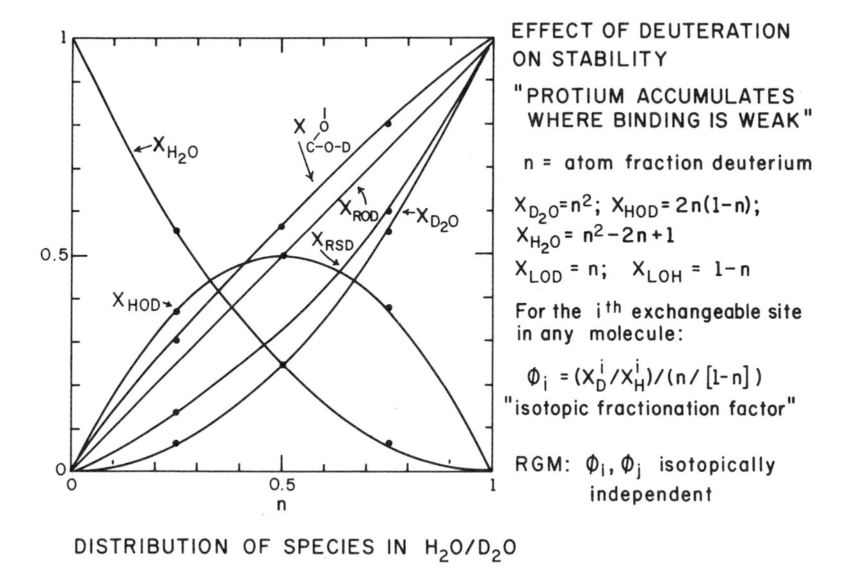

Figure 1. Plot of the relative abundances of various species in mixtures of H_2O and D_2O, atom fraction of deuterium n. Alcohols, with $\phi = 1$ (Table 2), increase in deuteration exactly with the atom fraction of deuterium (line labeled X_{ROD}). Thiols ($\phi = 0.40$–0.46) lag behind n in degree of deuteration while hemiacetals ($\phi = 1.23$–1.28) run ahead of n.

The factors are all easy to understand. Many are unity, indicating that net binding to the hydrogenic site in those functional groups is equal to that in bulk water. This is the case with the hydroxyl groups of alcohols and carboxylic acids and for amino and ammonium groups. The hydroxyl groups of *gem*-diols and hemiacetals, however, have $\phi > 1$, which means that deuterium accumulates in these groups and implies that binding is tighter there than in bulk water. This may be a result of "double bond, no bond" resonance in these compounds, which restricts rotation around the C—OH bonds, binding the hydrogen more tightly in place (Mata-Segreda et al., 1974). If a positive charge appears on the oxygen ($\overset{+}{\rangle O}$—H group), the bond to the proton is weakened and protium accumulates; thus $\phi < 1$ here. The same reasoning would have led to the expectation that $\phi < 1$ for the ammonium group (RHN_3^+), while it is found experimentally that $\phi \sim 1$. Perhaps the decreased binding along the N—H stretching coördinate is compensated by an increased resistance to bending in the stiff, tetrahedral ammonium ion; the net binding of the hydrogen may then be similar to that in bulk water and amines. The less-than-unit fractionation factor for the sulfhydryl group has been attributed to the low resistance offered to both

Table 2. Isotopic fractionation factors relative to water[a]

Functional group	ϕ	Approximate ϕ^{-1}
\ O—L	1.0	1.0
—CO_2—L	1.0	1.0
O—L \ / C / \ O |	1.23–1.28[b]	1/1.2–1/1.3
\ + O—L /	0.69	1.5
$^-$O—L	0.47–0.56[c]	2.0
\ N—L /	0.92	1.0
\ + —N—L /	0.97	1.0
\ S—L	0.40–0.46	2.0
\ —C—L /	0.62–0.64 (sp)[d] 0.78–0.85 (sp^2)[e] 0.84–1.18 (sp^3)[f]	1.6 1.2–1.3 0.85–1.2

[a] Unless otherwise indicated, taken from Schowen (1972).

[b] Mata-Segreda, Wint, and Schowen (1974).

[c] Probably a complex factor arising in part from the solvation shell (Gold and Grist, 1972).

[d] The measured K_{eq} of 0.473 in the gas phase at 25° for the reaction $C_2H_2 + D_2O \rightleftarrows C_2D_2 + H_2O$ (Pyper and Long, 1964), when corrected to liquid water by using the vapor pressure ratio for H_2O/D_2O of 1.152 (Jones, 1968), gives 0.64. The value of 0.74 quoted by More O'Ferrall (1975) is in error because the vapor pressure correction was applied the wrong way. The calculations of Hartshorn and Shiner (1972), when combined with the experimental values for the CH_2 of malate (Thomson, 1960) or the CH_3 of pyruvate (Meloche, Monti, and Cleland, 1977) give 0.62.

[e] Converted from the calculated values of Hartshorn and Shiner (1972), using 0.62 for acetylene. The lower value is for CH_2=, and the higher one for C—CH=.

[f] See the table in the Note on the Use of Fractionation Factors elsewhere in this volume, or Table 2 of Meloche, Monti, and Cleland (1977).

stretching and bending motions of hydrogen attached to sulfur (Lienhard and Jencks, 1966). Hydroxide ion exhibits a small fractionation factor that once was thought to originate in the extraordinarily free rotational motion of the diatomic ion relative to the corresponding, stiff angle-bending motion in water (Schowen, 1972). More probably the effect comes from loosening of the bonds to hydrogen in three solvent waters hydrogen-bonding to the lone-pair electrons of the hydroxide ion (Gold and Grist, 1972).

Rates and Equilibria in Isotopic Waters

In mechanistic studies, the utility of fractionation factors comes in their relationship to other equilibrium and kinetic isotope effects. This is readily appreciated from the scheme of equation 4.

$$\begin{array}{ccc} \text{LOD} + \text{RH} & \xrightarrow{k_H} & [\text{LOD} + \text{TH}]\rightarrow \\ \uparrow\downarrow\phi^R & & \uparrow\downarrow\phi^T \\ \text{LOH} + \text{RD} & \xrightarrow{k_D} & [\text{LOH} + \text{TD}]\rightarrow \end{array} \qquad (4)$$

Reactant RH, with one exchangeable hydrogen, is imagined to form activated complex TH, also with one exchangeable hydrogen, in a process with rate constant k_H (which is, of course, related to an equilibrium constant of activation $K_H{}^*$ by $k_H = (kT/h)K_H{}^*$). If the system were introduced into deuterium oxide, exchange would occur in the reactant state to generate RD (exchange equilibrium constant ϕ^R, the reactant isotopic fractionation factor) and in the transition state to generate TD (equilibrium constant ϕ^T). The process would now proceed with rate constant k_D. Because equation 4 consists of a closed thermodynamic cycle, $k_H\phi^T = \phi^R k_D$, or (equation 5) the isotope effect is given by the simple ratio of fractionation factors for transition and reaction states.

$$k_D/k_H = \phi^T/\phi^R; \quad k_H/k_D = (\phi^T)^{-1}/(\phi^R)^{-1} \qquad (5)$$

Because most of us are simultaneously accustomed to writing transition-state quantities in the numerator and to writing isotope effects in the form k_H/k_D, the second version of equation 5 may be specially convenient. It is for this reason that values of ϕ^{-1} are listed in Table 2.

If more than one exchangeable site exists in RH and TH, equation 5 may be generalized to obtain equation 6. Here the RGM is employed, which allows one to describe the effects of the different exchangeable sites by a simple product of fractionation factors.

$$k_D/k_H = \prod_i^{\nu} \phi_i{}^T / \prod_j^{\nu} \phi_j{}^R \qquad (6)$$

The kinetic isotope effect is thus determined by the reactant-state and transition-state fractionation factors which change on activation.

Any ϕ—regardless of its value—which is the same in reactant and transition states will cancel and can be neglected. The significant ϕ_j^R can frequently be measured or reliably estimated. If one then knew the ϕ_j^T (i.e., if one knew the nature of the transition state) one could predict k_D/k_H. Correspondingly, a knowledge of the (measurable) k_D/k_H sets limits on the possible structures of the transition state: acceptable models must generate ϕ_i^T which are consistent with k_D/k_H. Of course, these limits may be undesirably wide because a number of different sets $[\phi_i^T]$ may all produce the same product $\Pi_i^\nu \phi_i^T$. As shall be seen below, further limitations may be imposed by an examination not merely of k_D/k_H but of k_n in various mixed isotopic solvents. Nevertheless, much useful information may frequently be derived from values of k_D/k_H alone.

Equilibrium isotope effects are related to reactant and product fractionation factors (ϕ^R and ϕ^P, respectively) by the analog of equation 6, given as equation 7.

$$K_D/K_H = \prod_i^\nu \phi_i^P / \prod_j^\nu \phi_j^R \quad (7)$$

Fractionation Factors for Transition States

Needless to say, the ϕ_i^T cannot be measured by direct experiments and, in relating model transition-state structures to values of k_D/k_H, they have to be estimated by analogy to values obtained with stable molecules or by other indirect techniques (such as vibrational analysis with estimated force constants). The drawing of the appropriate analogy offers no difficulty when the transition-state site closely resembles one of the structures of Table 1. There are three common circumstances when this is not the case, and special approaches are required. These are cases in which 1) the hydrogenic site is a participant in the reaction coordinate, producing primary isotope effects; or 2) the hydrogenic site has an intermediate structure, with the group possessing a partial negative or positive charge; or 3) the hydrogenic site is part of a catalytic bridge in the transition state, although not necessarily a reaction-coordinate participant.

Fractionation factors for sites producing primary isotope effects frequently will be quite small numbers. For example, if a reactant hydroxyl group ($\phi^R = 1$) donates its proton to a base in a rate-determining process, producing a primary isotope effect of 5, ϕ^T will be $1/5$, or 0.2. Note that ϕ^T is always referred to a bulk water site as the standard of comparison so that if the hydrogen in question originates in a functional group with $\phi^R \neq 1$, then $(\phi^T)^{-1}$ will not give the "primary isotope effect" in the usual sense. Consider a sulfhydryl group ($\phi^R = 1/2$) which donates its proton, generating a primary isotope effect of 5 (that is, $k_H/k_D = 5 = (\phi^T)^{-1}/(\phi^R)^{-1}$). Obviously $(\phi^T)^{-1} = 10$, which is the equivalent isotope effect if the proton *had* originated in a site of $\phi^R = 1$. This point will rarely cause difficulty in interpreting ϕ^T values obtained by deduction from experiment but it ought to be kept in mind in generating values of k_H/k_D from transition-state models.

Sites that are of intermediate structural character in transition states can be expected to give rise to ϕ values of intermediate magnitude. Thus, a water molecule binding to a positive site (equation 8) and possessing a partial positive charge in the transition state should produce fractionation factors for both of its hydrogenic sites with values

$$H_2O + {>}C^{+}{-} \rightarrow \left[H_2\overset{\delta+}{O}\cdots C{<} \right] \rightarrow H_2\overset{+}{O}{-}C{\equiv} \quad (8)$$

between $\phi = 1$ (the reactant value) and $\phi = 0.69$ (the product value). A reasonable idea is to use an extension of the Brønsted hypothesis to arrive at a quantitative estimate. If we let α measure the degree of reaction progress at the transition state, then equation 9 yields an estimate for ϕ^T.

$$\phi^T = (\phi^R)^{1-\alpha}(\phi^P)^{\alpha} \quad (9)$$

In equation 8, $\alpha = \delta$ and $\phi^R = 1$ so that $\phi^T = (0.69)^{\delta}$. For a transition-state model in which $\delta = 0.75$, we thus obtain $\phi^T = (0.69)^{0.75} = 0.76$.

Many enzymic reactions will have hydrogenic sites involved in acid-base catalytic interactions where the hydrogen is bridging between oxygen, nitrogen, or sulfur atoms. Although these sites might have large amplitudes in the reaction coordinate, thus giving very small ϕ^T, this seems most frequently not to be the case (Schowen, 1972). Instead, such bridges commonly generate normal isotope effects of no more than 2–3, so that when $\phi^R = 1$ for the group involved, ϕ^T should be about 0.3–0.5 (Minor and Schowen, 1973). The reason for this is not certain, although it may be that unusually strong hydrogen bonds are formed in these transition states, producing reasonable isotope effects, but ones smaller than true primary effects.

Examples from Simple Systems

It should be instructive to consider a few examples in which these principles can be used to predict isotope effects or to understand isotope-effect observations. To avoid any concern about special complexities arising from protein structure (which is discussed next), the examples will be taken from simple nonenzymic systems.

Consider the ionization of an acid HA (equation 10).

$$\mathrm{H{-}A} + \mathrm{H_2O} \rightleftarrows \mathrm{H_2\overset{+}{O}{-}H} + \mathrm{A{:}^-} \quad (10)$$

Employing equation 7 and neglecting effects from solvation shells of the various species, $K_a^{D_2O}/K_a^{H_2O}$ is found to be given by equation 11.

$$K_a^{D_2O}/K_a^{H_2O} = (\phi_{LO^+})^3/\phi_{LA} = (0.69)^3/\phi_{LA} = 1/3\phi_{LA} \quad (11)$$

Acids with $\phi_{LA} \backsim 1$ (carboxylic acids and ammonium ions) therefore should be about three-fold stronger in H_2O than in D_2O, or $pK_a^D \backsim pK_a^H + 0.5$. In fact, nearly every simple carboxylic and ammonium acid with pK_a between 4 and 10 has $pK_a^D - pK_a^H = 0.5$–0.6 (see the compilation in Laughton and Robertson, 1969). When this is not true, it usually signifies an unusual effect such as intramolecular hydrogen bonding. Outside the pK_a range of 4–10, ϕ_{LA} probably becomes nonunit because of weakened binding in HA (less-than-unit factors on the acidic end of the scale) or because effects from the solvation shell of A^- enter (less-than-unit fractionation factors on the basic end of the scale). Table 2 shows that $\phi_{SL} \backsim 1/2$ for the sulfhydryl group; this implies that sulfhydryl acids should ionize only about 1.5 times more readily in H_2O or have $pK_a^D - pK_a^H \backsim 0.1$–$0.3$. These points will be of obvious use in our consideration of solvent isotope effects on the pH-rate behavior of enzymes.

There are many reactions that are subject to general base catalysis by buffer components but that show substantial rates in aqueous solution even in the absence of added catalyst (Johnson, 1965). Examples include the hydrolysis reactions of acetic anhydride (Batts and Gold, 1969) and alkyl trifluoroacetates (Winter and Scott, 1968; Mata-Segreda, 1975). It is reasonable to presume that the "water terms" for these reactions originate in the action of water itself as a general base catalyst of its own reaction (equation 12).

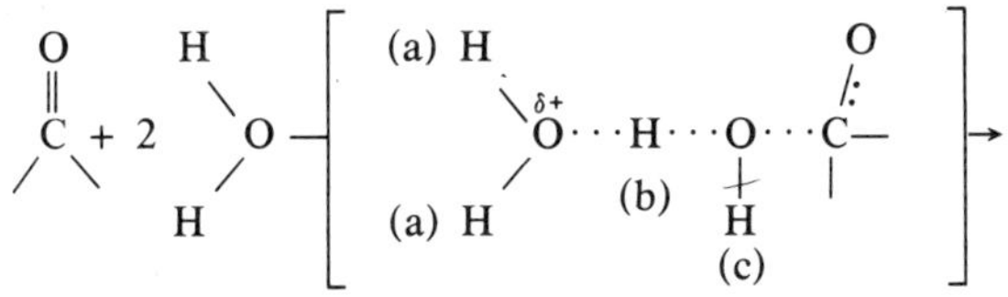

```
 O        H        ┌  (a) H                    O   ┐
 ‖          \      │       \ δ+               /:   │
 C  + 2      O  ───┤        O···H···O···C—         ├──→      (12)
/ \         /      │       /           ┼    |     │
          H        │  (a) H        (b)  H          │
                   └                   (c)        ┘
```

From equation 6 and the transition-state hypothesis of equation 12, we generate the expression of equation 13.

$$k_D/k_H = (\phi_a)^2\,\phi_b\phi_c/(1)^4 = (\phi_a)^2\,\phi_b\phi_c \qquad (13)$$

Let us assume that $\delta \simeq \frac{1}{2}$ in the transition state so that $\phi_a \simeq (0.69)^{1/2} \simeq 0.83$. The proton labeled (b) exists in a transition-state catalytic bridge between two oxygens so that we expect for it $\phi_b \simeq 0.4$. The proton labeled (c) should have a fractionation factor not very far from unity, so for simplicity we take $\phi_c \simeq 1$. This yields the prediction $k_D/k_H \simeq 1/3.6$. A large number of reactions in this category indeed show rates from 2–4 times faster in H_2O than in D_2O (see the tabulation by Johnson, 1965). Adjustments in the exact values of ϕ_a and ϕ_b can easily cover this range ($\phi_a \simeq 0.9$, $\phi_b \simeq 0.6$ yields $k_H/k_D \simeq 2$; $\phi_a \simeq 0.8$, $\phi_b \simeq 0.3$ yields $k_H/k_D \simeq 5$). More detailed studies employing mixtures of light and heavy water confirm the hypothesis of equation 12 (Batts and Gold, 1969; Barnes et al., 1972; Mata-Segreda, 1975).

SPECIAL CONSIDERATIONS WITH ENZYMES

Completeness of Isotope Exchange

It is well known that the exchange of some classes of enzyme hydrogenic sites with isotopic water is not instantaneous and, indeed, may occur at rates much slower than catalytic processes if the sites are deep within the protein structure or are otherwise not intermittently exposed to solvent by reversible conformation changes (Welch and Fasman, 1974). This clearly must be a matter of concern in studies of solvent isotope effects on enzyme action.

The most straightforward general procedure is to measure the rates of hydrogen exchange and then to equilibrate the enzyme with deuterium oxide (or a mixture of protium and deuterium oxides) before beginning the kinetic experiments. This may entail considerable experimentation, however, or may not be possible if the enzyme is unstable. Less thorough procedures frequently may suffice for a satisfactory result. It is common to initiate kinetic experiments by injection of small amounts of an enzyme stock solution into a much larger volume of a solution containing substrate, buffer, etc. A simple check for complications from incomplete exchange, therefore, can be made by employing three different enzyme stock solutions: one in pure protium oxide, one in pure deuterium oxide, and one in an equimolar mixture of the isotopic waters. If experiments employing all three of these give identical results (assuming a sufficiently large dilution factor on injection so that the stock solution leaves the isotopic composition of the larger volume unaffected), this constitutes a reasonable indication that exchange problems are not present. This does not mean, of course, that exchange is complete but only that it is not continuing at a significant rate under the storage conditions for the stock solution.

A second check is provided by the kinetic behavior itself. If simple kinetic behavior (e.g., first-order or zero-order time dependences) is observed throughout the period of a single kinetic experiment, this means either that exchange is not occurring during this period or that hydrogens that are exchanging during this period are not producing a significant kinetic isotope effect. Of course, the exchange characteristics of hydrogens that do not contribute to the isotope effect are not of interest in isotope-effect experiments. Thus, simple kinetics shows that all important hydrogenic sites (from the isotope-effect viewpoint) are already exchanged or else are exchanging very slowly relative to the period of the kinetic experiment.

Conformational Changes

Because of the importance of hydrogen bonds in maintaining the structural integrity of enzymes, it is reasonable to imagine that deuteration at the hydrogen-bonding sites would induce substantial conformational alterations. These alterations could affect the reaction rate for enzyme catalysis and, if the conformation change were undetected, incorrect mechanistic conclusions might be drawn from the observed kinetic isotope effect.

The most general solution to this problem is to conduct a thorough study of the enzyme conformation in deuterium oxide and in mixed isotopic solvents. Here again, this may constitute a project of considerable dimensions rather than an adjunct series of control experiments for a mechanistic investigation. Furthermore, it is possible that conformational changes too small to be detected by the normal techniques could still produce a substantial rate effect.

One source of comfort is that the most extreme form of concern—

that the cumulative effects of hydrogen-bond deuteration would universally produce drastically different enzyme structures—seems to be not at all justified. The x-ray crystallographic structure for ribonuclease completely exchanged with deuterium is identical to that for the protiated enzyme (Bello and Harker, 1961). Most studies of enzymes indicate small effects of deuterium oxide rather than very large effects. In the only case of an enormous structural isotope effect which has been studied in any detail (the subunit association of formyltetrahydrofolate synthetase, considered below) the approximately 60-fold stabilization of the tetramer by deuterium oxide appears to result from effects at only four structural sites, each generating a modest isotope effect.

The reason for this structural insensitivity may be that, contrary to one's initial expectation, isotope effects on hydrogen bonding are usually quite small (see, for example, Wolff et al., 1976). Perhaps the well known lowering of stretching frequencies on formation of hydrogen bonds (which should produce a normal isotope effect) is compensated by an increase in resistance to bending motions, leading to a near-equivalence of binding at the hydrogenic site between hydrogen-bonded and free species.

One check which can as a rule be made conveniently is whether the enzyme exhibits "normal" behavior of its rate/pH inflections on introduction to deuterium oxide (Bender and Hamilton, 1962). This is discussed in the next section.

The Question of pH

When the inflection points of the pH-rate profiles for enzymatic reactions originate in the association or dissociation of acidic groups of the enzyme or derived species, they can be expected to change "normally" when the solvent is made deuterium oxide if there is no gross structural change of the enzyme. "Normal" behavior means that equation 11 should be obeyed. For ammonium and carboxylic acids, $\phi_{LA} \frown 1$ so that the inflection points should move to more basic pHs by 0.5–0.6 units. If the ionization of a sulfhydryl function produces the inflection, ΔpK should be about 0.1–0.3 units. Failure of the pH-rate profile to behave in this fashion may be an indication that structural changes are being induced by deuteration, or that the inflections result from other mechanistic features than ionization, such as changes in the rate-determining step. Analysis of the results for an individual case may allow a distinction of these possibilities.

Figure 2 exhibits these points schematically. It emphasizes that the

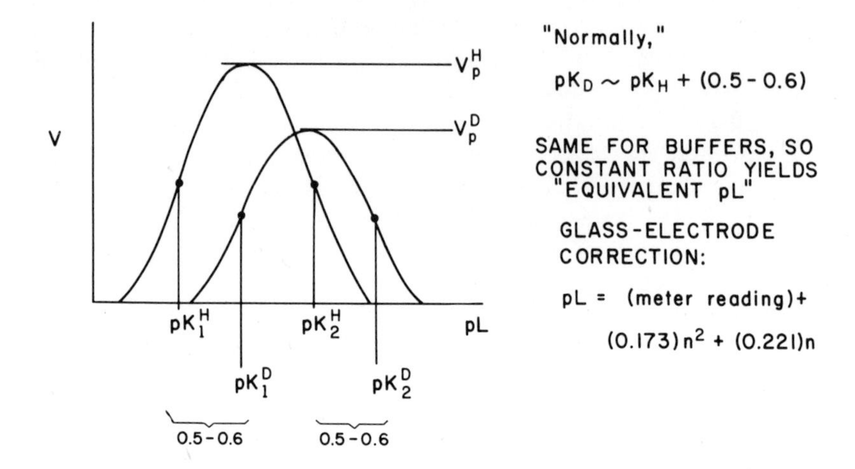

Figure 2. Schematic illustration of "normal" pK shifts (arising from ionization of carboxylic or ammonium acids) of a bell-shaped pH-rate profile. The change in peak velocity (V_p^H to V_p^D) represents an interpretable kinetic isotope effect. The glass electrode correction is from Schowen, Lee, and Schowen (1977).

pKs will shift "normally" in a way reflective of the isotope effect on ionization, while the amplitude of the curve will change in a way related to the kinetic isotope effect. A convenient feature results from the fact that buffer acids are also "normal" as a rule. Their pKs, therefore, will usually shift on deuteration by an amount equal to the shift in the pH-rate profile. If the buffer *ratio* in a kinetic experiment, therefore, is kept constant in a series of isotopically differing solvents, the rate obtained will always be at the same relative point on the pH-rate profile. Thus, although the pL will differ in the various solvents, it will be an "equivalent pL" in all of them.

The measurement with the glass electrode of pL in isotopic solvents requires a correction (Glasoe and Long, 1960) for the electrode is itself an acid for which the pK_a is different in deuterium oxide (Schowen et al., 1976). The formula for obtaining the pL is shown in Figure 2 (Schowen et al., 1976).

Outlook

Figure 3 emphasizes that the interpretation of all solvent isotope effects is a matter of model-building for comparison with observation. The

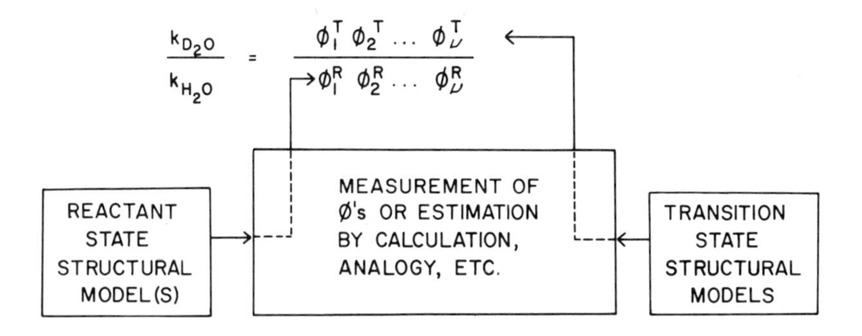

Figure 3. The comparison of structural models for reactants and transition states with measured isotope effects through estimation or measurement of fractionation factors for the various sites in each model. Models must include such effects as conformational reorganization, etc. Further information for model testing (and generation) is available through use of mixtures of isotopic waters, yielding $k_n(n)$.

potential complexity of enzyme behavior in light and heavy water needs always to be borne in mind when constructing models to explain observed solvent isotope effects. The entirety of this behavior, however, is almost surely susceptible of explanation within the framework which we have used in this treatment. As our experience of enzyme behavior in isotopic solvents grows, it is reasonably certain that the picture will simplify, rather than the contrary.

PROTON INVENTORIES: USE OF ISOTOPIC SOLVENT MIXTURES

As noted above, the solvent isotope effect k_H/k_D for the two pure isotopic solvents presents only a single number for comparison with calculated values based on models of the reactant state (each of which specifies a quantity $\Pi_j^\nu \phi_j^R$) and models of the transition state (each of which specifies a quantity $\Pi_i^\nu \phi_i^T$) (Schowen, 1977). Fortunately, the use of rates in mixtures of protium and deuterium oxides provides an opportunity for further testing of models and, indeed, will sometimes permit the dissection of the reactant-state and transition-state contributions to the isotope effect into the individual ϕ^R and ϕ^T values. Such studies have been called "proton inventories."

Rates and Equilibria in Mixed Isotopic Waters

The rate constant k_n in a mixed isotopic solvent with atom fraction n of deuterium is given by k_0, the protium-oxide value, multiplied by a series of corrections J_i^n, one for each hydrogenic site that contributes to the solvent isotope effect (equation 14). The RGM allows us to consider the net correction to be a simple product of the J_i^n. The form of these

$$k_n = k_0 \prod_i^{\nu} J_i^n \tag{14}$$

correction factors (which has been derived in a particularly lucid way by Kresge (1964)) is straightforward and reasonable. A plausible algebraic development (different from but equivalent to Kresge's) is given in equations 15 to 21.

$$J_i^n = f_{i,R,H}^n / f_{i,T,H}^n \tag{15}$$

$$f_{i,H}^n = [H]_i^n/([H]_i^n + [D]_i^n) = 1/(1 + [D]_i^n/[H]_i^n) \tag{16}$$

$$f_{i,H}^n = 1/(1 + [D]_i^n/[H]_i^n) = 1/(1 + \phi_i n/[1 - n]) \tag{17}$$

$$f_{i,H}^n = (1 - n)/(1 - n + n\phi_i) \tag{18}$$

$$J_i^n = \{(1 - n + n\phi_i^T)/(1 - n + n\phi_i^R)\} \tag{19}$$

$$k_n = k_0 \prod_i^{\nu} \{(1 - n + n\phi_i^T)/(1 - n + n\phi_i^R)\} \tag{20}$$

$$k_n = k_0 \prod_i^{\nu} (1 - n + n\phi_i^T)/ \prod_j^{\nu} (1 - n + n\phi_j^R) \tag{21}$$

In equation 15, the fs are the fraction of protium in the ith reactant-state hydrogenic site ($f_{i,R,H}^n$) and in the ith transition-state hydrogenic site ($f_{i,T,H}^n$), each at a solvent atom-fraction of deuterium n. The ratio of these quantities is the proper form for J_i^n for the following reason: If $f_{i,R,H}^n > f_{i,T,H}^n$, then protium accumulates in the ith reactant-state site (preferentially over the transition-state site). Thus, binding at this site is weaker in the reactant state and tighter in the transition state, which should lead to an *inverse* isotope effect contribution. Indeed, this is the case because $J_i > 1$, tending to make $k_n > k_0$. An entirely equivalent argument rationalizes the case when $J_i < 1$. Equation 21, in which the net contributions of the reactant state and of the transition state are separated, therefore should be a correct general description of rates in mixed isotopic waters, and equation 22 should correspondingly describe equilibria.

$$K_n = K_0 \prod_i^{\nu} (1 - n + n\phi_i^P) / \prod_j^{\nu} (1 - n + n\phi_j^R) \quad (22)$$

Note that when $n = 1$, these two equations (happily) dissolve into equations 6 and 7. The form of equation 22 has been verified repeatedly in cases where much is known about the ϕ_i^P and ϕ_j^R (such as the ionization of various acids). Its form, and by inference that of equation 21, therefore may be taken as reliably established (Gold, 1969). Both k_n and K_n are functions of n, the ϕ_i^T (or ϕ_i^P) and the ϕ_j^R, and experiments can be performed at many choices of n in order to establish the experimental shape of $k_n(n)$ or $K_n(n)$. The potential then exists in principle to arrive at a least squares set of ϕ values which are uniquely determined by the data. The precision of measurement required for such a procedure excludes it in practice at the present time for many cases of interest. Nevertheless, it is commonly possible to use the shape of $k_n(n)$ or $K_n(n)$ greatly to reduce the number of models consistent with solvent isotope-effect observations. Furthermore, when the solvent isotope-effect information is joined to all other accessible tests of the models (it goes without saying that this is imperative), its effect can be significant.

Significance of the Shape of $k_n(n)$

The use of $k_n(n)$ to learn about transition states is discussed here, but the similar applications of $K_n(n)$ to equilibrium circumstances follow the same rules. As an example, consider a solvent isotope effect $k_H/k_D = 4$, observed for a hypothetical enzymic rate. To be definite, we take the effect to be on k_{cat} so that the reactant state is the enzyme-substrate complex. We can readily generate (*inter alia*) four models consistent with the overall solvent isotope effect of 4. These are illustrated as (a) through (d) in Figure 4.

Model (a) A base (say a carbanion) abstracts a proton from a sulfhydryl group in the rate-determining step. The sulfhydryl site of the reactant contributes $\phi_{SL}^R = 1/2$, while a primary isotope effect is envisioned for the "in-flight" hydrogen; $\phi^T = 1/8$ will give $k_H/k_D = 4$. From equation 21 we have:

$$k_n = k_0(1 - n + n/8)/(1 - n + n/2)$$

This function is plotted in Figure 4. Note that it is nonlinear and bowed upward.

Model (b) The carbanion abstracts a proton from a hydroxyl group (for which $\phi^R = 1$). With $\phi^T = 1/4$, we now have:

$$k_n = k_0(1 - n + n/4)$$

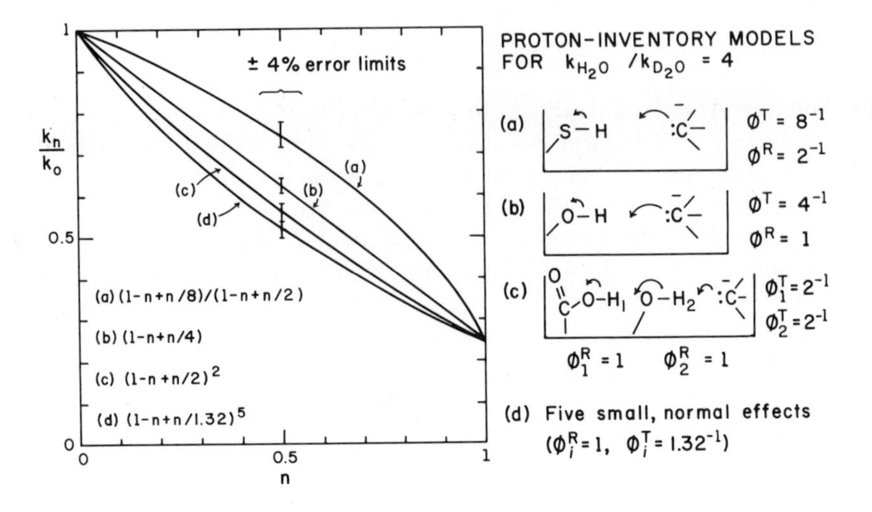

Figure 4. The shapes of $k_n(n)$ for various transition-state and reactant-state models, all of which yield the same overall isotope effect ($k_{H_2O}/k_{D_2O} = 4$).

This is a linear function, also plotted in the figure.

Model (c) Here the carbanion abstracts its proton from a hydroxyl group ($\phi_2{}^R = 1$), which is hydrogen-bonded to a carboxyl group ($\phi_1{}^R = 1$). The bridging hydrogen H_1 is imagined to function as a catalytic bridge between two oxygens, yielding (see above) $\phi_1{}^T \backsim \frac{1}{2}$. The "in-flight" proton H_2 is assigned $\phi_2{}^T \backsim \frac{1}{2}$. For $k_n n$, we obtain:

$$k_n = k_0(1 - n + n/2)^2$$

Again, we find nonlinearity, but now the sense of curvature is opposite to that for Model (a).

Model (d) The solvent isotope effect is considered to arise not from mechanistic sources in the usual sense but from rather small changes in the binding at five hydrogenic sites, say at structural hydrogen bonds. All the ϕ^R are taken as 1 and all the ϕ^T as $1/1.32$ ($= \frac{1}{4}{}^{1/5}$). Then:

$$k_n = k_0(1 - n + n/1.32)^5$$

This function, as the figure shows, is curved in the same sense as $k_n(n)$ for Model (c) but is slightly more curved.

As the figure shows, these four models, all of which agree equally well with the overall solvent isotope effect, give quite different pictures for $k_n(n)$. Thus, a proton inventory (even at the ± 4% precision level for which error limits are plotted on the curves) is capable of distinguishing these models. Of course, whichever curve (if any, of those shown) is found to fit the experimental data for a real case, there will be many other models also capable of generating the observed $k_n(n)$. The number of such alternatives is, in fact, infinite, as is true with every scientific experiment (Poincaré, 1952).

The gross distinctions in shape of these curves are of a general nature and can be useful in generating models likely to fit a particular set of observations (Albery and Davies, 1972). All the models correspond to an overall normal isotope effect ($k_0 > k_1$) so that the general "tilt" of all the curves is down. The opposite would, of course, be true for an overall inverse effect. Model (b) is one in which the entire isotope effect arises from a single transition-state site. Whenever this is true, $k_n(n)$ is linear. Indeed, whenever all or nearly all of the *net* effect arises from one ϕ^T (with a collection of other ϕ_i^T and ϕ_j^R exactly or roughly canceling) a linear $k_n(n)$ will be seen with data of common precision (Kresge, 1973). With Model (a), the net isotope effect results from partial cancellation of opposing contributions from one transition-state site and one reactant-state site. If the ϕ^T contribution had been unopposed by ϕ^R of ½, the net effect would have been 8. If the ϕ^R effect had not been canceled by ϕ^T but if the site had generated a $\phi^T = 1$ instead, then the net effect would have been inverse (½). The result is that increasing deuteration reduces the rate, but not as rapidly as with a single, unopposed contribution; the opposing factor "slows down" the fall-off and produces the upward bowing. In the case of the two-proton Model (c) and five-proton Model (d), the multiple-site effects are in the same direction. Now there is an opportunity, even at small n, for both sites to be deuterated. Thus, the rate drops off initially more rapidly than for a one-proton effect, and the result is the downward bowing seen for these curves. The larger the number of sites the greater the curvature. The distinction between the two-proton and five-proton models is not very great, however, and data of common precision, if the net isotope effect is modest, may frequently distinguish only one-proton from multiproton models.

Generation of Interpretative Models

Once one has in hand a set of $k_n(n)$ experimentally found for some enzymic reaction, the problem becomes one of generating various interpretative models for the reactant state and transition state. These can be related to sets ϕ_j^R and ϕ_i^T, respectively, and thus curves $k_n(n)$ can be calculated for each model from equation 21 and compared with the experimental data. The major source of clues for these models is usually other experimental data which suggest mechanistic features of the reaction. A second line of approach is visual inspection of the experimental $k_n(n)$ to see whether inverse isotope-effect contributions are indicated (upward bowing) or multiple normal effects (downward bowing), etc. Beyond these simple measures, we employ a polynomial regression method which is often helpful, and Albery (1975) has recently suggested a different technique (the "gamma method") for surveying the characteristics of models which will be consistent with a given data set.

The utility of the polynomial regression procedure is greatest if one has reason to suspect that one state (reactant or transition) does not contribute substantially to the isotope effect. Thus, if all the $\phi_j^R = 1$ (no reactant-state contribution), equation 21 simplifies to the form shown in equation 23, so that k_n is a polynomial in n, of order ν.

$$k_n = k_0 \prod_i^{\nu} (1 - n + n\phi_i^T) = k_0 + c_1 n + c_2 n^2 + \cdots + c_\nu n^\nu \quad (23)$$

Here ν is the number of "active protons," i.e., sites which contribute to the solvent isotope effect. The experimental $k_n(n)$ may now be fit by least squares procedures to polynomials in n of ever increasing order. Most computer centers have a program for this purpose; we use BMDO5R of the UCLA Health Sciences Computing Facility. Each term in n can then be evaluated for its statistical significance, for example, by the F test (Fisher and Yates, 1957). The exponent of n in the last statistically significant term gives the number of active protons justified by the data. Of course, this is a function not only of the physical system under study but also of the precision and abundance of the measurements. Therefore, at any given point, there may be further active protons hidden in the noise of the kinetic data and this needs to be remembered in formulating models.

In a case in which transition-state protons do not contribute (all $\phi_i^T = 1$), but the whole effect arises from reactant sites, the transition-state contribution in the numerator of equation 21 becomes unity. Then it is k_n^{-1} (equation 24) which becomes a polynomial in n.

$$k_n^{-1} = k_0^{-1} \prod_j^{\nu} (1 - n + n\phi_j^R) = k_0^{-1} + g_1 n + g_2 n^2 + \cdots + g_\nu n^\nu \quad (24)$$

The same regression technique as above can be used to explore values of ν.

Albery (1975) has introduced a procedure based on a quantity γ, which is related by equation 25 to the overall solvent isotope effect k_1/k_0 and the solvent isotope effect at $n = 0.5$, $k_{0.5}/k_0$.

$$\gamma = 8 \ln[(k_{0.5}/k_0)/(k_1/k_0)^{1/2}]/\{\ln(k_1/k_0)\}^2 \quad (25)$$

The point at $n = 0.5$ is of special utility because (as a glance at Figure 4 confirms) this is where the difference is greatest between the linear $k_n(n)$ and the various nonlinear possibilities. Although it is not obvious, $\gamma = 1$ when $k_n(n)$ is linear and γ tends toward zero as the number of sites contributing to the isotope effect becomes very large. It is even less obvious, but demonstrable by a stupendous algebraic exercise, that (for ϕs between 3.5 and 0.3, thus excluding some primary isotope effects) γ is also approximately given by equation 26.

$$\gamma = \sum_i \Lambda_i^2/a_i \quad (26)$$

In this equation, a_i is the number of sites of the ith class (positive for transition-state sites, negative for reactant-state sites) and Λ_i is the fraction of the overall isotopic difference in free energies of activation (for k_1/k_0) which is produced by sites of the ith class. Because equation 26 must be derived from models for reactant and transition states while equation 25 uses only experimental quantities, the two together constitute a method of comparing models with observations. In fact, nice plots can be constructed showing the spectrum of models consistent with experimental values of γ. Further details and extensions (to include ϕs outside the range 0.3–3.5) are provided by Albery (1975).

EXAMPLES OF ENZYMIC SOLVENT ISOTOPE EFFECTS

Catalysis of Amide and Ester Hydrolysis by Serine Proteases

Solvent isotope effects for enzymes in this class long ago were ascribed to acid-base catalysis by active-site functional groups (Bender and Hamilton, 1962; Bender and Kezdy, 1975). The more recent crystallographic discovery (Blow, Birktoft, and Hartley, 1969) of an interesting chain of two hydrogen bonds at the catalytic center, however, aroused further interest in the detailed interpretation of the effects, which are frequently (for these and other esterases and amidases) in the range $k_H/k_D \sim 2$–4. The hydrogen bond chain in question (the charge-relay system) links Ser-195 to His-57 to Asp-102 in α-chymotrypsin and the corresponding residues in other, similar enzymes (Blow, 1976). It is possible that both hydrogens of the chain undergo alterations in binding upon activation, thus contributing to the solvent isotope effect, or, alternatively, that only one proton (presumably that of His-57) engages in catalytic bridging, while the role of the aspartate is to orient the histidine or otherwise to assist without catalytic bridging. This is an obvious case in which the overall magnitude of the solvent isotope effect cannot distinguish the models but a proton inventory is required. If the solvent isotope effect arises from a single bridging proton (transition-state fractionation factor $\phi_1{}^T$) and this proton has $\phi_1{}^R \backsim 1$, then equation 21 becomes the linear function of equation 27.

$$k_n = k_0(1 - n + n\phi_1{}^T) \tag{27}$$

If both protons of the charge-relay chain contribute to the solvent isotope effect (with transition-state fractionation factors $\phi_1{}^T$ and $\phi_2{}^T$, and both $\phi_1{}^R$ and $\phi_2{}^R \backsim 1$), then the quadratic form of equation 28 is obtained.

$$k_n = k_0(1 - n + n\phi_1{}^T)(1 - n + n\phi_2{}^T) \tag{28}$$

Pollock, Hogg, and Schowen (1973) reported that the deacetylation of acetyl-α-chymotrypsin ($k_H/k_D = 2.4$) gave a linear $k_n(n)$, thus showing it to be a one-proton catalyst. Elrod (1975) has obtained confirmation of this conclusion with both acylation and deacylation reactions of various substrates and various serine proteases (see also Elrod et al., 1975). Figure 5 shows a few of his results (the caption gives the statistical information from the polynomial regression).

On the other hand, quantum-theoretical studies of hydrogen-bond chains suggest that rather small changes in the distances across these chains may strongly influence the coupling of bonding changes at the hydrogenic sites (Gandour, Maggiora, and Schowen, 1974). It is quite possible that long-chain polypeptide substrates would influence this distance more than the smaller molecules used in the studies cited above (cf. Maggiora and Schowen, 1977). It therefore seemed possible that coupled bonding changes (and thus a double contribution to the solvent isotope effect) might be observed with longer substrates. Recent work on trypsin by Quinn et al. (1976) shows that this indeed is the case (Figure 6). An exactly similar observation has been made by Hunkapillar, Forac, and Richards (1976) for α-lytic protease.

Catalysis of Amide Hydrolysis by Amidohydrolases

Glutaminases and asparaginases, which catalyze the conversion of glutamine to glutamic acid and asparagine to aspartic acid, show some

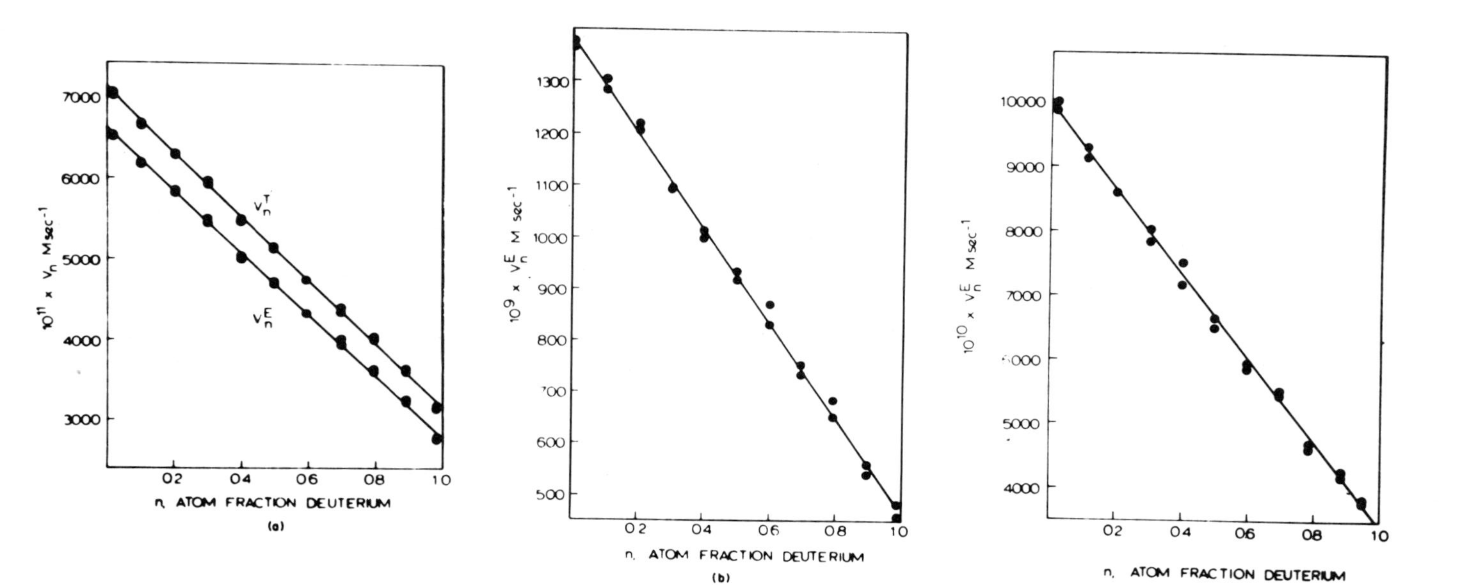

Figure 5. (a) Variation of the zero-order rate constants for the deacetylation of acetylelastase as a function of the atom fraction n of deuterium in binary mixtures of protium oxide and deuterium oxide. The rates were determined for the elastase-catalyzed hydrolysis of p-nitrophenyl acetate. $V_n{}^E$ represents the rates of the enzyme-catalyzed reaction, which were obtained by correction of the total hydrolysis rate $V_n{}^T$ for the small component of buffer-catalyzed reaction. The solid lines are linear least squares fits to the data (statistically significant with $P < 0.001$). A quadratic fit, although significant at $P < 0.01$, yields an isotope effect of 1.00 for the second proton. (b) A similar plot for deacylation of α-N-benzoyl-L-arginyltrypsin. Here the rates are for trypsin-catalyzed hydrolysis of α-N-benzoyl-L-arginyl ethyl ester and the background reaction is negligible. Again, the solid line represents a least squares fit (statistically significant with $P < 0.001$). In an attempted quadratic fit, the term in n^2 is not significant at $P < 0.2$. (c) The same experiment as in (b) with thrombin as catalyst. The linear fit again has $P < 0.001$, while the quadratic term is not significant at $P < 0.1$.

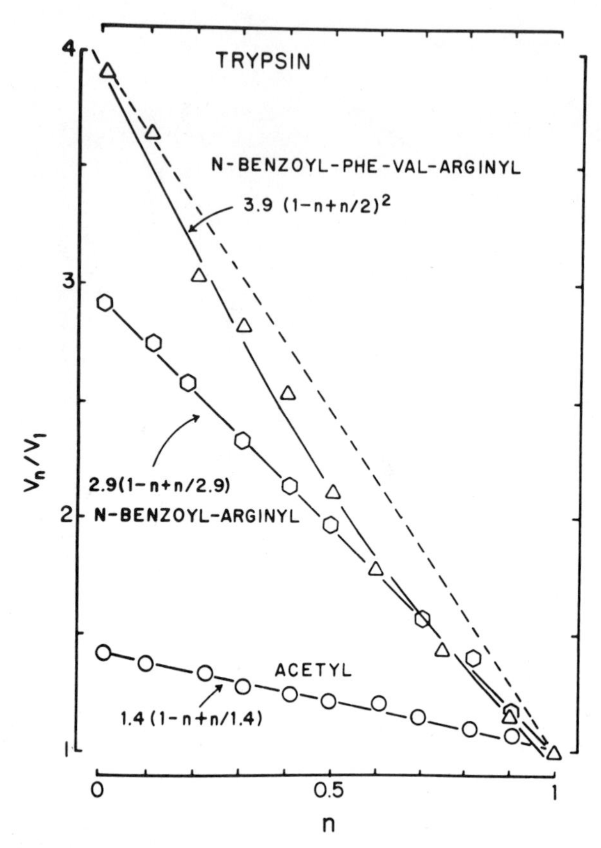

Figure 6. A series of proton-inventory results for three examples of trypsin catalysis. With a very small substrate-derived function (acetyl, lowest line), the enzyme is a one-proton catalyst with a small isotope effect. Increasing the size of this function to that of a dipeptide analog (*N*-benzoyl-arginyl, middle line) produces a larger isotope effect but the enzyme remains a one-proton catalyst. Finally, a tetrapeptide analog (*N*-benzoyl-phenylalanyl-valyl-arginyl) couples the charge-relay chain and the enzyme becomes a two-proton catalyst. The first result is for *p*-nitrophenyl acetate and the second for an ethyl ester as substrate so that deacylation is rate limiting. For the third case, a *p*-nitroanilide was used so that acylation may be rate limiting (Hunkapillar, Forgac, and Richards, 1976).

mechanistic similarities to the serine proteases, although mechanistic information is much sparser with the amidohydrolases and no detailed crystallographic results are yet available (Hartman, 1971; Wriston and Yellin, 1973). Therefore, we cannot be sure whether the active sites of these enzymes contain any special catalytic feature such as the charge-relay system. They do, however, frequently exhibit solvent isotope effects in the range $k_H k_D \sim 2$–4 (Venkatasubban, 1975). For the hydrolysis of glutamine by glutaminase-A of *Escherichia coli* and of asparagine by asparaginases of *E. coli* and *Erwinia carotovora,* the proton inventories are nonlinear, with $k_n(n)$ bowed downward. In fact, to a rather good approximation $\sqrt{k_n}$ is linear in n (Elrod et al., 1975). This suggests the simple model in which two transition-state protons contribute equally to the solvent isotope effect (equations 29 and 30).

$$k_n = k_0(1 - n + n\phi^T)^2 \tag{29}$$

$$\sqrt{k_n} = \sqrt{k_0}\,(1 - n + n\phi^T) \tag{30}$$

Venkatasubban (1975) noticed a curious fact with some of these enzymes: when a non-physiological substrate is employed, the solvent isotope effect is substantially reduced. "Thus it is a tempting hypothesis," he wrote, "to suggest that for the nonspecific substrate the mechanism has changed from a two proton catalysis to a one proton catalysis." This hypothesis seems to be correct. Figure 7 shows proton inventories for *Erwinia* asparaginase with asparagine and glutamine. With asparagine, a two-proton dependence is found, with glutamine, a one-proton dependence (Quinn et al., in preparation).

Thus here, as with the serine proteases, the results are consistent with a model in which the evolutionarily selected substrate, with a presumably excellent transition-state fit to the enzyme structure, gives multiproton catalysis. A less highly complementary transition state, for another substrate, uncouples the multiproton entity and the mechanism degenerates to one-proton catalysis.

Catalysis of Pyrophosphate Hydrolysis by Inorganic Pyrophosphatase

This enzyme catalyzes the hydrolysis of pyrophosphate with k_H/k_D = 1.9 when rates are compared at equivalent pL (see above). The pH-rate inflection is "normal," being shifted to higher pL by 0.5 units in deuterium oxide. The proton inventory of Konsowitz and Cooperman (1976) is shown in Figure 8. It is bowed upward, which shows that some form of inverse isotope-effect contribution is present. Konsowitz and Cooperman considered both reactant-state and transition-state origins for the inverse effect and discovered two dependences that fit the data about equally well. These are given as equations 31 and 32.

$$k_n = k_0(1 - n + 0.40n)/(1 - n + 0.90n)^2 \tag{31}$$

$$k_n = k_0(1 - n + 0.37n)(1 - n + 1.37n) \tag{32}$$

In equation 31, an inverse contribution arises from reactant-state fractionation factors of 0.90 (binding weaker than in bulk water) and in

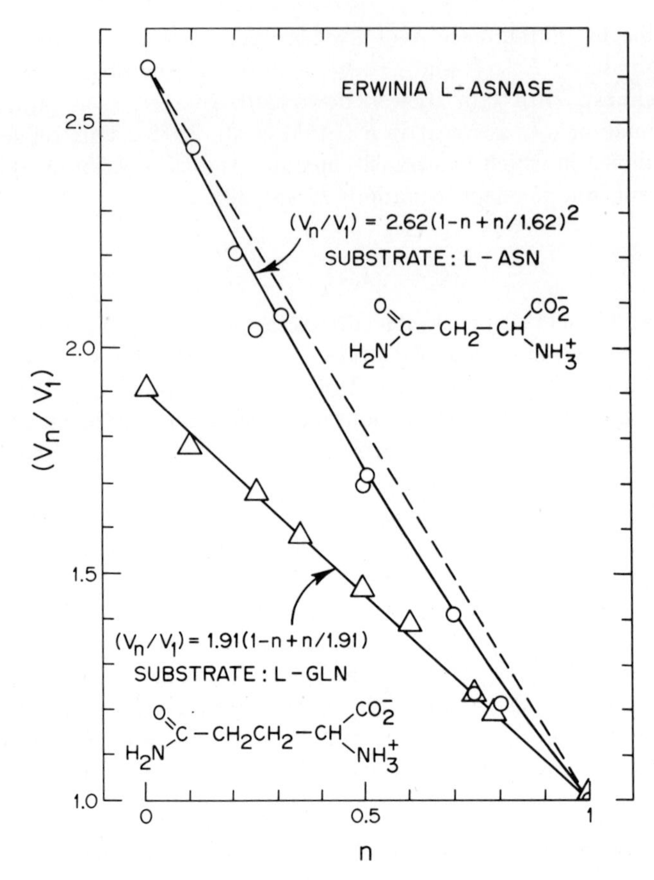

Figure 7. Proton inventories for hydrolytic action of asparaginase (*Erwinia carotovora*) upon its natural substrate asparagine, for which it is a two-proton (or multiproton) catalyst, and the unnatural substrate glutamine, for which it is a one-proton catalyst. The uncoupling of the enzyme catalytic entity is associated in this case with only a 30-fold rate reduction (see Elrod et al., 1975).

equation 32, the inverse contribution comes from a transition-state fractionation factor of 1.37 (binding tighter than in bulk water). Three models were adduced on the basis of these data and other mechanistic information and are shown in Figure 8 as Models A, B, and C. Models A and B correspond to equation 31. In both cases two hydrogens belong in the reactant state to a water molecule coordinated to Mg^{2+}. Because the water oxygen should have a partial positive charge, these hydrogens should have $\phi_1^R \sim \phi_2^R$ between 1.00 (water) and 0.69 ($H\overset{+}{O}\langle$). The assign-

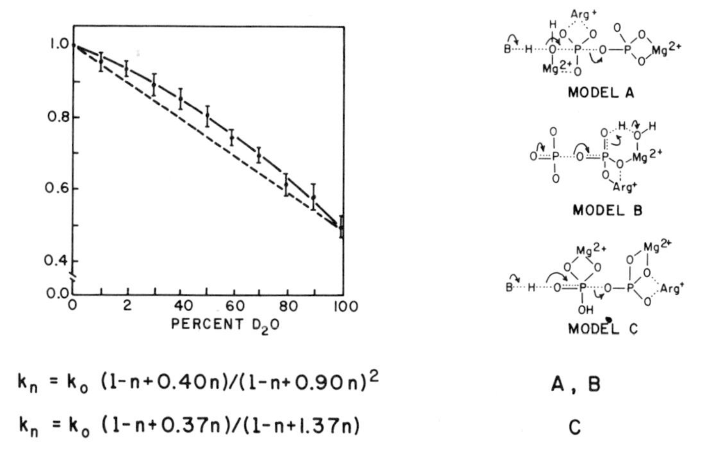

Figure 8. Proton inventory and interpretative models for hydrolysis of pyrophosphate by inorganic pyrophosphatase (Konsowitz and Cooperman, 1976).

ment $\phi_1{}^R = \phi_2{}^R = 0.90$, therefore, is made. In the transition state of Models A and B, one hydrogen is attached to an oxygen, which probably has little net charge and therefore, is assigned $\phi_2{}^T = 1$. The other hydrogen is in a transition-state bridge between oxygen and oxygen (B) or oxygen and the oxygen or nitrogen of B (A), which is consistent with $\phi_2{}^T = 0.40$.

Model C corresponds to equation 32. Both reactant-state hydrogens are here assigned $\phi^R = 1$. In the transition state, the nontransferring hydrogen is attached to an oxygen at a tetrahedral center to which lone-pair delocalization is doubtless significant. As in the case of *gem*-diols and hemiacetals (Mata-Segreda, Wint, and Schowen, 1974), restrictions to rotation about the single bond make $\phi^T > 1$. Thus, it is logical to associate this center with $\phi_1{}^T = 1.37$. Here again the other hydrogen bridges between two oxygens in a transition state, substantiating the assignment $\phi_2{}^T = 0.37$.

Conformational Change in Ribonuclease-A

The enzyme ribonuclease-A undergoes a rapid conformational change near neutral pH which can be observed by the temperature-jump technique (Wang et al., 1975). The isomerization interconverts two forms of the enzyme, one with $pK_a > 8$, the other with pK_a 6.1 (H_2O) so that an indicator can follow the reaction through the uptake or release of protons. The rate constants for interconversion were found to be reduced by about five-fold in deuterium oxide, in accord with some form of protolytic involvement in the rate-determining step (French and Hammes, 1965).

Figure 9 shows, at lower right, the variation of the rate constant k_{12}^n (for conversion of the less acidic to the more acidic conformer) with n. The dependence is strongly nonlinear with downward bowing. The most likely models thus will be ones in which more than one transition-state site contributes a normal isotope effect. Polynomial regression analysis confirms that at least two sites are required and suggests three: the solid line on the plot is for the best-fit cubic polynomial.

One attractive model for the transition state is that shown at upper right. Here it is imagined that the less acidic enzyme conformer is stabilized by an interaction between the imidazolium moiety of a histidine and some other structural feature of the enzyme. The interaction shifts the pK_a of the histidine to above 8. The rate-determining step in the isomerization is then imagined as the transfer of the "outer" imidazolium proton to an adjacent water molecule to produce a hydronium ion. The now-neutral imidazole no longer enjoys its stabilizing interaction and, in a fast step, moves to its new conformational position where its pK_a becomes 6.1. We take the ϕ_j^R to be 1 (two water sites and two imidazolium sites), and we let $\phi_i^T = \phi_2^T = 0.69$ for the outer water sites of the transi-

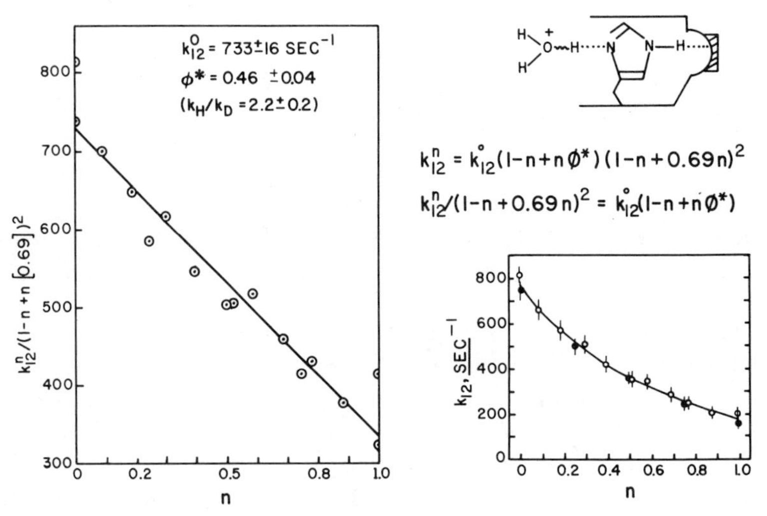

Figure 9. Proton inventory data and interpretative model for conformational isomerization of ribonuclease (Wang et al., 1975).

tion state. This is because proton transfer from an acid of $pK_a \sim 8.5$ to the very weak base water should have a very product-like transition state, with the hydronium ion nearly fully formed. For the moment, we do not specify the fractionation factor for the bridging proton, calling it ϕ^*, and we consider $\phi_3{}^T = 1$ for the other imidazole proton. This leads to equation 33.

$$k_n = k_0(1 - n + 0.69n)^2(1 - n + \phi^* n) \quad (33)$$

Notice that this equation is convertible to equation 34.

$$k_n/(1 - n + 0.69n)^2 = k_0(1 - n + \phi^* n) \quad (34)$$

A plot of the left side of equation 34 versus n will be linear if the model is consistent with the data and furthermore will determine the appropriate value of ϕ^*. At the left side of Figure 9 the plot is shown. It yields the very reasonable value $\phi^* = 0.46$.

Subunit Association in Formyltetrahydrofolate Synthetase

The inactive monomer of formyltetrahydrofolate synthetase (*Clostridium cylindrosporum*) is induced by certain monovalent cations, such as potassium or cesium ions, to form a tetrameric aggregate incorporating two of the inorganic cations (Harmony, Himes, and Schowen, 1975). This tetramer is the enzymically active species. Both the rate and the equilibrium constant of tetramer formation are increased in deuterium oxide, the rate about three-fold and the equilibrium constant at least 50-fold.

At a higher ionic strength than was used for the other experiments, it was shown that the rate isotope effect is wholly localized in a rapid, reversible step in which monomer binds cation. The second step of the mechanism, in which the monomer-cation complex combines with another monomer unit, shows no isotope effect. A plot of rate constant against n is nonlinear, but the reciprocal rate constant is linear in n (cf. equation 24). This is consistent with the view that the equilibrium cation-binding to the monomer produces its inverse isotope effect by causing a single, loosely bound proton ($\phi \sim 1/3$) to be expelled from the cation binding site and to enter a more ordinary binding state ($\phi \sim 1$).

This model receives nice support from observations on the equilibrium constant for tetramer formation, K_{A_n}, which is plotted versus n in Figure 10. The isotope effect is in the same direction as that on the rate but much larger. In fact, if it is supposed that each of the two cations which bind the tetramer together occupy simultaneously two binding sites (say one on each monomer) and must displace from each

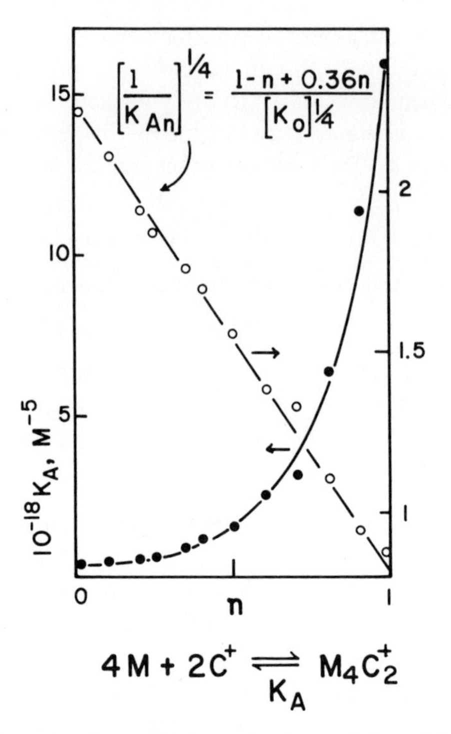

Figure 10. Proton inventory for equilibrium subunit association of formyltetrahydrofolate synthetase (Harmony, Himes, and Schowen, 1975).

binding site one of the weakly bound protons mentioned above, then the isotope effect on equilibrium tetramer formation would arise from the conversion of four reactant-state (monomer) protons with $\phi^R < 1$ to ordinary protons with $\phi^P \sim 1$ in the tetramer. This model generates equation 35,

$$K_{An} = K_{A0}/(1 - n + n\phi^R)^4 \tag{35}$$

which readily goes over to equation 36.

$$(1/K_{An})^{1/4} = (1/K_{A0})^{1/4}(1 - n + n\phi^R) \tag{36}$$

The latter is plotted in Figure 10 and is consistent with the data. The value of ϕ^R obtained from Figure 10 using equation 36 is 0.36, not far from the value of 0.33 estimated for cation binding from the kinetic data.

These results do not in any way prove that the subunit-association isotope effects arise from the binding-site structure mentioned above, but they are consistent with the view that the very large effect of D_2O should be understandable in reasonably simple, chemical terms. As structural and enzymological studies progress, it will be interesting to see what picture of the binding site eventually emerges.

CONCLUSION

Although the study of enzymic reactions in deuterated solvents is effectively at a primitive stage of development and surprises can be expected from future work, it seems not too early to conclude that the effects are interpretable as a general rule in terms of moderately straightforward concepts and should be useful in mechanisms research. As Figure 11 emphasizes, the most powerful use of solvent isotope effects will be as part of a general iterative scheme for the development of interpretative models which incorporate the results of all kinds of experiments and which themselves suggest new experiments to produce further refinement of the models.

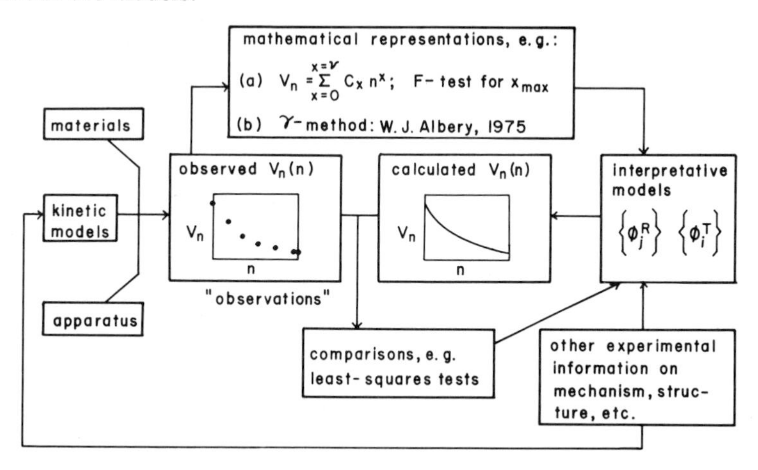

Figure 11. The role of solvent isotope effects and their interpretation in the iterative development of mechanistic models.

ACKNOWLEDGMENTS

It is a pleasure to thank all my collaborators, many of whose papers are cited, for their manifold contributions to this article, and to express our joint gratitude to the National Science Foundation and National Institutes of Health for their support of our research. Professors B. S. Cooperman and J. H. Richards kindly sent unpublished papers, and the organizers of the Steenbock Symposium provided the occasion and the impetus.

LITERATURE CITED

Albery, W. J. 1975. Solvent isotope effects. In: E. Caldin and V. Gold (eds.), Proton-Transfer Reactions, pp. 263-316. Chapman and Hall, London.

Albery, W. J., and Davies, M. H. 1969. Effect of the breakdown of the rule of the geometric mean on fractionation factor theory. Trans. Faraday Soc. 65:1059-1065.

Albery, W. J., and Davies, M. H. 1972. Mechanistic conclusions from the curvature of solvent isotope effects. J. Chem. Soc. Faraday Trans. I: 167–181.

Arnett, E. M., and McKelvey, D. R. 1969. Solvent isotope effect on thermodynamics of nonreacting solutes. In: J. F. Coetzee and C. D. Ritchie (eds.), Solutes-Solvent Interactions, pp. 344–399. Marcel Dekker, New York.

Barnes, D. J., Cole, M., Lobo, S., Winter, J. G., and Scott, J. M. W. 1972. Studies in solvolysis. Part V. Further investigations concerning the solvolysis of primary, secondary and tertiary trifluoroacetates. Can. J. Chem. 50: 2175–2181.

Batts, B. D., and Gold, V. 1969. Rates of water-catalyzed reactions in water-deuterium oxide mixtures. J. Chem. Soc. A. 984–987.

Bello, J., and Harker, D. 1961. Crystallization of deuterated ribonuclease. Nature (London) 192:756.

Bender, M. L., and Hamilton, G. A. 1962. Kinetic isotope effects of deuterium oxide on several α-chymotrypsin-catalyzed reactions. J. Am. Chem. Soc. 84:2570-2577.

Bender, M. L., and Kézdy, F. J. 1975. Proton-transfer in enzymatic catalysis. In: E. Caldin and V. Gold (eds.), Proton-Transfer Reactions, pp. 385–408. Chapman and Hall, London.

Blow, D. M. 1976. Structure and mechanism of chymotrypsin. Accts. Chem. Res. 9:145–152.

Blow, D. M., Birktoft, J. J., and Hartley, B. S. 1969. Role of a buried acid group in the mechanism of action of chymotrypsin. Nature (London) 221: 337-340.

Elrod, J. P. 1975. Comparatively mechanistic study of a set of serine hydrolases using the proton inventory technique and beta-deuterium probe. Unpublished doctoral dissertation, University of Kansas, Lawrence, Kansas.

Elrod, J. P., Gandour, R. D., Hogg, J. L., Kise, M., Maggiora, G. M., Schowen, R. L., and Venkatasubban, K. S. 1975. Proton bridges in enzyme catalysis. Faraday Symp. Chem. Soc. 10:145–153.

Fisher, R. A., and Yates, F. 1957. Statistical Tables for Biological Agricultural and Medical Research, 5th Ed. Oliver and Boyd, Edinburgh and London.

French, T. C., and Hammes, G. G. 1965. Relaxation spectra of ribonuclease. II. Isomerization of ribonuclease at neutral pH values. J. Am. Chem. Soc. 87:4669-4673.

Friedman, L., and Shiner, V. J. 1966. Experimental determination of the disproportionation of hydrogen isotopes in water. J. Chem. Phys. 44:4639-4640.
Gandour, R. D., Maggiora, G. M., and Schowen, R. L. 1974. Coupling of proton motions in catalytic activated complexes. Model potential-energy surfaces for hydrogen-bond chains. J. Am. Chem. Soc. 96:6967-6979.
Glasoe, P. K., and Long, F. A. 1960. Use of glass electrodes to measure acidities in deuterium oxide. J. Phys. Chem. 64:188-190.
Gold, V. 1963. The fractionation of hydrogen isotopes between hydrogen ions and water. Proc. Chem. Soc. 141-143.
Gold, V. 1968. Rule of the geometric mean: Its role in the treatment of thermodynamic and kinetic deuterium solvent isotope effects. Trans. Faraday Soc. 64:2770-2779.
Gold, V. 1969. Protolytic processes in H_2O—D_2O mixtures. Adv. Phys. Org. Chem. 7:259-331.
Gold, V., and Grist, S. 1972. Deuterium solvent isotope effects on reactions involving the aqueous hydroxide ion. J. Chem. Soc. Perkin II:89-95.
Halevi, E. A. 1971. Solvent isotope effects—second thoughts about an old problem. Israel J. Chem. 9:385-395.
Harmony, J. A. K., Himes, R. H., and Schowen, R. L. 1975. The monovalent cation-induced association of formyltetrahydrofolate synthetase subunits: A solvent isotope effect. Biochemistry 14:5379-5386.
Hartman, S. C. 1971. Glutaminase and γ-glutamyltransferases. In: P. D. Boyer (ed.), The Enzymes, ch. 4, pp. 79-100. 3rd Ed. Academic Press, New York.
Hartshorn, S. R., and Shiner, Jr., V. J. 1972. Calculation of H/D, $^{12}C/^{13}C$, and $^{12}C/^{14}C$ fractionation factors from valence force fields derived for a series of simple organic molecules. J. Am. Chem. Soc. 94:9002-9012.
Hunkapillar, M. W., Forgac, M. D., and Richards, J. H. 1976. Mechanism of action of serine proteases: Tetrahedral intermediate and concerted proton transfer. Biochemistry 15:5581-5588.
Johnson, S. L. 1965. General base and nucleophilic catalysis of ester hydrolysis and related reactions. Adv. Phys. Org. Chem. 5:237-330.
Jones, W. M. 1968. Vapor pressure of tritium oxide and deuterium oxide. Interpretation of the isotope effects. J. Chem. Phys. 48:197-214.
Katz, J. J., and Crespi, H. L. 1970. Isotope effects in biological systems. In: C. J. Collins and N. S. Bowman (eds.), Isotope Effects in Chemical Reactions, pp. 286-363. Van Nostrand-Reinhold, New York.
Konsowitz, L. M., and Cooperman, B. S. 1976. Solvent isotope effect in inorganic pyrophosphatase-catalyzed hydrolysis of inorganic pyrophophate. J. Am. Chem. Soc. 98:1993-1995.
Kresge, A. J. 1964. Solvent isotope effect in H_2O—D_2O mixtures. Pure Appl. Chem. 8:243-258.
Kresge, A. J. 1973. Solvent isotope effects and the mechanism of chymotrypsin action. J. Am. Chem. Soc. 95:3065-3067.
Kresge, A. J., and Allred, A. L. 1963. Hydrogen isotope fractionation in acidified solutions of protium and deuterium oxide. J. Am. Chem. Soc. 85: 1541.
Laughton, P. M., and Robertson, R. E. 1969. Solvent isotope effects for equilibria and reactions. In: J. F. Coetzee and C. D. Richie (eds.), Solute-Solvent Interactions, pp. 400-538. Marcel Dekker, New York.

Lienhard, G. E., and Jencks, W. P. 1966. Thiol addition to the carbonyl group. Equilibria and kinetics. J. Am. Chem. Soc. 88:3982-3995.

Maggiora, G. M., and Schowen, R. L. 1977. The interplay of theory and experiment in bioorganic chemistry: Three case histories. In: E. E. van Tamelen (ed.), A Survey of Contemporary Bioorganic Chemistry. Academic Press, New York.

Mata-Segreda, J. F. 1975. Chemical models for aqueous biodynamical processes. Unpublished doctoral dissertation, University of Kansas, Lawrence, Kansas.

Mata-Segreda, J. F., Wint, S., and Schowen, R. L. 1974. Tight binding of hydroxyl protons in *gem*-diols and hemiacetals. J. Am. Chem. Soc. 96:5608.

Meloche, H. P., Monti, C. T., and Cleland, W. W. 1977. Magnitude of the equilibrium isotope effect in carbon-tritium bond synthesis. Biochim. Biophys. Acta. 480:517–519.

Minor, S. S., and Schowen, R. L. 1973. One-proton solvation bridge in intramolecular carboxylate catalysis of ester hydrolysis. J. Am. Chem. Soc. 95: 2279-2281.

More O'Ferrall, R. A. 1975. Substrate isotope effects. In: E. Caldin and V. Gold (eds.), Proton-transfer reactions, p. 216. John Wiley and Sons, New York.

Poincaré, H. 1952. Science and Hypothesis, ch. IX. Dover Publications, New York.

Pollock, E., Hogg, J. L., and Schowen, R. L. 1973. One-proton catalysis in the deacetylation of acetyl-α-chymotrypsin. J. Am. Chem. Soc. 95:968.

Pyper, J. W., and Long, F. A. 1964. Equilibrium in the deuterium exchange of acetylene and water. J. Chem. Phys. 41:1890-1896.

Quinn, D. M., Patterson, M., Jarvis, R., Ranney, G., and Schowen, R. L. Changes in catalytic coupling in the action of amidohydrolases and serine hydrolases with changes in substrate structure. In preparation.

Schowen, K. B. 1977. Solvent hydrogen isotope effects. In: R. D. Gandor and R. L. Schowen (eds.), Transition States of Biochemical Processes. Plenum Press, New York.

Schowen, K. B., Lee, J. K., and Schowen, R. L. 1977. A polybasic-acid model for the glass electrode. In preparation.

Schowen, R. L. 1972. Mechanistic deductions from solvent isotope effects. Prog. Phys. Org. Chem. 9:275-332.

Thomson, J. F. 1960. Fumarase activity in D_2O. Arch. Biochem. Biophys. 90: 1-6.

Van Hook, W. A. 1972. Disproportionation of HOD in condensed phases. J. Chem. Soc. Chem. Comm. 479-480.

Venkatasubban, K. S. 1975. Proton bridges in enzymic and nonenzymic amide solvolysis. Unpublished doctoral dissertation, University of Kansas, Lawrence, Kansas.

Wang, M.-S., Gandour, R. D., Rodgers, J., Haslam, J. L., and Schowen, R. L. 1975. Transition-state structure for a conformation change of ribonuclease. Bioorg. Chem. 4:392-406.

Welch, W. H., Jr., and Fasman, G. D. 1974. Hydrogen-tritium exchange in polypeptides. Models of α-helical and β conformations. Biochemistry 13: 2455–2466.

Winter, J. G., and Scott, J. M. W. 1968. Studies in solvolysis. Part I. The neutral hydrolysis of some alkyl trifluoroacetates in water and deuterium oxide. Can. J. Chem. 46:2887–2894.

Wolff, H., Bauer, O., Goetz, R., Landeck, H., Schiller, O., and Schimpt, L. 1976. Association and vapor pressure isotope effect of variously deuterated methanols in *N*-hexane. J. Phys. Chem. 80:131-138.

Wriston, J. C., Jr., and Yellin, T. O. 1973. L-Asparaginase: A review. Adv. Enzymol. 39:185-248.

Secondary Kinetic Isotope Effects

J. F. Kirsch

Richards (1970) reviewed the subject of secondary kinetic isotope effects (KIEs) in enzyme-catalyzed reactions for the third edition of *The Enzymes.* At the time of his writing significant results obtained from this technique were available only from two enzymes: fumarase and lysozyme. Several additional examples have appeared since then, many but not all of which will be discussed here. The fact remains, however, that secondary KIEs are investigated by enzymologists far less frequently than are primary. There are two substantial reasons why this is so. First, Secondary KIEs are far smaller in magnitude than are primary. For example, primary KIEs for hydrogen normally lie in the range, $k_H/k_D = 2\text{–}10$, while secondary hydrogen isotope effects usually have limiting values of 1.02–1.40. Thus, the demands upon experimental technique are considerably more stringent. Second, the theoretical basis of the secondary KIE is intuitively more difficult to grasp. Although it may not be absolutely correct, the presence of a significant primary KIE is generally attributed to a rate-determining transition state in which the bond to the isotope is "being more or less broken." Such a simple qualitative rationalization for secondary KIEs is lacking. The manifestation of the latter effect implies only that the force constants to the isotope have changed in proceeding from the ground state to the transition state of the reaction.

Nonetheless, although again it is not strictly correct, one can glean an analogous, intuitive understanding for secondary KIEs in terms of the relative geometries of the ground and transition states.

The vast majority, if not all, of the secondary hydrogen KIEs of interest to biochemists arise from experiments typified by the general reaction shown in equation 1

$$\underset{\text{I}}{\mathrm{H^{*}{-}\overset{\overset{\displaystyle X}{\|}}{C}{-}Y} \atop {+ \atop \mathrm{ZH}}} \rightleftarrows \underset{\text{II}}{\left[\mathrm{H^{*}{-}\overset{\overset{\displaystyle X}{\vdots}}{\underset{\underset{\displaystyle ZH}{\vdots}}{C}}\cdots Y}\right]^{\ddagger}} \rightleftarrows \underset{\text{III}}{\mathrm{H^{*}{-}\overset{\overset{\displaystyle XH}{|}}{\underset{\underset{\displaystyle Z}{|}}{C}}{-}Y}} \qquad (1)$$

In this reaction trigonally hybridized carbon is converted to its tetrahedral form or vice versa. The transition state ordinarily will lie between these two extremes. The measured values of k_H/k_D are termed the α-deuterium isotope effects. Substitution at more remote positions often will also give rate differences, but the number of examples presently available which are of interest to biochemists precludes their discussion here.

THEORY

Streitwieser Formulation

A semi-empirical formulation of the origin of secondary isotope effects was provided by Streitwieser et al. (1958) (Melander, 1960) and is shown in equations 2 to 5.

$$\frac{k_D}{k_H} = \exp\left\{-\frac{hc}{2kT}\left[\Sigma\,(\bar{\nu}_D{}^{\ddagger} - \bar{\nu}_H{}^{\ddagger}) - \Sigma\,(\bar{\nu}_D - \bar{\nu}_H)\right]\right\} \qquad (2)$$

$$\frac{\nu_H}{\nu_D} \approx \sqrt{\frac{m_D}{m_H}} = 1.41 \quad \text{(Empirically, 1.35 works better)} \qquad (3)$$

$$\frac{k_D}{k_H} = \exp\left\{-\frac{hc}{2kT}\left(\frac{1}{1.35} - 1\right)\Sigma\,(\bar{\nu}_H{}^{\ddagger} - \bar{\nu}_H)\right\} \qquad (4)$$

$$\frac{k_D}{k_H} = \exp\left\{\frac{0.187}{T}\,\Sigma\,(\bar{\nu}_H{}^{\ddagger} - \bar{\nu}_H)\right\} \qquad (5)$$

The isotope effect, to a very good approximation, arises from the difference in zero-point energies between the labeled and unlabeled transition states minus the corresponding values for the ground states (equation 2). The ratio of the hydrogen to deuterium frequencies is given by the square root of the mass ratio (equation 3). This would be an exact value if the mass of the atom to which the hydrogen is bonded were infinite. Because the mass of carbon is not that much greater than that of hydrogen, it is necessary to take into account the reduced mass of the oscillating couple, which gives the somewhat smaller value of 1.35. This substitution leads to equation 4, and the values of the universal constants to equation 5. Thus, the value of the isotope effect can in principle be obtained from a knowledge of the vibrational frequencies affecting the isotope in the transition state compared with those of the ground state. Of course, the former are not measurable but, as will be shown, they can sometimes be estimated with reasonable precision from infrared spectra of model compounds chosen to resemble the transition states.

Limitations of Interpretation

Most of this discussion is centered upon the relatively straightforward case in which there is a single rate-determining transition state for a reaction sequence. If one or more intermediates exist along the reaction pathway whose rates of formation and decomposition are both kinetically significant, then the assignment of the isotope effect to a particular transition state may well be obscured (e.g., Northrop, 1975). Consider, for example, the relatively common instance in which a substrate with a tetrahedral ground state forms an intermediate with trigonal geometry, which then collapses to form a different tetrahedral compound. In such a case, the observation of a small but significant kinetic isotope effect cannot be assigned with certainty to either an early transition state leading to the formation of the intermediate or to a late transition state leading from the intermediate to the final product. Both enzymatic and nonenzymatic examples will be discussed where this ambiguity arises. The observation of a large, i.e., nearly maximal, kinetic isotope effect, however, is unambiguously indicative of a transition-state structure very closely resembling the intermediate; in such a case, the question of a transition-state structure representing the late formation of the intermediate or its early decomposition is much less important.

Magnitudes

The above considerations lead to the question of the estimation of the magnitude of the KIE for the formation of a given transition-state structure. This is ordinarily very difficult. In most cases of enzymatic interest, the lack of appropriate force constants or even good infrared spectral assignments for molecules resembling transition states, and even for most substrates, are unavailable. There is a general tendency in the literature to accept a figure of $k_H/k_D \approx 1.25$ as being more or less the limiting value for sp^3 to sp^2 conversion. This stems from the extensive investigations of Shiner (Shiner, 1970) and others of solvolysis reactions. Unfortunately, this figure cannot be generalized, as is shown by the equilibrium data in Table 1. All of the reactions reported in this table involve the addition of a nucleophile to a formyl or a carbonyl compound to form a tetrahedral adduct. These and other data discussed by Bilkadi, de Lorimier, and Kirsch (1975) bring out the importance of, among other things, the nucleophilic element (e.g., reactions 1 to 3) in determining the values of the equilibrium isotope effects for addition to the formyl or carbonyl group. It is seen from the table that the addition of hydroxide ion to methyl formate gives a value of $K_D/K_H = 1.14$. Where nitrogen is the nucleophile the limiting values are much larger, between 1.32 and 1.38. Therefore, it seems that without knowledge of the nature of the nucleophilic element on the enzyme, one cannot draw very definitive conclusions about the transition-state structure in the catalyzed reaction.

REACTIONS OF METHYL FORMATE AND FORMAMIDE

Some of the above considerations are illustrated by investigations of nonenzymatic acyl transfer reactions of methyl formate and formamide undertaken by Dr. Bilkadi in our laboratory (Bilkadi, de Lorimier, and Kirsch, 1975; Bilkadi and Kirsch, in preparation).

Table 1. Limiting values of equilibrium α-deuterium isotope effects (K_D/K_H) for addition of nucleophiles to formyl and carbonyl groups

Reaction	K_D/K_H	Reference
1) R—O—R′ + $HCO_2Me \rightleftarrows$ HC(OR)(OR′)(OMe) R,R′ = H, alkyl	1.21	a
2) $OH^- + HCO_2Me \rightleftarrows HC(O^-)(OMe)(OH)$	1.14	a
3) $NH_2NH_2 + HCO_2Me \rightleftarrows HC(OH)(OMe)(NHNH_2)$	1.32-1.38	a
4) *p*-Methoxybenzaldehyde + HCN $\rightleftarrows$ $pMeOC_6H_4CH(OH)CN$	1.28	b
5) Benzaldehyde + $NH_2OH \rightleftarrows C_6H_5CH(OH)NHOH$	1.36	b
6) $CH_3CHO + 2C_2H_5SH \rightleftarrows CH_3CH(SC_2H_5)_2$	ca. 1.28	c

a) Bilkadi, de Lorimier, and Kirsch, 1975; b) do Amaral, Bull, and Cordes, 1972; c) calculated from spectral assignments of Wladislaw, Olivato, and Sala, 1971, using the Streitwieser equation.

The hydrolysis of methyl formate almost certainly involves the formation of a tetrahedral intermediate which is the hydrate of the ester. The relevant IR stretching and bending frequencies of the ester have been assigned and are shown in Table 2. Those of the hydrate are not known but can be predicted with considerable confidence from a knowledge of the IR spectra of analogous compounds such as the ortho ester (Bilkadi, de Lorimier, and Kirsch, 1975). The equilibrium α-deuterium isotope effect arises primarily because of the 340 cm^{-1} change in the out-of-plane bending frequency of the C—H bond between the ester and its hydrate. We do not have corresponding infrared data for the addition of nitrogen to acyl or carbonyl carbon, but experimental data obtained by Bilkadi and Kirsch (in preparation) and by Cordes (do Amaral, Bull, and Cordes, 1972) provide clear evidence for a much larger isotope effect in this case (Table 1). Our investigation of the KIE for the hydrazinolysis of methyl formate

$$\mathrm{L{-}C({=}O){-}OMe} + \mathrm{NH_2NH_2} \underset{}{\overset{K_1}{\rightleftharpoons}} \mathrm{L{-}C(O^-)(NHNH_2){-}OMe} \xrightarrow{k_2} \mathrm{L{-}C({=}O){-}NHNH_2} + \mathrm{MeOH} \quad (6)$$

was stimulated by the important discovery of Blackburn and Jencks (1968) that this reaction proceeds through the formation of a tetrahedral intermediate and that the rate-limiting step changes from formation of this intermediate at high pH to its decomposition at a lower value of pH. Furthermore, our earlier work on primary oxygen-18 leaving group kinetic isotope effects indicated that the transition state for the decom-

Table 2. C—H stretching and bending frequencies (cm^{-1}) in methyl formate and its hypothetical hydrate $HC(OH)_2OCH_3$, and the resulting α-D isotope effect at 25°

Mode	$HCOOCH_3$	$HC(OH)_2OCH_3$	$\Delta\nu$	$\ln(K_D/K_H)$
C—H stretch	2943	2900	−43	−0.02
C—H bend	1032 (o.p.)	1370	+338	+0.21
C—H bend	1371 (i.p.)	1370	—	—
				+0.19

From Bilkadi, de Lorimier, and Kirsch, 1975, reproduced by permission.

position of this intermediate must be very late (Sawyer and Kirsch, 1973). The reaction measured at high pH is, in fact, the equilibrium formation of the intermediate from methyl formate and hydrazine. The measured α-deuterium isotope effect under these conditions is equal to 1.38. At lower values of pH where decomposition of the intermediate is the rate-determining step, the isotope effect virtually vanishes, indicating that the transition state has nearly trigonal geometry in the decomposition reaction. Thus, the following isotope effects can be derived:

$$\text{For equilibrium step 1;} \quad K_D/K_H = 1.38$$

$$k_{\text{overall}} = k_2K_1; \quad k_D/k_H = 1.02$$

$$\text{For } k_2; \quad k_D/k_H = \frac{1.02}{1.38} = 0.74$$

In addition to delineating the transition-state structure, this investigation illustrates how a simple variation in pH can change the value of an observed kinetic isotope effect from a very large to a very small figure. Clearly, investigations at intermediate values of pH, where both formation and decomposition of the intermediate are partially rate determining, would have led to intermediate values of the observed KIE which could not have been unambiguously assigned to a given transition-state structure.

The alkaline hydrolysis reaction of formamide, in common with those for other amides and anilides (Bender and Thomas, 1961; Biechler and Taft, 1957; Schowen, Jayaramann, and Kershner, 1966), exhibits an observed rate constant that is dependent upon terms both first and second order in hydroxide ion concentration (Figure 1). The value of the α-deuterium KIE varies smoothly with hydroxide ion concentration, as shown in Figure 2.

A mechanism embodying these observations as well as those of other workers in this field is

$$OH^- + LCONH_2 \underset{k_{-a}}{\overset{k_a}{\rightleftharpoons}} L\text{—}\underset{OH}{\overset{O^-}{C}}\text{—}NH_2 \overset{K'[OH^-]}{\rightleftharpoons} L\text{—}\underset{O^-}{\overset{O^-}{C}}\text{—}NH_2 \xrightarrow{k_2} LCO_2^- + NH_3 + OH^- \quad (7)$$

$$L\text{—}\underset{OH}{\overset{O^-}{C}}\text{—}NH_2 \overset{K}{\rightleftharpoons} L\text{—}\underset{O^-}{\overset{O^-}{C}}\text{—}NH_3^+ \xrightarrow{k_1} LCO_2^- + NH_3$$

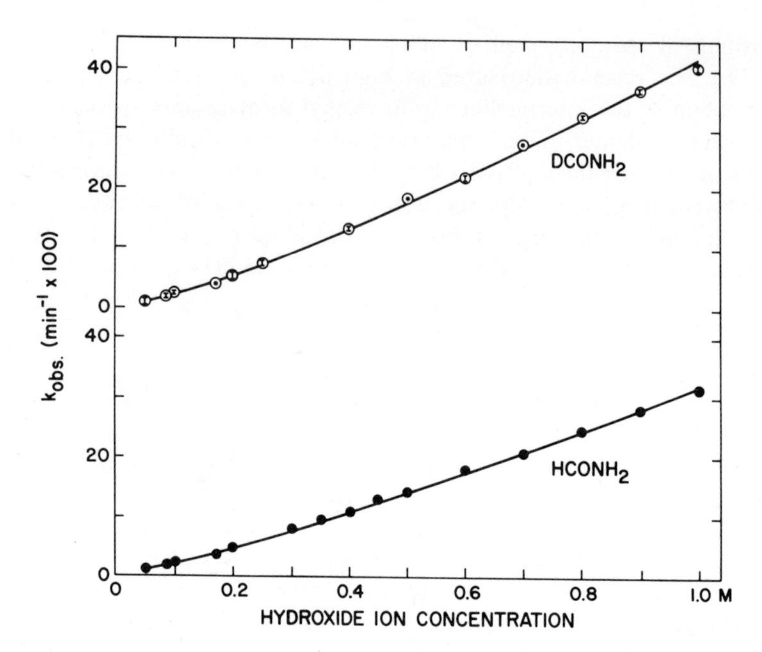

Figure 1. The dependence of the observed rate constants for the alkaline hydrolysis of $HCONH_2$ and $DCONH_2$ upon hydroxide ion concentration. (Data of Bilkadi and Kirsch, in preparation.)

The important feature distinguishing the alkaline hydrolysis of amides from esters arises from the fact that NH_2^-, unlike RO^-, cannot be expelled directly from the tetrahedral intermediate. In the reaction pathway which is dependent on the first power of hydroxide ion concentration, a proton switch of undefined mechanism would form the highly dipolar intermediate $[HC(O^-)_2(NH_3^+)]$, which decomposes to products in an uncatalyzed or water-catalyzed reaction. At higher concentrations of hydroxide ion the initially formed intermediate is trapped by a second molecule of base to produce the doubly negatively charged species, which then decomposes to products in a reaction that is probably general acid catalyzed by water. The steady-state expression for this mechanism involves three adjustable parameters whose values were obtained by nonlinear least squares regression analysis of 70 sets of data points. The value of the isotope effect, k_D/k_H (Table 3), is most precisely determined for the first step in the reaction sequence, the addition of hydroxide ion to formamide. The figure of 1.53 ± 0.23 indicates a very late transition state for this step. Similarly, the low value

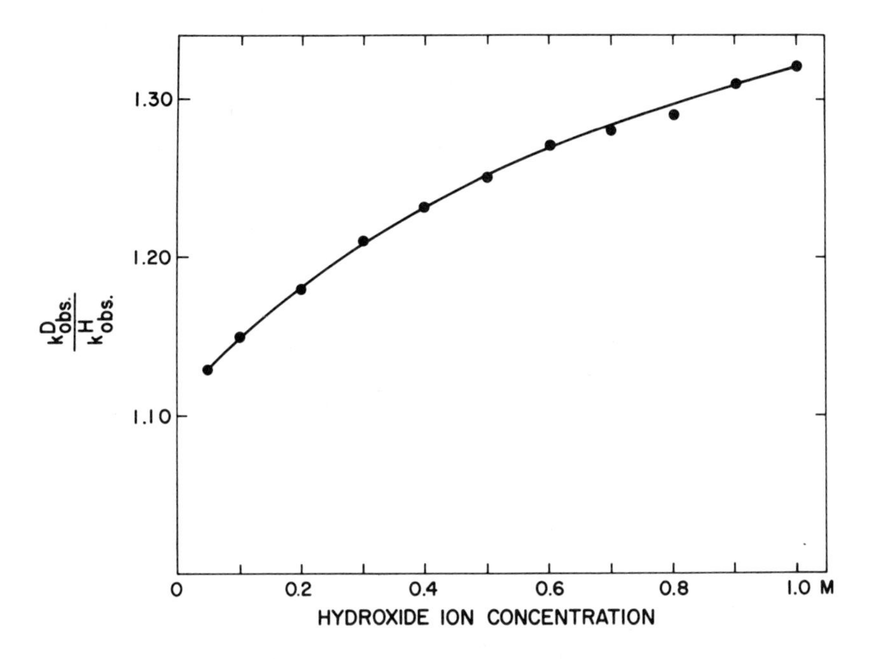

Figure 2. The dependence of the α-deuterium isotope effect for the hydrolysis of formamide upon hydroxide ion concentration. (Data of Bilkadi and Kirsch, in preparation.)

Table 3. Secondary deuterium isotope effects in the alkaline hydrolysis of formamide[a]

Constant	k_D/k_H or K_D/K_H
k_a	1.53 ± 0.23
k_{-a}	Ca. 1.0
K	1.0
K'	0.9
k_2	0.71 ± 0.31
k_1	0.63 ± 0.1

[a] Isotope effects on k_a, k_1, and k_2 were obtained by nonlinear regression analysis for the steady-state expression for the mechanism shown in equation 7. Isotope effects on K and K' are estimates based on known α-deuterium isotope effects on ionization constants.

of the isotope effect on k_1 implies a late transition state for the decomposition of the dipolar intermediate to formate ion plus ammonia. The value of k_2 is too imprecisely determined to permit a statement much stronger than that the transition state for this step is not early.

As in the case of the hydrazinolysis of methyl formate already discussed, it is clear that no single measurement of the kinetic isotope effect at a given hydroxide ion concentration would have provided a meaningful result insofar as the interpretation of the reaction mechanism is concerned. Interpretation was made possible only by the extensive series of measurements throughout the hydroxide ion concentration range and with the aid of the results from earlier investigations, which pointed to the existence of the singly and doubly negatively charged tetrahedral intermediates. It will be well to keep the possibility of these potential pitfalls in mind as the enzymatic results are considered.

GLYCOSIDASES

The most extensive use of secondary KIEs in the determination of enzyme mechanism has been made with glycoside lysing enzymes. The following mechanism would seem to apply to the most thoroughly investigated, lysozyme and β-galactosidase.

```
  _O  OR                     _O+                     _O   OH
 {   \/       Enz           {   \\                  {   \ /
 {___/\      ----> Enz···   {     C—H    ---->      {    C
      H                     {____/       H2O        {___/ \
                                                           H
                              ?↓↑                                  (8)

                             _O   Enz
                            {   \ /
                            {    C
                            {___/ \
                                   H
```

The available results for these enzymes indicate that a carbonium ion, or more precisely an oxocarbonium ion, is formed along the reaction route. Distinguishing differences arise depending upon whether or not the trapping of the carbonium ion by the enzyme is on the normal pathway for the substrates that have been investigated and on the nature of the rate-determining step.

Lysozyme

The paper published by Dahlquist, Rand-Meir, and Raftery (1969) was pioneering both in the technical aspects of the application of the double labeling technique to the determination of KIEs and for the important information with respect to transition-state structure for the lysozyme mechanism obtained. The reaction studied is

$$\text{RO-(glucopyranoside, CH}_2\text{OH, OH, H*, OH)–O–Ph*} \xrightarrow[\text{H}^+]{\text{lysozyme or}} \left[\text{(oxocarbonium ion, CH}_2\text{OH, O}^+\text{, OH, H*, O–R, OH)}\right] \xrightarrow{H_2O} \text{RO-(CH}_2\text{OH, OH, OH)}\sim\text{(H, OH)} \quad (9)$$

R = NAG, H

and some of the kinetic isotope effect results are summarized in Table 4. The unique technical feature of these experiments was the use of ^{14}C and ^{3}H labels in the phenyl moiety to follow the deuterium to hydrogen ratio in the C—1 position of the sugars. The kinetic isotope effect was therefore simply determined by measuring $^{14}C/^{3}H$ ratios of the phenol produced as a function of the extent of reaction. Most evidence indicates that the acid-catalyzed hydrolysis of glycopyranosides proceeds through an A—1 mechanism involving rate-determining carbonium ion formation (Capon, 1969). The large secondary α-deuterium KIEs observed for the acid-catalyzed hydrolysis of this phenyl glycoside and for

Table 4. Hydrogen isotope effects on the hydrolysis of phenyl-β-glucosides 1-D

Catalyst	k_H/k_D
Acid	1.13
Sodium methoxide	1.03
β-Glucosidase	1.01
Lysozyme	1.11

From Dahlquist, Rand-Meir, and Raftery, 1969.

other acetals (Bull et al., 1971) strongly support a carbonium ion mechanism for this reaction type. Because the value of k_H/k_D for lysozyme is similar to that observed for acid-catalyzed hydrolysis, it is very reasonable to conclude that a carbonium ion intermediate is formed in the rate-determining step for the lysozyme mechanism as well.

Responding to the potential criticism arising from the special nature of the phenol leaving group, Smith, Mohr, and Raftery (1973) prepared α-tritium labeled chitotriose from in vivo labeled insect chitin

OH O OH RO— H* O CH$_3$CONH — OH O OH ~(H, OH) CH$_3$CONH —Lysozyme or H$^+$→ OH O OH RO— ~(H*, OH) CH$_3$CONH + OH O OH HO— ~(H, OH) CH$_3$CONH (10)

R = NAG, H

They observed a $k_H/k_T = 1.19$, which is equivalent to $k_H/k_D = 1.14$. Curiously, in this case, unlike that of the phenyl glycoside discussed above, the isotope effect for the acid-catalyzed reaction is only equal to 1.08, which is significantly lower than that found for the enzyme (Mohr, Smith, and Raftery, 1973). Such a result may indicate participation by the neighboring acetamido group in the acid-catalyzed but not in the enzyme-catalyzed reaction.

β-Galactosidase

Although trapping of the carbonium ion intermediate of lysozyme by the side chain residue Asp-52 to form a covalent enzyme substrate complex seems a reasonable possibility and has been proposed (e.g., Atkinson and Bruice, 1974) (equation 8), supportive evidence is lacking. In contrast, rather substantial evidence exists for the formation of a covalent enzyme-substrate complex in the β-galactosidase reaction.

First, D-galactal is a potent inhibitor of β-galactosidase and was thought by virtue of its suspected half-chair conformation to act as a transition-state analog. Wentworth and Wolfenden (1974) showed, however, that the rate constant for the association of this inhibitor with the enzyme is only 270 $M^{-1}sec^{-1}$, a value that seems much too low for a simple noncovalent interaction. The rate of association is further reduced in solvent D_2O by a factor of 2, and the inhibitor dissociates from the enzyme primarily as 2-deoxygalactose. Thus, it appears likely that the inhibitor forms a covalent complex with the enzyme which hydrolyzes in a subsequent step.

D-galactal + E $\rightleftharpoons$ [galactosyl enzyme] $\overset{H_2O}{\rightleftharpoons}$ 2-deoxygalactose (11)

Additional evidence for a galactosyl enzyme was provided by Fink and Angelides (1975), who investigated the rate of reaction of this enzyme with *o*-nitrophenylgalactoside at sub-zero temperatures and obtained evidence for a pre-steady-state burst of *o*-nitrophenol equivalent to about 61% of the calculated number of active sites. This result indicates that under these conditions the reaction of the enzyme with *o*-nitrophenylgalactoside to produce a galactosyl enzyme is fast and that the slower steady-state reaction is controlled by the rate of degalactosylation. Last, Sinnott and Viratelle (1973) demonstrated that the addition of the nucleophile methanol increased the steady-state rate of substrate consumption only when very rapidly reacting substrates were used and not for the more slowly reacting ones.

These observations are consistent with a rate-limiting degalactosylation reaction in which the galactosyl enzyme is formed rapidly from very reactive substrates and with some prior step being rate limiting for the less reactive substrates.

In what is probably the most thorough investigation of secondary deuterium isotope effects in enzymology, Sinnott and Souchard (1973) (Figure 3) showed that the α-deuterium KIE on k_{cat} increases more or less linearly with the value of k_{cat}. The pK_a of the leaving group correlates only very poorly with k_{cat} or k_{cat}/K_m, indicating that the chemical reactivity, at least as measured by the acidity of the leaving group, is not an important determinant of reaction rate for most of the substrates investigated. It was further noted that the addition of methanol increases the value of k_H/k_D for 3,5-dinitrophenyl β-galactoside to 1.34 but does not change the value of the KIE for the more slowly reacting substrates. The mechanism proposed by Sinnott and his colleagues is

$$\mathrm{E} + \text{GAL-X} \overset{K_s}{\rightleftharpoons} \mathrm{E}\cdots\text{GAL-X} \underset{k_{-2}}{\overset{k_2}{\rightleftharpoons}}$$

$$\mathrm{E}^*\cdots\text{GAL-X} \xrightarrow[\searrow \mathrm{XH}]{k_3} \mathrm{E}^*\cdots\mathrm{GAL}^+ \xrightarrow[k_5]{\left[\begin{smallmatrix}\mathrm{MeOH} \\ \text{or} \\ \mathrm{H_2O}\end{smallmatrix}\right]} \mathrm{E} + \text{GAL-OR} \qquad (12)$$

$$k_4 \downarrow\uparrow k_{-4}$$

$$\text{E—GAL}$$

Sinnott and his colleagues argue that the Michaelis complex undergoes a conformational change (k_2), which constitutes the rate-determining step for most of the aryl galactosides investigated. The substituted phenol is expelled in the k_3 step leading to the formation of an enzyme-bound carbonium ion. This step is not rate determining for the most slowly reacting aryl glycosides, but it becomes partially so for the more reactive members of the series. It is interesting that galactosyl pyridinium salts are also substrates for this enzyme, but they react at much slower rates than do the aryl glycosides (Sinnott and Withers, 1974). The values of k_{cat} for these compounds are highly correlated with the pK_as of the leaving groups, which indicates that the k_3 step is rate determining. Consistent with this conclusion is the observation of rather large secondary KIEs for the galactosyl pyridinium salts. Sinnott further

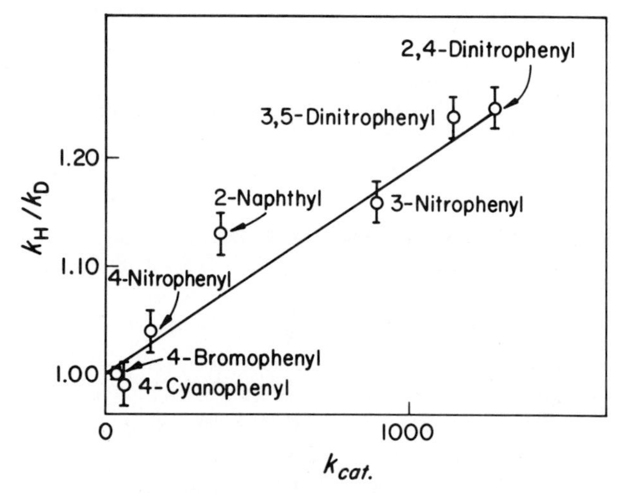

Figure 3. α-Deuterium kinetic isotope effects for the β-galactosidase-catalyzed hydrolysis of aryl galactosides as a function of k_{cat}. (From Sinnott and Souchard, 1973, by permission.)

postulates that for the most highly reactive aryl glycosides, both the conformational change (k_2) and carbonium ion formation steps (k_3) are fast and that the enzyme bound carbonium ion, once formed, is largely trapped by a nucleophilic group of the enzyme to form the covalent complex. The k_{-4} and k_5 steps are fast relative to the preceding steps for all but the most reactive substrates and are, therefore, not of kinetic significance. This postulated sequence would explain the methanol effect, which presumably acts solely on the k_5 step. The k_{-4} and k_5 steps are, however, rate determining for the most reactive substrates, and the large KIE observed for them indicates that carbonium ion formation is far advanced in the transition state. In fact, the extremely large value observed for k_H/k_D of 1.34 for the more reactive substrates in the presence of added methanol would lead me to predict that the nucleophilic element on the enzyme when identified will turn out to be nitrogen, rather than oxygen or sulfur (Table 1).

Phosphorylase

Some of the hazards manifest in attempting to measure the expected small values of secondary deuterium KIEs become apparent when one considers the current literature available on the phosphorylase reaction.

CH$_2$O(H), H*, OH, HO, OH, —O—(P) + glycogen $\xrightleftharpoons{\text{phosphorylase}}$ CH$_2$O(H), H*, OH, HO, OH, —O—glycogen + P_i (13)

Tu, Jacobson, and Graves (1971) attempted a direct measurement of the α-deuterium KIE for rabbit muscle phosphorylase-*b* by comparing the rates of release of inorganic phosphate from C-1(H) and C-1(D) substrates. Their data give $k_H/k_D = 1.10$. They also obtained a value of $k_H/k_D = 1.13$ for acid-catalyzed hydrolysis of the same compounds, as expected, and a value of 1.00 for the alkaline phosphatase reaction, which is also as predicted because this reaction involves cleavage of the P—O rather than the C—O bond. On the other hand, Firsov, Bogacheva, and Bresler (1974), in a much more extensive set of experiments, used the double label technique to measure both the α-tritium and α-deuterium effects in both directions, and they report that the KIEs are no larger than 1%. They conclude that the formation of a carbonium ion is not a rate-determining step in this reaction and suggest an S$_n$2-

type displacement. But clearly any rate-determining step not involving the changing of the force constants at C-1, such as a conformational change, would produce the same result. These experiments serve to illustrate how small differences (i.e., 10–15%) in sensitive kinetic determinations can lead to qualitatively different mechanistic explanations and emphatically point out how small deterministic errors may far outweigh in importance the statistically determined standard errors obtained in replications of the same experiment.

ALDOLASE

An important early application of the secondary KIE method to enzymes is the work of Biellmann, O'Connell, and Rose (1969) on aldolase, which catalyzes the reaction

$$\begin{array}{c} CH_2O\textcircled{P} \\ | \\ C{=}O \\ | \\ HO{-}C{-}H^* \\ | \\ H{-}C{-}O{-}H \quad B \\ | \\ H{-}C{-}OH \\ | \\ CH_2O\textcircled{P} \end{array} \underset{k_r}{\overset{k_f}{\rightleftharpoons}} \begin{array}{c} CH_2O\textcircled{P} \\ | \\ C{=}O \\ | \\ HO{-}C{-}H^* \\ | \\ H \end{array} + \begin{array}{c} CHO \\ | \\ HC{-}OH \\ | \\ CH_2O\textcircled{P} \end{array} \qquad (14)$$

These workers investigated both the yeast and muscle enzymes, whose mechanisms differ in that the electron sink at the 2-position of fructose diphosphate is provided in the first case by a metal and in the second by Schiff base formation through an enzyme lysine residue. The secondary tritium isotope effect was measured in both directions. In the synthesis of fructose diphosphate, stereospecifically labeled (1-R) (1-^{3}H) dihydroxyacetone phosphate was used. The isotope effects are recorded in Table 5. Earlier exchange experiments with the yeast enzyme had established that the formation of the stabilized carbanion is rate determining at pH 8; therefore, the large kinetic isotope effect observed for k_r at this pH most probably is a measure of the extent of rehybridization at the 1-position of dihydroxyacetone phosphate from sp^3 to sp^2. The isotope effect in the synthesis reaction catalyzed by the yeast enzyme vanishes at pH 6, indicating that another step has become rate determining.

Table 5. Secondary hydrogen isotope effects on reactions catalyzed by aldolase (equation 14)

Enzyme source	pH	Rate constant	k_H/k_T	Calculated k_H/k_D
Yeast	8.0	k_f	1.43	1.28
Yeast	8.0	k_r	1.28	1.19
Muscle	6.0	k_f	1.33	1.22
Muscle	6.0	k_r	0.99	0.99

From Biellmann, O'Connell, and Rose, 1969.

ASPARTASE

This enzyme catalyzes the deamination of L-aspartate to fumarate and ammonia. The reaction might proceed by either a carbanion mechanism

$$^{-}O_2C{-}C(H^*)(H){-}C(H^*)(NH_3^+){-}CO_2^- \xrightleftharpoons{\text{fast}} {}^{-}O_2C{-}\overset{-}{C}(H){-}C(H^*)(NH_3^+){-}CO_2^- \xrightarrow[-NH_3]{\text{slow}} {}^{-}O_2C(H)C{=}C(H^*)CO_2^- \qquad (15)$$

or by a carbonium ion mechanism

$$^{-}O_2C{-}C(H^*)(H){-}C(H^*)(NH_3^+){-}CO_2^- \xrightarrow[-NH_3]{\text{slow}} {}^{-}O_2C{-}C(H^*)(H){-}\overset{+}{C}(H^*){-}CO_2^- \xrightarrow{\text{fast}} {}^{-}O_2C(H)C{=}C(H^*)CO_2^- \qquad (16)$$

No primary hydrogen KIE was observed for aspartate labeled in the 3-position. A secondary effect of $k_H/k_D = 1.11 \pm 0.02$ was obtained for substrate labeled in the 2-position (Dougherty, Williams, and Younathan, 1972). Limiting carbanion and carbonium ion mechanisms for this reaction are shown in equations 15 and 16. The absence of a primary hydrogen KIE indicates that if the carbanion mechanism is correct, proton removal must be fast and the equilibrium isotope effect for this reaction is near 1. The secondary KIE arises from the developing sp^2 character in the deamination step. The fact that hydrogen exchange at the 3-position is not seen provides some argument against the carbanion mechanism. The rate-limiting carbonium ion mechanism is consistent with both the lack of a primary isotope effect and the presence of a secondary kinetic isotope effect. As noted by Dougherty, Williams, and Younathan (1972), however, this latter value is somewhat smaller than that expected for a limiting carbonium ion mechanism; it may, in fact, represent an equilibrium isotope effect with the rate-limiting step being the dissociation of fumarate. This last possibility becomes more attractive when one considers the reverse reaction, which would have to involve the equilibrium protonation of fumaric acid to give the carbonium ion followed by its slow collapse to aspartate if the limiting carbonium ion mechanism were to obtain.

THYMIDYLATE SYNTHETASE

This enzyme catalyses the reductive methylation of deoxyuridylate (dUMP) to thymidylate (dTMP) using 5-10-methylenetetrahydrofolate as the methyl and hydrogen donor. The enzyme also catalyses the reductive debromination of 5-BrdUMP to dUMP (Wataya and Santi, 1975).

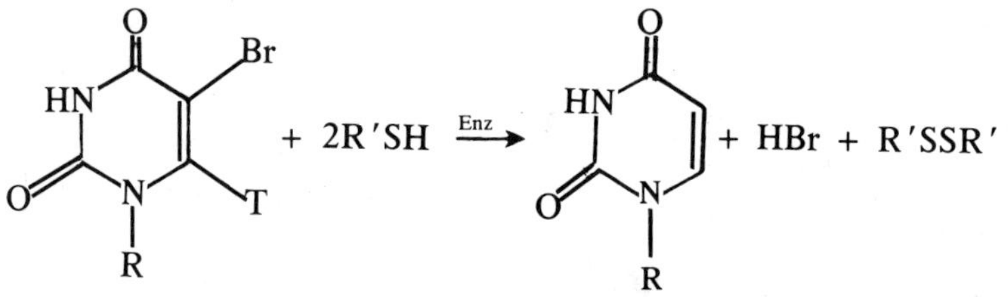

+ 2R′SH $\xrightarrow{\text{Enz}}$ + HBr + R′SSR′ (17)

An inverse tritium KIE for the compound labeled in the 6-position, corresponding to a $k_D/k_H = 1.10$, has also been observed (Garrett and Santi (unpublished data)). This observation supports a transition state in which the C-6 atom is at least partially rehybridized toward a tetrehedral transition state

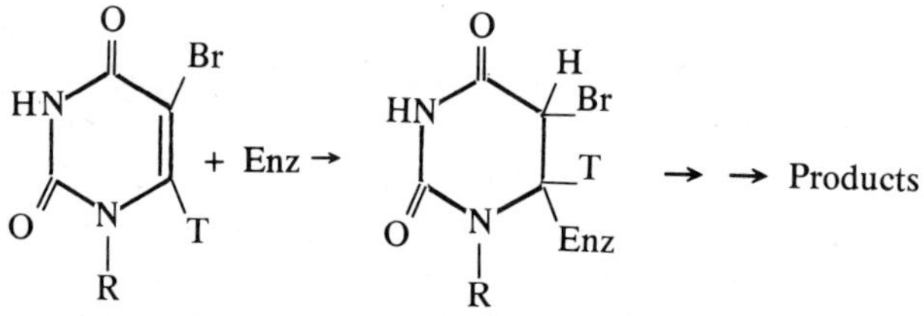

+ Enz → → → Products (18)

The subsequent steps leading from the hypothetical enzyme-BrdUMP adduct most likely involve attack of thiol on the bromine atom to produce an unstable sulfenyl bromide which decomposes to final products. Further evidence for addition at the 6-position emanates from the isolation of an enzyme-6 adduct after incubation of 5-FdUMP and 5-10-methylenetetrahydrofolate with the enzyme and from the observation of a normal $k_H/k_T = 1.23$ ($k_H/k_D = 1.15$) for the rate of dissociation of (6-^{3}H) FdUMP from the complex (Santi, McHenry, and Sommer, 1974).

FORMYLTETRAHYDROFOLATE SYNTHETASE

This enzyme catalyses the synthesis of *N*-10-formyltetrahydrofolate from formate, ATP, and tetrahydrofolate

$$\mathrm{H^*{-}C}\begin{matrix}{}^{*}\mathrm{O} \\ \mathrm{O}^{*}\end{matrix} + \mathrm{ATP} \rightleftharpoons \mathrm{H^*{-}\overset{\overset{\displaystyle O^*}{\|}}{C}{-}Ⓟ} \xrightleftharpoons{\mathrm{THF}} \mathrm{H^*{-}\overset{\overset{\displaystyle O^*}{\|}}{C}{-}N\langle} \quad + \mathrm{ADP} \tag{19}$$

Although direct evidence is lacking, it seems likely that formyl phosphate is an intermediate in this reaction. We have determined secondary deuterium and mixed primary and secondary oxygen-18 kinetic isotope effects for formate in this reaction. A small but significant value of $k_D/k_H = 1.045 \pm 0.005$ was measured. The oxygen-18 isotope effect is $k_{16}/k_{18} = 1.005 \pm 0.005$ (Rosenberg, Rabinowitz, and Kirsch, in preparation). The reaction sequence is complicated, and complete interpretation of the KIE is unwarranted at this time. It does seem clear from the secondary kinetic isotope effect data that there is a rate-determining transition state involving at least some sp^3 character introduced into the formate ion. Moreover, the absence of a significant oxygen-18 primary KIE implies that if formyl phosphate formation is rate determining, then the transition state is quite early, rather than late, with respect to carbon-oxygen bond cleavage. Transition states involving extensive cleavage of this bond would be expected to exhibit much larger isotope effects (Sawyer and Kirsch, 1973).

CONCLUSIONS

Secondary and primary KIEs constitute the only widely applicable approach capable of quantitatively delineating transition-state structures for enzyme-catalyzed reactions. Technical problems arising from difficulties in preparing isotopically labeled compounds in sufficiently pure form and in carrying out kinetic measurements with the necessary degree of precision have severely limited the number of attempts to measure secondary KIEs in enzymatic systems. While the observation of a large secondary KIE can often be interpreted in terms of transition-state structure with confidence, smaller values can only be rationalized in an ambiguous manner because of problems arising from intermediate partitioning or steps not involving the isotope being partially rate determining. Finally, sufficient data are not presently available for the enzymologist to estimate precisely the limiting values of the secondary KIE for many of the reactions of interest. These caveats should be taken in a cautionary rather than in a prohibitive sense, because it does seem likely that rewards in terms of precisely defined transition state structures for enzyme-catalyzed reactions will be realized as these obstacles are overcome.

DISCUSSION

V. J. Shiner. It might be helpful, although from my point of view perhaps a little dangerous, for me to illustrate the use of some of the fractionation factors I showed earlier to try to get a value for the hydration of an aldehyde.

Previously we have looked at aldehyde hydrations in a very elementary way as simply the addition of an oxygen leaving group, which, by analogy with solvolytic reactions, suggests an effect of about 1.23, whereas the experimental results are between 1.35 and 1.38. Warren Buddenbaum has now calculated the fractionation factor for formaldehyde, which turns out to be very close to that for methane (see Table 1 in Buddenbaum and Shiner, this volume). Our solvolytic results suggest that the effects of α-oxygen and α-fluorine are similar, so that methylene fluoride should be a reasonable model for formaldehyde hydrate; the fractionation factor between methane and methylene fluoride is $(1.465/1.248)^2$ or 1.38, in agreement with the experimental values for aldehyde hydration. The earlier estimate of 1.23 is too small apparently because the fractionation factor for formaldehyde is significantly lower than that of, e.g., an ethyl cation.

J. F. Kirsch. It's interesting that for acetaldehyde we calculated the hydration isotope effect using the Streitwieser equation and got 1.37. At least in that case the three-frequency approximation works very well.

T. I. Kalman. I'd like to make a short comment on thymidylate synthetase. I will show some material during the poster session. It includes one aspect which should be considered occasionally when one calculates deuterium isotope effects from tritium isotope effects rather than directly determining them. In the case of thymidylate synthetase, for example, if one measures the reactivation of the inhibited enzyme under circumstances when the sixth position of 5-fluorodUMP is labeled with deuterium, one finds a secondary deuterium isotope effect of 1.24 (Kalman, 1975). The techniques of measuring secondary deuterium isotope effects are different, obviously, from those of measuring tritium isotope effects. In determining a secondary tritium isotope effect, one uses a mixture of tritium and ^{14}C labeled 5-fluorodUMP to inactivate the enzyme—^{14}C representing the lighter (hydrogen) and tritium, the heavier isotope—and one measures the release of *all* radioactive label from the drug-protein complex under dissociation conditions. Because there is multiple binding of the ligand to two sites of the enzyme, one may observe a mixed value representing the average of two reactions, one which is going through sp^3—sp^2 rehybridization and one which does not (non-covalent binding). Therefore, one may actually find a lower secondary tritium isotope effect than the corresponding deuterium isotope effect, because the latter is caused by *only* the covalent interaction, fully expressing the effect of the configurational change at position 6 of the pyrimidine ring.

W. P. Jencks. I have a question for anyone who would be willing to try to answer it. I have wondered whether the interaction of a tightly bound substrate with an enzyme could restrict the motion of hydrogen atoms enough so as to introduce a secondary isotope effect or to alter a secondary isotope effect. We have heard how interactions with neighboring atoms on the same carbon atom can do this, and I wonder how strong an interaction through van der Waals interactions with the enzyme itself would be required to change these motions enough to be detectable.

W. W. Cleland. We may have an example of just that. With fumarase, erythro-L-malate-3-D (synthesized enzymatically from fumarate in D_2O) gives an inverse isotope effect of 8 or 9% in equilibrium perturbation experiments at neutral pH. Apparently there is a rapid catalytic equilibrium on the enzyme between malate and fumarate, with the proton that is removed being held by a carboxyl group which (as Boyer has shown) loses its proton to solvent more slowly than fumarate is released. Release of fumarate or malate from the enzyme is rate limiting, and when deuterium substitution alters the catalytic equilibrium in favor of fumarate and deuterated carboxyl, fumarate is initially formed more rapidly from deuterated malate than it is hydrated to form malate. All the kinetic evidence we have agrees with this proposal.

We have derived the equations for this model and discover that what corresponds to commitments to catalysis in the usual equations (see chapters by Cleland and Northrop, this volume) is V_1/V_2, the ratio of maximum velocities. Because this ratio is experimentally available, one can calculate the intrinsic isotope effect, which is predicted by these equations to be $1/\beta$, the equilibrium deuterium isotope effect on the reaction

$$\text{Malate} + \text{E—COO}^- \rightleftarrows \text{Fumarate} + \text{E—COOH}$$

We get a calculated $1/\beta$ that is 12–14% inverse, while the effect on the overall fumarase equilibrium, which has been accurately measured by Thomson, is only inverse by 7%. Thus, the carboxyl group on the enzyme shows a fractionation factor relative to water of 1.05–1.07. For a normal carboxyl group this number is unity, so it appears we have an OH bond that is stiffer by 5–7% than it is in water. Perhaps this is an example of restricted motion caused by the tight geometry of the active site.

R. L. Schowen. Dan Quinn has measured distribution of *p*-$NO_2C_6H_4O_2CCD_3$ and *p*-$NO_2C_6H_4O_2CCH_3$ between chlorocyclohexane and water, and there appears to be a substantial effect, conceivably as large as 12 or 13%. It favors the view that the binding is looser to the isotopic atoms in the organic phase than in water. If you imagine that as the carbonyl compound is extracted into chlorocyclohexane, you desolvate the carbonyl group and that increases the hyperconjugation of the CH bonds, that will weaken the binding, and it will be weaker in chlorocyclohexane. The size of the effect is what is startling and a little bit worrisome, but I have a strong suspicion that it's correct. As many people will also know, Thornton and his group found phase transfer isotope effects using high-pressure liquid chromatography to measure them, and they are of the order, I believe, of about 0.5–1% per deuterium. They are doubtless not connected with phenomena such as hyperconjugation, because they occur with simple alkanes being distributed between the stationary phase and the moving phase.

LITERATURE CITED

Atkinson, R. F., and Bruice, T. C. 1974. Ring strain and general acid catalysis of acetal hydrolysis: Lysozyme catalysis. J. Am. Chem. Soc. 96:819–825.

Bender, M. L., and Thomas, R. J. 1961. The concurrent alkaline hydrolysis and isotopic oxygen exchange of a series of *p*-substituted acetanilides. J. Am. Chem. Soc. 83:4183–4189.

Biechler, S. S., and Taft, Jr., R. W. 1957. The effect of structure on kinetics and mechanism of the alkaline hydrolysis of anilides. J. Am. Chem. Soc. 79:4927–4935.

Biellmann, J. F., O'Connell, E. L., and Rose, I. A. 1969. Secondary isotope effects in reactions catalyzed by yeast and muscle aldolase. J. Am. Chem. Soc. 91:6484–6488.

Bilkadi, Z., de Lorimier, R., and Kirsch, J. F. 1975. Secondary α-deuterium kinetic isotope effects and transition-state structure for the hydrolysis and hydrazinolysis reactions of formate esters. J. Am. Chem. Soc. 97:4317–4322.

Bilkadi, Z., and Kirsch, J. F. Secondary α-deuterium isotope effects and transition-state structures for acyl transfer reactions of formamide. In preparation.

Blackburn, G. M., and Jencks, W. P. 1968. The mechanism of the aminolysis of methyl formate. J. Am. Chem. Soc. 90:2638–2645.

Bull, H. G., Koehler, K., Pletcher, T. C., Ortiz, J. J., and Cordes, E. H. 1971. Effects of α-deuterium substitution, polar substituents, temperature, and salts on the kinetics of hydrolysis of acetals and ortho esters. J. Am. Chem. Soc. 93:3002–3011.

Capon, B. 1969. Mechanism in carbohydrate chemistry. Chem. Revs. 69:407–498.

Dahlquist, F. W., Rand-Meir, T., and Raftery, M. A. 1969. Application of secondary α-deuterium kinetic isotope effects to studies of enzyme catalysis, glycoside hydrolysis by lysozyme and β-glucosidase. Biochemistry 8:4214–4221.

do Amaral, L., Bull, H. G., and Cordes, E. H. 1972. Secondary deuterium isotope effects for carbonyl addition reactions. J. Am. Chem. Soc. 94:7579–7580.

Dougherty, T. B., Williams, V. R., and Younathan, E. S. 1972. Mechanism of action of aspartase. A kinetic study and isotope rate effects with ^2H. Biochemistry 11:2493–2498.

Fink, A. L., and Angelides, K. J. 1975. The β-galactosidase-catalyzed hydrolysis of *o*-nitrophenyl-β-galactoside at subzero temperatures: Evidence for a galactosyl enzyme intermediate. Biochem. Biophys. Res. Commun. 64:701–708.

Firsov, L. M., Bogacheva, T. I., and Bresler, S. E. 1974. Secondary isotope effect in the phorphorylase reaction. Eur. J. Biochem. 42:605–609.

Kalman, T. I. 1975. Molecular aspects of mechanism of action of 5-fluorodeoxy-uridine. Annu. N.Y. Acad. Sci. 255:326–331.

Melander, L. 1960. Isotope Effects on Reaction Rates, chs. 2 and 5. Ronald Press, New York.

Mohr, L. H., Smith, L. E. H., and Raftery, M. A. 1973. α-Secondary tritium isotope effects in the aqueous hydrolysis of glycopyranosides of *N*-acetyl-β-D-glucosamine. Arch. Biochem. Biophys. 159:505–511.

Northrop, D. B. 1975. Steady-state analysis of kinetic isotope effects in enzymic reactions. Biochemistry 14:2644–2651.

Richards, J. H. 1970. Kinetic isotope effects in enzymic reactions. In: P. D. Boyer (ed.), The Enzymes, pp. 321–333. Vol. II. Academic Press, New York.

Rosenberg, S., Rabinowitz, J. C., and Kirsch, J. F. Primary oxygen-18 and secondary α-deuterium isotope effects in the formyl tetrahydrofolate synthetase reaction. In preparation.

Santi, D. V., McHenry, C. S., and Sommer, H. 1974. Mechanism of interaction of thymidylate synthetase with 5-fluorodeoxyuridylate. Biochemistry 13:471.

Sawyer, C. B., and Kirsch, J. F. 1973. Kinetic isotope effects for reactions of methyl formate-*methoxyl*-^{18}O. J. Am. Chem. Soc. 95:7375–7381.

Schowen, R. L., Jayaramann, H., and Kershner, L. 1966. Catalytic efficiencies in amide hydrolysis. The two-step mechanism. J. Am. Chem. Soc. 88:3373–3375.

Shiner, V. J., Jr. 1970. Deuterium isotope effects in solvolytic substitution at saturated carbon. In: C. J. Collins and N. S. Bowman (eds.), Isotope Effects in Chemical Reactions, pp. 90–159. Van Nostrand Reinhold, New York.

Sinnott, M. L., and Souchard, I. J. L. 1973. The mechanism of action of β-galactosidase. Biochem. J. 133:89–98.

Sinnott, M. L., and Viratelle, O. M. 1973. The effect of methanol and dioxane on the rates of the β-galactosidase-catalyzed hydrolyses of some β-D-galacto-pyranosides: Rate-limiting degalactosylation. Biochem. J. 133:81–87.

Sinnott, M. L., and Withers, S. G. 1974. The β-D-galactosidase-catalyzed hydrolyses of β-D-galactopyranosyl pyridinium salts. Biochem. J. 143:751–762.

Smith, L. E. H., Mohr, L. H., and Raftery, M. A. 1973. Mechanism for lysozyme-catalyzed hydrolysis. J. Am. Chem. Soc. 95:7497–7500.

Streitwieser, A., Jr., Jagow, R. H., Fahey, R. C., and Suzuki, S. 1958. Kinetic isotope effects in the acetolysis of deuterated cyclopentyl tosylates. J. Am. Chem. Soc. 80:2326–2332.

Tu, J.-I., Jacobson, G. R., and Graves, D. J. 1971. Isotopic effects and inhibition of polysaacharide phosphorylase by 1,5-gluconolactone. Relationship to the catalytic mechanism. Biochemistry 10:1229–1236.

Wataya, Y., and Santi, D. V. 1975. Thymidylate synthetase catalyzed dehalogenation of 5-bromo and 5-iodo-2′-thymidylate. Biochem. Biophys. Res. Commun. 67:818–823.

Wentworth, D. F., and Wolfenden, R. V. 1974. Slow binding of D-galactal, a "reversible" inhibitor of bacterial β-galactosidase. Biochemistry 13:4715–4720.

Wladislaw, B., Olivato, P. R., and Sala, O. 1971. The infrared and raman spectra of some thioacetals. J. Chem. Soc. B. pp. 2037–2040.

Determining the Absolute Magnitude of Hydrogen Isotope Effects

D. B. Northrop

This chapter discusses recent developments in the steady-state kinetic theory of hydrogen isotope effects in enzyme-catalyzed reactions. Some of the theory is still in the formative stage, and comments or criticisms from members of this gathering will be highly welcome. I should say that I am a recent convert to this field; I was encouraged to take a look at isotope effects by Professor Sih of the School of Pharmacy just 2 years ago. Upon examining the literature, I was struck by two basic problems: first, there was confusion as to exactly what kinetic parameters should be measured in isotopic experiments with enzymes; and second, how the observed values of isotope effects should be compared and evaluated. These are fundamental problems. A basis for comparison of different results of any kind of experimental data requires a common point of reference and some common scale of values, both of which appeared inadequate.

To illustrate this latter point, consider an enzymatic reaction having an intrinsic isotope effect of $k_H/k_D = 15$. (By intrinsic isotope effect I mean the full effect originating from the single isotopically sensitive step of catalysis, exclusive of all interference from isotopically insensitive steps.) Obviously, an observed isotope effect with a value of 2 has a quite difference meaning in this reaction than it has in a second enzymatic reaction where the intrinsic isotope effect is only 3. Also, comparisons to values of isotope effects observed in chemical reactions are of limited use because the bond-breaking step is frequently the rate-limiting step in chemical reactions, and one observes the intrinsic value directly. This is seldom the case in enzymatic reactions. Instead, one observes greatly suppressed values resulting from additional rate limitations imposed by the release of products (Cleland, 1975), and release of substrates as well, as will be seen shortly.

Because the intrinsic isotope effect may vary from reaction to reaction, a comparison between observed and intrinsic isotope effects within a single enzymatic reaction is clearly what is desired in order to make the kinds of mechanistic interpretations normally associated with isotope effects on enzyme-catalyzed reactions. Of course, such a comparison requires that one somehow be able to measure or estimate the intrinsic isotope effect. A method of calculating the absolute magnitude of isotope effects (intrinsic) has now been found, based on a fixed relationship between deuterium and tritium isotope effects (Northrop, 1975). The method provides the desired point of reference and the necessary scale on which to analyze data from different experiments and also to compare results from different enzymes. In the course of developing the kinetic equations for this method, answers to the other problem, that of what to measure, presented themselves.

The present discussion begins with an overly simplified enzymatic mechanism in order to focus on the fundamental concepts necessary for evaluating apparent effects and understanding the basis for the method of calculating intrinsic effects, then moves to a second level of complexity which incorporates reversibility and a multiplicity of reaction steps more characteristic of real enzymes, and third, addresses complications arising from the presence of equilibrium isotope effects. From there the discussion shifts to more practical problems of what to measure and how to design laboratory experiments in order to enhance apparent isotope effects. Enhancement is necessary for the application of the method because many observed effects are disappointingly small, and this results in an extreme lack of precision in the calculation of intrinsic values. Throughout the discussion, comments are limited to enzymatic reactions involving direct transfer of deuterium and tritium between substrates and it is assumed that only primary hydrogen isotope effects are present and that they affect only a single step of catalysis. Solutions to these excluded problems must await another time, but at present they are not considered to be major complications in most situations.

COMPONENTS OF A KINETIC MECHANISM

Scheme 1 shows a minimal enzymatic mechanism illustrating the three fundamental components of any mechanism:

$$E \underset{k_2}{\overset{k_1 S}{\rightleftharpoons}} ES \overset{k_3}{\rightarrow} EP \overset{k_5}{\rightarrow} E + P$$

Scheme 1

The first is the substrate binding component, governed by k_1 and k_2. The second is the catalytic component, governed by k_3, considered to be an irreversible step in this example. The third is the product release component, governed by k_5, also an irreversible step because the concentration of product is assumed to be zero during initial velocity measurements. The kinetic expressions describing the mechanism of Scheme 1 are:

$$V = \frac{k_3 k_5 E_t}{k_3 + k_5} \tag{1}$$

$$\frac{V}{K} = \frac{k_1 k_3 E_t}{k_2 + k_3} \tag{2}$$

$$K = \frac{k_5(k_2 + k_3)}{k_1(k_3 + k_5)} \tag{3}$$

$$v = \frac{k_1 k_3 k_5 [S] E_t}{k_5 (k_2 + k_3) + k_1 (k_3 + k_5) [S]} \tag{4}$$

where V represents the maximal velocity, K the Michaelis constant, and v the initial velocity at one concentration of substrate, S. Considered from the point of view of fundamental components, equations 1 and 2 reveal that V and V/K are each determined by only two of the three components, but not the same two components. By definition, V represents the rate at saturating substrate concentrations, which eliminates the binding component and leaves the second and third components. V/K always includes all rate constants up to and including the first irreversible step, which in this example eliminates the product release component and leaves the first and second components. In contrast, equations 3 and 4 reveal that K and v are each determined by all three components. As a result, isotope effects on Michaelis constants or single-rate measurements are difficult or impossible to interpret. The maximum utility of isotope effects should be found in their expression on V and V/K, because these depend on the fewest variables. Furthermore, because V and V/K are dependent on different components of a reaction mechanism, interpretations of isotope effects on V and V/K must also be different.

DEPENDENCE OF APPARENT ISOTOPE EFFECTS ON COMPONENTS

Having established that V and V/K are the kinetic parameters of interest in measurements of isotope effects on enzyme-catalyzed reactions, let us consider what determines the level of expression of the intrinsic isotope effect in the apparent isotope effects one observes on V and V/K. Dividing equation 1 (representing a reaction with a hydrogen-containing substrate) by an analogous equation representing a reaction with a deuterium-containing substrate, one can obtain the following equation:

$$\frac{V_H}{V_D} = \frac{k_{3H}/k_{3D} + k_{3H}/k_5}{1 + k_{3H}/k_5} \tag{5}$$

This equation shows that the size of the apparent isotope effect on the maximal velocity (V_H/V_D) relative to the size of the intrinsic isotope effect (k_{3H}/k_{3D}) is dependent upon the ratio of the rate of catalysis to the rate of product release (k_{3H}/k_5).

Similarly, from equation 2 one can obtain the following equation for V/K isotope effects:

$$\frac{(V/K)_H}{(V/K)_D} = \frac{k_{3H}/k_{3D} + k_{3H}/k_2}{1 + k_{3H}/k_2} \tag{6}$$

This equation shows that the size of the apparent isotope effect on V/K relative to the intrinsic isotope effect is dependent upon the ratio of the rate of catalysis to the rate of substrate release (k_{3H}/k_2).

NOMENCLATURE AND INTERPRETATION OF APPARENT ISOTOPE EFFECTS

The equations for apparent isotope effects tend to be confusing, particularly in more complex mechanisms. One tends to get lost in the algebra and fails to grasp the overall form of these equations. For this reason, I wish to suggest a new nomenclature. I do this with some hesitation, because the usual result of such an effort is an introduction of new symbols which, by their newness, are more abstract than the old, whereas my purpose is to move in the opposite direction. When we speak of a "deuterium isotope effect on V_{max}," both the hydrogen rate and its ratio to a deuterium rate are implied. Therefore, it seems logical and consistent to incorporate the same unstated implications in the symbolism and to identify only the isotope and the kinetic parameter of interest, preferably in a form compatible with the spoken reference.

With this in mind, I propose that equations 5 and 6 be written in a general form:

$$^{D}V = \frac{^{D}k + R}{1 + R} \quad (7)$$

$$^{D}(V/K) = \frac{^{D}k + C}{1 + C} \quad (8)$$

The intrinsic isotope effect on catalysis (^{D}k) may be written without the subscript denoting the reactive step involved because this, too, is implied by a primary isotope effect. Two new symbols are required, *R* and *C*. *R* stands for "ratio of catalysis" and represents the ratio of the rate of the catalytic step to the rate of the other forward steps contributing to the maximal velocity. *C* stands for "commitment to catalysis" and represents the tendency of the enzyme-substrate complex to go forward through catalysis as opposed to its tendency to break down to free enzyme and substrate.

The functions represented by *R* and *C* have been referred to as "partitioning factors" by others in the field. I prefer "ratio of catalysis" and "commitment to catalysis" because these make a distinction between the different components which contribute to *V* and *V/K* effects. These definitions convey a sense of directionality to the functions they represent, and they convey a clearer mental image of an otherwise abstract collection of rate constants. These points aid in the interpretation of isotope effects on enzyme-catalyzed reactions, which will be illustrated later in detail. For the moment, it is important to note that isotope effects on *V* are in reality a measure of the ratio of catalysis, which is the basis for the familiar usage of isotope effects as a means of identifying a rate-determining step. On the other hand, isotope effects on *V/K* are similarly a measure of the commitment to catalysis, a notion not generally recognized. In fact, erroneous attempts have been made to identify rate-determining steps from *V/K* data, primarily because of a failure to recognize discrimination isotope effects as being dependent on *V/K* effects (see below).

CALCULATION OF INTRINSIC ISOTOPE EFFECTS

Interpretations of isotope effects based on the ratio of, or commitment to, catalysis are, of course, hampered if the magnitude of the intrinsic isotope effect is not known. Useful limits can be calculated in some instances (Northrop, 1975). A comparison between deuterium and tritium isotope effects would appear to provide a means of obtaining the intrinsic isotope effects, because *R* and *C* should not be perturbed by the alternate isotope. The secret to making such a comparison turns out to be the subtraction of 1 from both sides of an equation for an apparent isotope effect. This was unexpected, but by hindsight it appears obvious because current conventions assign a value of 1 to the absence of an isotope effect, while the mathematics of comparisons requires zero as the reference point.

Subtracting 1 from both sides of equation 8 yields

$$
\begin{aligned}
{}^{D}(V/K) - 1 &= \frac{{}^{D}k + C}{1 + C} - 1 \\
&= \frac{{}^{D}k + C - 1 - C}{1 + C} \\
&= \frac{{}^{D}k - 1}{1 + C} \qquad (9)
\end{aligned}
$$

The effect of this maneuver is to remove C from the numerator of the equation. Dividing equation 9 by a similar expression for tritium isotope effects yields

$$
\begin{aligned}
\frac{{}^{D}(V/K) - 1}{{}^{T}(V/K) - 1} &= \frac{\dfrac{{}^{D}k - 1}{1 + C}}{\dfrac{{}^{T}k - 1}{1 + C}} \\
&= \frac{{}^{D}k - 1}{{}^{T}k - 1} \qquad (10)
\end{aligned}
$$

Cancellation of the common denominators of the deuterium and tritium expressions removes *C* from the equation and leaves only the apparent isotope effects as a function of the intrinsic isotope effects.

Swain et al. (1958) have calculated a fixed relationship between intrinsic isotope effects of deuterium and tritium which allows us to express either one in terms of the other. Incorporating the Swain relationship into equation 10 yields

$$
\frac{{}^{D}(V/K) - 1}{{}^{T}(V/K) - 1} = \frac{{}^{D}k - 1}{{}^{D}k^{1.44} - 1} \qquad (11)
$$

This equation provides the apparent deuterium and tritium isotope effects on V/K, which are measurable, as a function of a single intrinsic isotope effect, which therefore is calculable. The calculation cannot be done in a direct fashion because of the exponential function involved. Rather, it is necessary to assume values for the intrinsic effect and generate a series of values for ($V/K - 1$) ratios, which can then be compared to real data. This comparison may be done through the use of the graph shown in Figure 1, or more easily by use of the table given at the end of this volume.

A similar calculation theoretically could be accomplished using apparent isotope effects on V (starting with equation 7) but tritium isotope effects on V require total substitution of tritium in substrates, which is not technically feasible.

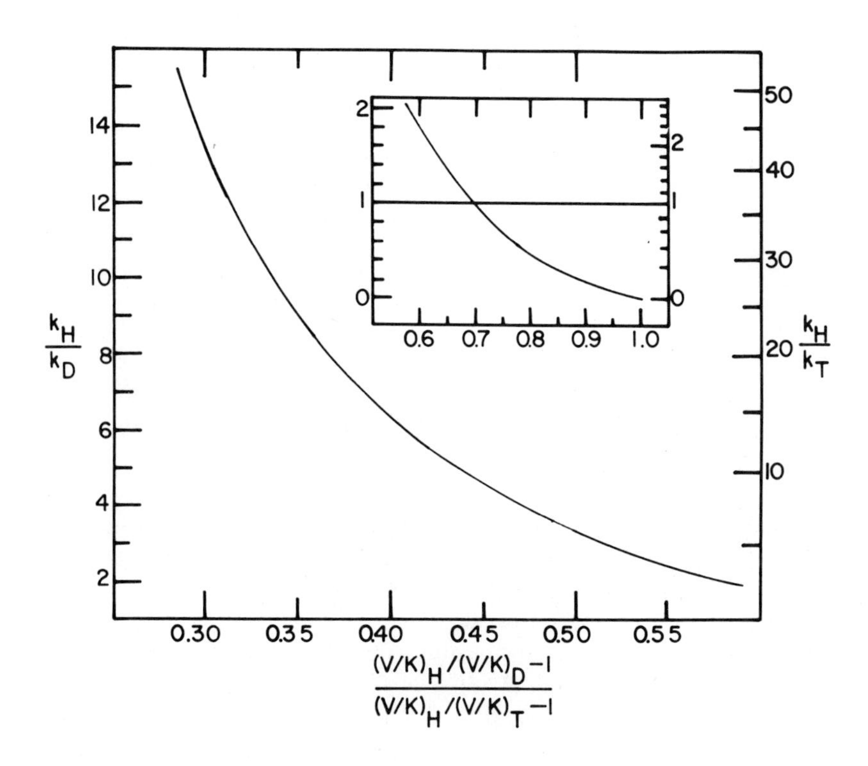

Figure 1. Intrinsic isotope effects as a function of apparent deuterium and tritium isotope effects on V/K.

THE SWAIN RELATIONSHIP

The validity of equation 11 is clearly dependent upon the validity of the Swain relationship, so it is important to consider the basis and reliability of this function. Swain's group employed a simple quantum-mechanical model in their theoretical investigation which considered only stretching vibrations, assumed that these are in harmonic oscillation, and further assumed isotopic differences arose only from differences in zero-point vibrational energies, i.e., tunneling effects were excluded. They found that the relative magnitude of deuterium versus tritium isotope effects could be expressed as the exponential function of the isotopic masses:

$$\frac{k_H}{k_T} = \left(\frac{k_H}{k_D}\right)^{\left[\frac{1/\sqrt{m_1} - 1/\sqrt{m_2}}{1/\sqrt{m_1} - 1/\sqrt{m_3}}\right]} \quad (12)$$

Using values of $m_1 = 1.0081$, $m_2 = 2.0147$, and $m_3 = 3.0170$:

$$\frac{k_H}{k_T} = \left(\frac{k_H}{k_D}\right)^{1.442} \quad (13)$$

Swain predicted this equation to hold as long as the temperature remained sufficiently low to maintain X—H bonds in their lowest vibrational state, noting a temperature range of 0–100 °C.

Most of the subsequent investigations of the Swain relationship have focused on the possible effects of transition-state vibrational energy differences or tunneling. Bigeleisen (1962) evaluated a more complete quantum-mechanical model that included tunneling and proposed a range of 1.33–1.56 for the mass exponent, but suggested that more extensive tunneling would lead to lower values. The model was further elaborated by More O'Ferrall and Kouba (1967) who found deviations of only 2% from Swain's value if nonhydrogenic bonds were treated as rigid to stretching motion. Then Lewis and Robinson (1968) demonstrated that tunneling need not produce significant variations in the value, citing experimental examples having deuterium isotope effects of 16 and 23. Finally, Stern and Vogel (1971) and Stern and Weston (1974) considered 180 model-reaction systems, incorporated secondary isotope effects, extended the temperature range to 20–1000 °K, and concluded that the limits set by Bigeleisen are not likely to be exceeded.

The only troublesome area seems to be the possibility of steric isotope effects, which can be both significant and inverse (Carter and Melander, 1973). The model studies above were restricted to normal isotope effects that reflected only force-constant changes at the isotopic positions between reactants and transition states. Stern and Weston (1974) acknowledged many deviations in the mass exponent when these conditions were not met. It is generally agreed that thermodynamic isotope effects of steric origin, such as an effect on substrate binding, are probably insignificant. However, an isotope effect on the conformational changes leading to the enzyme-substrate transition state is conceivable when a tight fit between the substrate and the active site "pocket" of the enzyme is considered. Nevertheless, the significance of a kinetic isotope effect of steric origin is reduced as catalysis becomes more rate limiting, because slower catalysis causes conformational changes to approach an equilibrium state. Hence, steric effects may be a problem only when apparent isotope effects are small, which gives an additional reason to try to enhance apparent effects (see below).

Equally significant to the model studies is the absence of any experimental documentation of a deviation from the Swain relationship. This, plus the model studies, the narrow temperature range available to the enzymologist, and the accepted notion that transition-state chemistry of enzymes is similar to that in nonenzymatic catalysis, leads me to believe that errors caused by deviations from the Swain relationship are not likely to exceed the limits of experimental error of determinations of apparent isotope effects.

COMPARISON OF APPARENT AND INTRINSIC ISOTOPE EFFECTS

Once the intrinsic isotope effect has been calculated, a comparison of this value with the apparent isotope effect on V suggests a measure of the contribution of the bond-breaking step to the overall rate. A direct comparison is misleading, however, and once again it is necessary to subtract a value of 1 from each of the values in order to obtain a common reference point. The comparison then takes the form

$$f_v = \frac{{}^{D}V - 1}{{}^{D}k - 1} \qquad (14)$$

f_v is called the "fractional reduction of the maximal velocity." This term represents a measure of the rate limitation imposed by catalysis on the overall rate. In an extreme situation where catalysis is totally rate limiting, ${}^{D}V = {}^{D}k$ and $f_v = 1$. At the other extreme, for example, if product release is rate limiting, $^{D}V = 1$ and $f_v = 0$. One can anchor one's thinking about f_v at either extreme. I prefer the latter, considering f_v a measure of the observed maximal velocity relative to what one would expect if catalysis had not contributed to the overall rate.

The meaning and use of f_v are not easy concepts, and I can best convey what they mean by a series of hypothetical examples. Table 1 lists possible apparent and intrinsic isotope effects plus calculated f_v values for a series of alternate substrates for an enzyme. The first three examples represent conditions in which the intrinsic isotope effect remains the same. Apparent isotope effects on V are different because of changes either in the rate of catalysis or the rates of product release. The degree of rate limitation imposed by the rate of catalysis is suggested by the magnitude of the apparent isotope effect. Clearly, the larger isotope effects indicate that catalysis is more rate limiting with S_2 ($^{D}V = 3$) than S_1 ($^{D}V = 2$) and even more so with S_3 ($^{D}V = 5$). The question is: How much more? The answer is given by the f_v values. From these one can say that catalysis is twice as rate limiting with S_2 as S_1 ($f_v = 0.2$ vs. 0.1) and four times as much with S_3 ($f_v = 0.4$ vs 0.1). Such quantitative comparisons cannot be made from ^{D}V values alone. This is particularly true if the intrinsic isotope effect is also changed by the alternate substrate, as in the last two examples. Although the apparent isotope effect with S_4 is the same as with S_1, the effect obviously has a different meaning because the intrinsic isotope effects are drastically different. Similarly, S_5 yields a different value for ^{D}V than the rest of the examples, and this might be thought to indicate a unique degree of rate limitation imposed by catalysis. However, S_2, S_4, and S_5 all yield $f_v = 0.2$, which means that the contribution of catalysis to the overall rates is the same for these three substrates.

The above discussion illustrates the meaning and use of f_v in a relative sense, but it can also be thought of in an absolute sense. Multiplied by 100, the f_v values yield the percentage of rate limitation of the max-

Table 1. Comparisons of apparent and intrinsic isotope effects

Substrate	^{D}V	^{D}k	f_v
S_1	2	11	0.1
S_2	3	11	0.2
S_3	5	11	0.4
S_4	2	6	0.2
S_5	4	16	0.2

imal velocity imposed by catalysis. For example, using data obtained by Bright and Gibson (1967), I previously calculated $f_v = 0.28$ for glucose oxidase (Northrop, 1975). This means that the observed maximal velocity of glucose oxidase is 28% less than one would expect under conditions in which the isotopically sensitive step is infinitely fast.

A similar approach may be applied to V/K isotope effects, using equation 15.

$$f_{v/k} = \frac{^{D}(V/K) - 1}{^{D}k - 1} \tag{15}$$

$f_{v/k}$ is the fractional reduction of V/K attributed to the bond-breaking step.

Although this illustration assumed hypothetical results with a single enzyme, it could just as well have been applied to a series of enzymes. Here, then, in f_v and $f_{v/k}$ is the "scale" on which we can compare and evaluate isotopic data on enzyme-catalyzed reactions.

EQUILIBRATION STEPS AND REVERSIBILITY

In order to extend this analysis to real enzymes, a more general mechanism than depicted in Scheme 1 must be considered. Two features must be added: equilibration steps between substrate binding, catalysis, and product release, and reversibility of catalysis. Scheme 2 illustrates such a mechanism.

$$E_1 \underset{k_2}{\overset{k_1 B}{\rightleftharpoons}} E_2 \underset{k_4}{\overset{k_3}{\rightleftharpoons}} E_3 \underset{k_6^*}{\overset{k_5^*}{\rightleftharpoons}} E_4 \underset{k_8}{\overset{k_7}{\rightleftharpoons}} E_5 \underset{k_{10}P}{\overset{k_9}{\rightleftharpoons}} E_6 \overset{k_{11}}{\longrightarrow} E_1$$

Scheme 2

Step 1 represents substrate binding; steps 2 and 4 represent conformational changes of the enzyme and/or additional catalytic steps; step 3 is the isotopically sensitive catalytic step (denoted by asterisks); step 5 is the release of the first product; and step 6 represents the sum of all additional steps necessary to regenerate E_1. The mechanism is general; it may be considered as one half of a ping-pong mechanism, with k_{11} representing the net rate constant for the other half; or the inner pair of reactants in an ordered mechanism, with k_{11} being the net rate constant for the release of other products and binding of other substrates; or a similar portion of a random mechanism at saturating concentrations of other substrates.

The apparent V/K isotope effect with respect to substrate B takes the form:

$$^{D}\left(\frac{V}{K_b}\right) = \frac{^{D}k_5 + C_f + C_r}{1 + C_f + C_r} \tag{16}$$

where

$$C_f = \frac{k_5}{k_4}\left(1 + \frac{k_3}{k_2}\right) \tag{17}$$

$$C_r = \frac{k_6}{k_7}\left(1 + \frac{k_8}{k_9}\right) \tag{18}$$

Three new features are present in equation 16. First, the C term now consists of two parts: C_f, the forward commitment to catalysis, and C_r, the reverse commitment to catalysis.[1] This feature ensues from the presence of reversibility and leads to an important insight into V/K effects: to the extent that the distribution of enzyme forms in the steady-state favors a flow toward catalytic forms E_3 and E_4, C_f and C_r will be large and the isotope effect will be suppressed; to the extent that the distribution favors free enzyme on both the substrate and product sides, C_f and C_r will be small and the apparent isotope effect will be enhanced. Second, the terms of the commitment to catalysis are more complex and contain one ratio of rate constants for each step preceding catalysis. Consider the case in which k_7, rather than k_5, represents catalysis. The expression for the forward commitment to catalysis becomes[2]

$$C_f = \frac{k_7}{k_6}\left(1 + \frac{k_5}{k_4}\left(1 + \frac{k_3}{k_2}\right)\right) \tag{19}$$

Comparison of equations 17 and 19 with the earlier definition of C reveals the pattern of this series, which, once grasped, makes it possible to write kinetic expressions for V/K isotope effects by simple inspection of a proposed mechanism (O'Leary, 1977). Third, because both C_f and C_r are present, in the absence of an equilibrium isotope effect it is ne-

[1] The term "commitment to catalysis" is algebraically defined differently here than when it was first introduced (Northrop, 1975). Its conceptual meaning, however, remains the same.

[2] The commitment to catalysis is as much as a property of the isotope as of the substrate or enzymatic mechanism. For a given enzyme, equation 17 might apply for a hydrogen isotope, while at the same time equation 19 might describe a carbon isotope effect. This situation ensues from the carbon and hydrogen isotopes affecting different steps of catalysis, as is the case with isocitrate dehydrogenase (O'Leary and Limburg, 1977).

cessary that ${}^D(V/K_b)$ equals ${}^D(V/K_p)$, thus providing a useful check on the internal consistency of one's data.

The apparent isotope effect on V takes the form

$$ {}^DV = \frac{{}^Dk_5 + R_f + C_r}{1 + R_f + C_r} \tag{20} $$

where

$$ R_f = \left(\frac{k_5}{k_3} + \frac{k_5}{k_7}\left(1 + \frac{k_8}{k_9}\right) + \frac{k_5}{k_9} + \frac{k_5}{k_{11}}\right)\left(\frac{k_3}{k_3 + k_4}\right) \tag{21} $$

The presence of reversibility brings C_r into the expression, and the addition of equilibration steps increases the complexity of the forward ratio of catalysis, R_f. This term retains its earlier meaning, however, noted by the separate ratios of catalysis to each of the other forward rate constants. Hence, the individual ratios are additive, causing an increased level of suppression of ${}^DV$ simply as a function of the number of steps in the mechanism.

Despite the increased complexity of the kinetic equations for Scheme 2, the method of calculating intrinsic isotope effects still applies. Subtracting 1 from each side of equation 16 removes the commitment factors from the numerator.

$$ {}^D(V/K) - 1 = \frac{{}^Dk_5 + C_f + C_r}{1 + C_f + C_r} - 1 $$

$$ = \frac{{}^Dk_5 + C_f + C_r - 1 - C_f - C_r}{1 + C_f + C_r} $$

$$ = \frac{{}^Dk_5 - 1}{1 + C_f + C_r} \tag{22} $$

Division by the tritium expression removes the remaining commitment factors.

$$ \frac{{}^D(V/D) - 1}{{}^T(V/K) - 1} = \frac{\dfrac{{}^Dk_5 - 1}{1 + C_f + C_r}}{\dfrac{{}^Tk_5 - 1}{1 + C_f + C_r}} $$

$$ = \frac{{}^Dk_5 - 1}{{}^Tk_5 - 1} $$

$$ = \frac{{}^Dk_5 - 1}{{}^Dk_5{}^{1.44} - 1} \tag{23} $$

Thus, the addition of extra steps and reversibility to an enzymatic mechanism does not influence the calculation of intrinsic isotope effects. Because C_r is present in equations 16 and 20, however, subsequent calculation of f_v and $f_{v/k}$ are altered. The fractional reductions are no longer dependent solely upon the rate of catalysis, but they are influenced by the reverse commitment to catalysis as well. Nevertheless, the scales represented by f_v and $f_{v/k}$ remain more appropriate for comparisons than direct comparisons of apparent isotope effects. What is now needed in this area is an independent means of determining C_r.

ISOTOPE EFFECTS ON THE EQUILIBRIUM CONSTANT

Isotope effects on catalysis need not be equal in both directions of a reaction and, when unequal, appear as an isotope effect on the equilibrium constant. From our initial assumptions, we can define:

$$^DK_{eq} = \frac{^Dk_f}{^Dk_r} \tag{24}$$

where $^DK_{eq}$ is the deuterium isotope effect on the equilibrium constant, and ^Dk_f and ^Dk_r are the forward and reverse intrinsic isotope effects. The general equations for apparent isotope effects and calculations of intrinsic effects now take the following forms (Cleland, 1977). For the reaction in the forward direction:

$$^DV_f = \frac{^Dk_f + R_f + C_r \cdot {^DK_{eq}}}{1 + R_f + C_r} \tag{25}$$

$$^D(V/K)_f = \frac{^Dk_f + C_f + C_r \cdot {^DK_{eq}}}{1 + C_f + C_r} \tag{26}$$

$$\frac{^D(V/K)_f - 1}{^T(V/K)_f - 1} = \frac{^Dk_f - 1 + C_r\,(^DK_{eq} - 1)}{^Dk_f{}^{1.44} - 1 + C_r\,(^DK_{eq}{}^{1.44} - 1)} \tag{27}$$

For the reverse reaction:

$$^DV_r = \frac{^Dk_r + R_r + C_f/^DK_{eq}}{1 + R_r + C_f} \tag{28}$$

$$^D(V/K)_r = \frac{^Dk_r + C_f/^DK_{eq} + C_r}{1 + C_f + C_r} \tag{29}$$

$$\frac{{}^{D}(V/K)_r - 1}{{}^{T}(V/K)_r - 1} = \frac{{}^{D}k_r - 1 + C_f\left(\frac{1}{{}^{D}K_{eq}} - 1\right)}{{}^{D}k_r^{\,1.44} - 1 + C_f\left(\frac{1}{{}^{D}K_{eq}^{\,1.44}} - 1\right)} \tag{30}$$

These equations reveal that the influence of equilibrium isotope effects is dependent upon the values of the commitment factors. Because these are not known, the errors introduced into calculations of intrinsic isotope effects cannot be determined. A set of limits of error, however, can be obtained by comparing the results of calculations of the forward and reverse intrinsic isotope effects. To understand and obtain these limits, it is important to note that the influence of ${}^{D}K_{eq}$ on equations 27 and 30 is not symmetrical. First, the expression of ${}^{D}K_{eq}$ is determined by C_r in equation 27 and by C_f in equation 30, but C_r need not equal C_f. Second, given a normal (i.e., greater than one) isotope effect on K_{eq}, a fraction of C_r is added in equation 27, whereas a fraction of C_f is subtracted in equation 30. Third, the sizes of the fractions added or subtracted are not equal.

This lack of symmetry can best be conveyed by a hypothetical example. Figure 2 is a plot of calculated values of intrinsic isotope effects as a function of commitments to catalysis, assuming true values of ${}^{D}k_f = 6$ and ${}^{D}k_r = 5$, which yields ${}^{D}K_{eq} = 1.2$. The first lack of symmetry requires that separate curves be drawn for each intrinsic effect, indicated by the separate axes for C_f and C_r. It is possible for C_f to be large and C_r to be very small, in which case the error in the calculation of ${}^{D}k_r$ would be insignificant while that in ${}^{D}k_r$ would be large. The second lack of symmetry causes the curves to have slopes of opposite sign. As C_f increases, calculated ${}^{D}k_r$ values also increase, but as C_r increases, calculated ${}^{D}k_f$ values decrease. The third lack of symmetry results in different magnitudes of slopes for the two curves. Given equal commitments to catalysis (i.e., $C_f = C_r = 5$) the two calculated intrinsic isotope effects have different amounts of error (i.e., ${}^{D}k_r$ is too small by 1 unit, whereas ${}^{D}k_f$ is too large by more than 1.6 units). Hence, the direction of error can be predicted, but the amount of error in each intrinsic effect cannot.

The first indication of error can be obtained by comparing a measured equilibrium isotope effect to the ratio of the calculated intrinsic isotope effects. Because the direction of error is predictable and given a normal ${}^{D}K_{eq}$, then

$$^{D}K_{eq} \geqslant (^{D}k_f/^{D}k_r)_{\mathrm{CALC}} \tag{31}$$

Furthermore, by cross-multiplying this function, it becomes clear that the true value for $^{D}k_f$ is less than $^{D}K_{eq}(^{D}k_r)_{\mathrm{CALC}}$ but greater than $(^{D}k_f)_{\mathrm{CALC}}$. Similarly, the outside limits of $^{D}k_r$ are less than $(^{D}k_r)_{\mathrm{CALC}}$ but greater than $(^{D}k_r)_{\mathrm{CALC}}/^{D}K_{eg}$.

It is not necessary to determine experimentally the apparent isotope effects on V/K in the reverse reaction in order to calculate these limits, because a comparison of equations 26 and 29 yields:

$$\frac{^{D}(V/K)_f}{^{D}(V/K)_r} = {}^{D}K_{eq} \tag{32}$$

From equation 32, the reverse V/K isotope effects may be calculated from the equilibrium isotope effect and kinetic data from the forward direction only. $^{D}k_r$ can most easily be so obtained by incorporating the relationship of equation 32 into equation 30, because

$$\frac{^{D}(V/K)_f \cdot {}^{D}K_{eq} - 1}{^{T}(V/K)_f \cdot {}^{D}K_{eq}{}^{1.44} - 1} = \frac{^{D}(V/K)_r - 1}{^{T}(V/K)_r - 1} \tag{33}$$

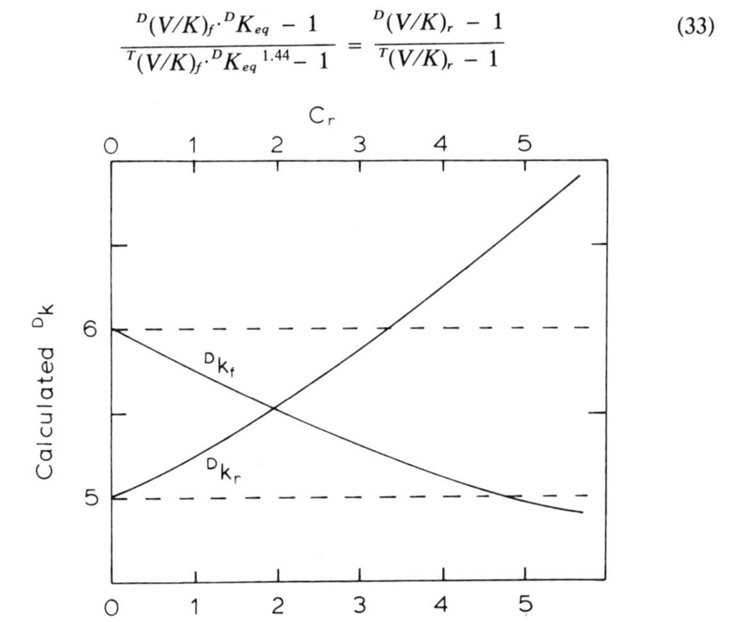

Figure 2. Errors imposed on calculated values of intrinsic isotope effects due to an equilibrium isotope effect. The figure assumes the true $^{D}k_f$ = 6, $^{D}k_r$ = 5, and $^{D}K_{eq}$ = 1.2.

Nevertheless, it is desirable to obtain experimental data in both directions of a reaction because equation 32 provides a useful check on the internal consistency of experimental results.

ENHANCEMENT OF APPARENT ISOTOPE EFFECTS

Experimental determinations of apparent isotope effects frequently have given very small values having large percentages of error. The percentage of error is expanded during the calculation of ($V/K - 1$) functions, expanded again during the calculation of intrinsic isotope effects, and further expanded when the limits of confidence because of equilibrium isotope effects are taken into consideration. The net result is that the larger the apparent isotope effect, the more precise the calculation of the intrinsic isotope effect. While an intrinsic isotope effect should be a constant, an apparent one need not be, and examples of variation with experimental conditions are known. These experimental conditions, therefore, may be manipulated to enhance the apparent isotope effects and increase the reliability of these isotopic methods.

Variation of Second Substrate

Preliminary experiments to detect isotope effects in enzymatic reactions are frequently conducted at saturating substrate concentrations. Because increased substrate concentrations increase the commitment to catalysis, saturation may have suppressed many interesting and useful isotope effects.

A good example of this phenomenon can be found in the data of Bush, Shiner, and Mahler (1973) in studies of NADH/NADD oxidation by liver alcohol dehydrogenase. Table 2 shows apparent isotope effects as a function of acetaldehyde concentrations. At 5 mM, only a

Table 2. NADH/NADD oxidation by liver alcohol dehydrogenase

CH_3CHO (mM)	^{D}V
→ ∞	1.0
5.0	1.2
0.5	1.6
0.05	2.2
→ 0.0	2.6

Adapted from Bush, Shiner, and Mahler, 1973.

small effect was detected, but at lower concentrations the effect became significant. Extrapolation to saturating concentrations abolished the isotope effect, indicating no effect on the maximal velocity. Extrapolation to zero concentration reveals a *V/K* isotope effect of 2.6.

Earlier it was pointed out that the desired kinetic parameters to be characterized regarding isotope effects are *V* and *V/K*, which implies a variation in the isotopically labeled substrate. But now it is clear that in multi-substrate cases, other substrates and/or cofactors should also be varied in order to obtain both measurable and limiting values of these effects.

Use of Alternate Substrates or Cofactors

Alternate substrates often display a wide variation in kinetic parameters. Those that are "less sticky" i.e., have higher Michaelis or dissociation constants, have the lowest commitment to catalysis, and therefore should display the largest *V/K* isotope effects. Because isotope effects on *V* are controlled by two terms, alternate substrates may enhance the apparent isotope effect in two ways. A substrate giving rise to a less sticky product will have a lower reverse commitment to catalysis. Alternatively, substrates displaying low velocities may do so as a result of a shift in the rate-limiting step from product release to catalysis which reduces the ratio of catalysis.

Several examples of alternate substrate enhancement of isotope effects exist, but perhaps the most dramatic and best documented case comes from recent studies on glutamate dehydrogenase (Cleland, this volume). Deuterated glutamate yields an isotope effect on *V/K* of 1.2 and an insignificant effect on *V*. When glutamate is replaced by norvaline, an alternate substrate having a K_m six-to-seven-fold higher than glutamate, the *V/K* isotope effect is enhanced to 3.3 and the effect on *V* enhanced to 6.0. That these enhancements result in part from a reduction of the reverse commitment to catalysis is supported by initial velocity studies on the reverse reaction. The pattern with α-ketoglutarate versus ammonia is the normal intersecting pattern of a sequential mechanism. However, the pattern with α-ketovalerate versus ammonia intersects on the vertical axis, indicating a rapid equilibrium ordered mechanism.

It is possible that the use of alternate substrates which alter the rate of catalysis may alter the magnitude of the intrinsic isotope effects as well. This will have to be checked using a series of alternate substrates giving significant but different apparent isotope effects. The results should prove interesting regardless of whether or not the intrinsic isotope effect is altered. A particularly attractive approach would be to vary the choice of cofactor, metal ions for example, because these frequently affect the kinetic parameters of substrates and could be varied within a single experimental design using natural substrates.

Variation of pH

Both V and V/K are known to vary as a function of pH, and frequently they vary independent of each other. Few data appear to be available at present concerning possible variation of isotope effects on V and V/K as a function of pH, and the theory relating pH to isotope effects is still in the developmental stage. Nevertheless, there is a reason to predict that extremes of pH should enhance apparent isotope effects. O'Leary (1977) has noted that in order for isotope effects to be pH dependent, alternate reaction pathways leading up to, but not necessarily including, the catalytic step must exist whose relative contribution to the overall reaction is pH dependent. To illustrate how this would work and how it relates to these kinetic equations, Figure 3 shows a pH dependent reaction mechanism. The mechanism assumes that the isoelectric form of the enzyme is the most active and is the only form that undergoes catal-

$$
\begin{array}{cccccccccccc}
 & E_1^{-3} & & & & & & & & & & E_6^{-3} \\
 & \updownarrow & & & & & & & & & & \updownarrow \\
 & E_1^{-2} & \overset{A}{\rightleftharpoons} & E_2^{-2} & & & & & & E_5^{-2} & \rightarrow & E_6^{-2} \\
 & \updownarrow & & \updownarrow & & & & & & \updownarrow & & \updownarrow \\
 & E_1^{-1} & \overset{A}{\rightleftharpoons} & E_2^{-1} & \rightleftharpoons & E_3^{-1} & & E_4^{-1} & \rightleftharpoons & E_5^{-1} & \rightarrow & E_6^{-1} \\
 & \updownarrow & & \updownarrow & & \updownarrow & & \updownarrow & & \updownarrow & & \updownarrow \\
\text{pH}\uparrow & E_1 & \overset{A}{\rightleftharpoons} & E_2 & \rightleftharpoons & E_3 & \underset{k}{\overset{k}{\rightleftharpoons}} & E_4 & \rightleftharpoons & E_5 & \rightarrow & E_6 \\
 & \updownarrow & & \updownarrow & & \updownarrow & & \updownarrow & & \updownarrow & & \updownarrow \\
 & E_1^{+1} & \overset{A}{\rightleftharpoons} & E_2^{+1} & \rightleftharpoons & E_3^{+1} & & E_4^{+1} & \rightleftharpoons & E_5^{+1} & \rightarrow & E_6^{+1} \\
 & \updownarrow & & \updownarrow & & & & & & \updownarrow & & \updownarrow \\
 & E_1^{+2} & \overset{A}{\rightleftharpoons} & E_2^{+2} & & & & & & E_5^{+2} & \rightarrow & E_6^{+2} \\
 & \updownarrow & & & & & & & & & & \updownarrow \\
 & E_1^{+3} & & & & & & & & & & E_6^{+3}
\end{array}
$$

Figure 3. Effect of pH on commitments to catalysis. The reaction mechanism diagrammatically illustrates the shift in the distribution of enzyme forms away from the catalytic complexes at the extremes of pH.

ysis. At acid or alkaline pH, the enzyme is titrated into other forms. Some of these will not bind substrate (i.e., E^{-3}), some can bind but are unable to undergo the conformational change preceding catalysis (i.e., E^{-2}), and others can reach but not undergo the catalytic step (i.e., E^{-1}). The net effect of these additional enzyme forms and partial pathways which cannot undergo catalysis is to shift the distribution of enzyme away from catalysis in the steady state and, hence, lower the overall commitment to catalysis. In turn, this should enhance apparent isotope effects.

Does this actually work? It should work better on V/K than V isotope effects, because ${}^{D}V$ is dependent on the ratio of catalysis in addition to the commitment to catalysis, and variation of the ratio of catalysis as a function of pH is not predictable, nor necessarily favorable, at extremes of pH. However, the only examples I know of show enhancement of ${}^{D}V$ with ${}^{D}(V/K)$ being either insignificant or independent of pH within the range of pH examined. Studies in Cleland's laboratory with malic enzyme gave deuterium isotope effects on V near 3 at pH 3.9 and 9.5 but no detectable effect at neutral pH (Schimerlik, Grimshaw, and Cleland, 1977). Similarly, in O'Leary's laboratory, isocitrate dehydrogenase gave an effect on V of 1.5 at pH 5.5 and none at neutral pH (O'Leary and Limburg, 1977).

The problem again arises as to whether the sizes of intrinsic isotope effects will be altered by attempts to enhance apparent isotope effects. Changes in pH are not likely to have any direct effect on catalysis, as enzymes probably have to achieve a single and optimum state of protonation for the transition state of catalysis. This, too, awaits verification.

Variation of Temperature

General differences between the reactive steps comprising an enzymatic reaction mechanism (i.e., binding and release of reactants, enzyme conformational changes, and catalysis) suggest that these probably have different temperature coefficients. If so, then variation in temperature should alter the ratio of catalysis and commitments to catalysis, giving rise to changes in apparent isotope effects. Isotope effects on enzymatic reactions are rarely studied as a function of temperature, particularly with enough data to calculate V and V/K effects, but I have found two thorough studies and both support this hypothesis. Bright and Gibson (1967) performed steady-state experiments at 25°C and transient-state experiments at 3°C on the oxidation of glucose-1-D by glucose oxidase.

From their data, I calculate a ${}^{D}V$ of 4.92 at 3 °C and 7.83 at 25 °C, accompanied by a change in ${}^{D}(V/K)$ from 1.98 to 3.84. Similarly, studies by Bardsley, Craffe, and Shindler (1973) on diamine oxidase give ${}^{D}V$ and ${}^{D}(V/K)$ values of 1.6 and 2.3 at 30 °C which increase to 2.75 and 4.83, respectively, at 50 °C.

Intrinsic isotope effects also change with temperature, but these decrease with increasing temperature. Within the temperature range available to the enzymologist, the changes are small and may be considered insignificant (Vogel and Stern, 1971). Hence, the changes in apparent isotope effects observed with glucose oxidase and diamine oxidase can be attributed to changes in ratios of catalysis and/or commitments to catalysis.

EXPERIMENTAL DESIGNS

The development of the theory for calculating intrinsic isotope effects was accompanied by frustrations resulting from a lack of data in the literature to which these ideas might be applied and tested. Of course, without foreknowledge of these methods, one could not expect investigations to expend extra effort to obtain limiting values for V/K isotope effects. But, aside from the requirements for determining intrinsic isotope effects, many earlier studies were poorly designed in other ways which greatly devalued their findings.

Competitive versus Noncompetitive Experiments

Simon and Palm (1966) clearly and rigorously drew sharp distinctions between competitive and noncompetitive isotope experiments in enzyme-catalyzed reactions. Unfortunately, these distinctions have not become part of the conceptual tools of working enzymologists interested in isotope effects. It is important that they do.

Noncompetitive experiments employ total or near-total isotopic substitution of substrates and are usually associated with deuterium, because only trace labeling is possible with tritium. One varies the concentration of the normal substrate and measures the rates of the enzymatic reaction in one set of experiments, then repeats this procedure with deuterated substrate. The term noncompetitive derives from the fact that the enzyme does not encounter labeled and unlabeled substrate at the same time and is somewhat misleading because of preconceptions associated with the more common noncompetitive inhibition of steady-state kinetics. However, when the data are plotted on a double-reciprocal plot, they do look like noncompetitive inhibition and should be analyzed in much the same way, in that one looks for slope and intercept effects: ${}^{D}V$ is determined from the ratio of intercepts, while ${}^{D}(V/K)$ is obtained from the ratio of slopes of the two double-reciprocal plots. Graphical analysis, however, has severe statistical bias, and these calculations are best conducted by properly weighted least squares computer analysis. Computer programs are given and described at the end of this volume.

These computer programs perform a second operation as well. It is not always possible to prepare labeled compounds with total isotopic substitutions, nor is it necessary if one is dealing with large deuterium isotope effects. In such cases a correction must be made to obtain values for full deuterium isotope effects. This correction is incorporated into the programs. Obviously, a separate experimental determination of the fraction of deuterium label in the labeled substrates is required.

Two important precautions must be taken in noncompetitive measurements of isotope effects. First, the concentrations of labeled and unlabeled substrates must be accurately known. Concentration errors will appear as systematic errors in ${}^{D}(V/K)$ but will not affect ${}^{D}V$. Second, the purity of labeled and unlabeled substrate must be equivalent. Because one usually has to prepare the labeled substrate oneself, a common approach is to prepare unlabeled substrate by the same methods and compare its kinetic behavior to commercial material. Impurities can affect slopes and/or intercepts of double-reciprocal plots (Cleland, Gross, and Folk, 1973), thus distorting ${}^{D}V$ and/or ${}^{D}(V/K)$.

Competitive experiments employ trace labeling of substrates and usually are associated with tritium and heavy atom isotope effects, although deuterium isotope effects can also be determined in this way. In the presence of a trace-labeled substrate, enzymes encounter labeled and unlabeled substrate simultaneously, hence the term competitive, and the isotope effect appears in the form of isotopic discrimination. Given a normal isotope effect, the unlabeled substrate reacts with the enzyme at a faster rate than the labeled substrate, leaving behind isotopically enriched substrate. The kinetics of the initial velocities of this process follow the equation:

$$\frac{v_H}{v_T} = \frac{(V/K)_H \cdot [S_H]}{(V/K)_T \cdot [S_T]} \tag{34}$$

which reveals that competitive experiments measure only V/K isotope effects. Therefore, competitive experiments have the disadvantage of yielding less information than noncompetitive experiments and cannot be used to detect the rate-limiting step in the overall reaction. Competitive experiments have the advantage, however, that the measured isotope effect is independent of the total concentration of substrate, which therefore need not be determined accurately, and independent of the presence of impurities, because both labeled and unlabeled substrate will be equally affected.

Isotope effects are obtained from competitive experiments by measuring isotope enrichment in substrate or product over the course of the reaction, rather than by measuring velocities directly (Bigeleisen and Wolfsberg, 1958). Integration and rearrangement of equation 34 yield

$$^{T}(V/K) = \frac{\log(1 - f)}{\log(1 - f \cdot (SA)_0/(SA)_p)} \tag{35}$$

where f is the fractional conversion of substrate to product, $(SA)_0$ is the initial specific radioactivity of substrate, and $(SA)_p$ is the specific radioactivity of product at a fractional conversion f. To determine $^{T}(V/K)$ experimentally, one incubates an enzyme with tritiated substrate of accurately known specific activity, stops the reaction after a short period of time, determines the extent of the reaction, isolates the tritiated product, and determines accurately its specific activity. Corrections for, or precautions against, quenching during counting procedures are obviously extremely important. With some enzymatic reactions, it may be more convenient or more accurate to determine $(SA)_0$ from the specific activity of products after driving the reaction to completion, because identical procedures can then be used in the determination of both $(SA)_0$ and $(SA)_p$.

With highly reversible reactions, significant error may be introduced if the reaction is allowed to proceed too far, because equations 34 and 35 do not account for any reversal of the reaction and its subsequent additional isotope effect. This error may be avoided by keeping the fractional conversion very low (i.e., less than 5% of the conversion necessary to reach chemical equilibrium). Under these conditions, the procedures are simplified in that the fractional conversion need not be determined, because at low values of f, equation 35 reduces to

$$^{T}(V/K) = \left| \frac{(SA)_0}{(SA)_p} \right|_{f \to 0} \tag{36}$$

Low levels of conversion present an experimental problem of isolating and determining accurately the specific activity of very small amounts of product. This, in turn, can be solved by incorporating a ^{14}C label into the experimental design to yield a doubly labeled product. Cold carrier product can then be added during isolation and the tritium specific activity determined by channels-ratio methods of scintillation counting, provided substrate contamination is not a problem. Further refinements of experimental design are a function of the reaction mechanisms of individual enzymes.

Ping-Pong Mechanisms

An enzymatic reaction obeying a ping-pong mechanism is perhaps the least difficult to study in terms of obtaining limiting values for V/K isotope effects. Consider the mechanism in Scheme 3:

$$E + A \underset{k_2}{\overset{k_1}{\rightleftharpoons}} EA \underset{k_4}{\overset{k_3}{\rightleftharpoons}} FP \xrightarrow{k_5} F + P$$

$$F + B \underset{k_8}{\overset{k_7}{\rightleftharpoons}} FB \underset{k_{10}}{\overset{k_9}{\rightleftharpoons}} EQ \xrightarrow{k_{11}} E + Q$$

Scheme 3

The V/K isotope effect with respect to substrate A is governed by the equation:

$$^D(V/K_a) = \frac{^Dk_3 + C_f + C_r}{1 + C_f + C_r} \tag{37}$$

where

$$C_f = k_3/k_2 \tag{38}$$
$$C_r = k_4/k_5 \tag{39}$$

Thus $^D(V/K_a)$ is independent of the second half-reaction and independent of the concentration of substrate B. The kinetic pattern obtained with a ping-pong mechanism consists of two sets of parallel lines: one set with labeled A, and the other set with unlabeled A. Only the determination of $^D(V)$, and $^D(V/K_b)$ if desired, requires a variation of the concentration of substrate B.

Ordered Mechanisms

Sequential mechanisms yield isotopic data dependent upon all substrates. Consider the ordered mechanism of Scheme 4.

```
     A     B     P     Q
     ↓     ↓     ↑     ↑
  ─────────────────────────
  E    EA    (X)    EQ    E
```

Scheme 4

Analysis of Scheme 4 reveals the utility of the concept of commitment to catalysis. Increasing the concentration of substrate B drives EA into the central complex and, in doing so, raises the commitment to catalysis of EA to infinity. Hence, as [B] → ∞, observed $^{D}(V/K_a) \rightarrow 1$. Similarly, as [B] → 0, observed $^{D}v \rightarrow {}^{D}(V/K_b)$. This is the situation encountered earlier with alcohol dehydrogenase, as described in Table 2. The desired deuterium *V/K* isotope effect to be used in calculating intrinsic isotope effects, therefore, is $^{D}(V/K_b)$, which is readily obtained from the slopes of double reciprocal plots with varied B, regardless of which substrate carries the deuterium label, and independent of the concentration of A.

In experiments with tritium discrimination in ordered mechanisms, it does make a difference which substrate carries the tritium label. If B is labeled then $^{T}(V/K_b)$ is obtained directly. If A carries the label, then the apparent isotope effect varies as a function of the concentration of B, and this concentration must be extrapolated to zero. Discrimination as a function of the concentration of substrate B follows the equation:

$$^{T}v = {}^{T}\left(\frac{V}{K_b}\right) \frac{1 + \dfrac{K_a{}^t\,[B]}{K_{ia}{}^t\,K_b{}^t}}{1 + \dfrac{K_a\,[B]}{K_{ia}\,K_b}} \tag{40}$$

where K_{ia} is the inhibition constant of substrate A. Subtraction of 1 from both sides of the equation, followed by a rearrangement which assumes no isotope effect on $K_{ia}{}^t$ (which in Scheme 4 is reasonable because K_{ia} is the dissociation constant of A) yields

$$^{T}v - 1 = \frac{^{T}(V/K_b) - 1}{1 + \dfrac{K_a\,[B]}{K_{ia}\,K_b}} \tag{41}$$

which in reciprocal form becomes:

$$\frac{1}{^{T}v - 1} = \frac{1}{^{T}(V/K_b) - 1} + \frac{1}{^{T}(V/K_b) - 1} \quad \frac{K_a\,[B]}{K_{ia}\,K_b} \tag{42}$$

Thus, plotting the reciprocal of the apparent isotope effect minus one *versus* the concentration of B gives a straight line. The *V/K* isotope effect can be obtained from the intercept (Cleland, 1977).

RANDOM MECHANISMS

Random mechanisms are similar to ordered mechanisms in that the concentration of co-substrate influences the size of the apparent V/K isotope effect, but the mechanisms differ in that saturation by one substrate will not abolish the V/K isotope effect of the other. This is apparent from examination of the mechanism of Scheme 5, because saturation with substrate B does not drive EA into the central complex, as in an ordered mechanism, but, instead, shifts the course of the reaction to the lower pathway by combining with free E, thus raising the commitment to catalysis to a finite, not infinite, value.

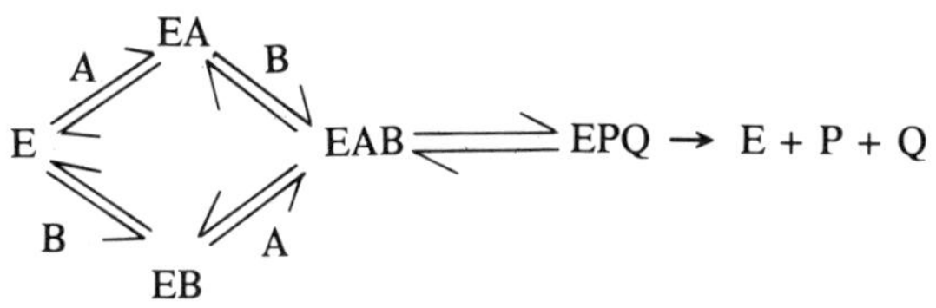

$$E \overset{A}{\rightleftharpoons} EA \overset{B}{\rightleftharpoons} EAB; \quad E \overset{B}{\rightleftharpoons} EB \overset{A}{\rightleftharpoons} EAB; \quad EAB \rightleftharpoons EPQ \rightarrow E + P + Q$$

Scheme 5

Limiting values for deuterium isotope effects on V/K may be obtained in a straightforward manner from double-reciprocal plots, with provisions for varying each substrate while holding the others constant. Tritium discrimination is again more complex, and the methods for extracting limiting values for ${}^{T}(V/K)$ are not fully resolved. With substrate A labeled with tritium, it can be shown that as $[B] \rightarrow \infty$, ${}^{T}\nu \rightarrow {}^{T}(V/K_a)$. The extrapolations should be performed in much the same manner as described for an ordered mechanism. The problem is that plots of $1/({}^{T}\nu - 1)$ versus substrate concentration are not linear as in ordered mechanisms. Whether the deviation from linearity is large or whether the extrapolation requires curve-fitting on a computer remains to be determined.

A limiting case of a random mechanism leads to some useful generalizations. A rapid equilibrium random mechanism has, by definition, a very low commitment to catalysis. Consequently, ${}^{D}V = {}^{D}(V/K_a) = {}^{D}(V/K_b)$. To the extent that this is not true, one is not dealing with a fully rapid equilibrium mechanism. Furthermore, the substrate with the lowest V/K isotope effect is the one with the highest commitment to catalysis, i.e., the "stickiest" substrate, which reveals the preferred pathway of the random mechanism. If ${}^{D}V$ is larger than either ${}^{D}(V/K)$, then both substrates are sticky, whereas lower ^{D}V values indicate the presence of a slow step following the release of the first product.

CONCLUDING COMMENTS

Clearly, in isotopic studies of enzyme-catalyzed reactions one should measure the isotope effects on limiting values of *V* and *V/K*. This may require extrapolation of co-substrate concentrations. The collection of necessary data is straightforward with deuterium, but tritium discrimination is trickier and yields only *V/K* values. A comparison of deuterium and tritium isotope effects on *V/K* may provide the absolute magnitude of the isotope effect on the bond-breaking step. This is true regardless of the number of discrete steps in the enzyme mechanism. The only limitation arises from the presence of equilibrium isotope effects, but nevertheless one can still calculate maximum and minimum values, and I believe that ultimately the means will be found to overcome this limitation.

To date there have been few applications of these methods. In fact, only one example exists that is free from complications such as secondary isotope effects, and that is the study of malic enzyme by Schimerlik, Grimshaw, and Cleland (1977). An equilibrium isotope effect of 1.2 was present; consequently only maximum and minimum values could be obtained. For the oxidation of malate, the limiting values for the intrinsic isotope effect were 5 and 8, while the reverse reaction gave 4 and 6.5. The range is large because of the small apparent isotope effects on *V/K* of 1.5 for deuterium and 2.0 for tritium.

It will be interesting to look for more application of these techniques in the next few years. I am still somewhat amazed that so simple an operation as subtracting the number 1 from an equation could lead to such a torrent of useful algebra and insight into enzyme mechanisms, leading ultimately to the intrinsic isotope effect. But it is important to note that, as exciting and interesting as it may seem, the determination of intrinsic isotope effects is not the ultimate goal in these studies; rather, it provides a new tool for probing further into enzyme catalysis. The absolute magnitude of isotope effects on catalysis provides clues about chemical mechanisms and limitations surrounding transition states, and the level of expression of the intrinsic values in apparent isotope effects on *V* and *V/K* under different conditions should provide a wealth of information about kinetic mechanisms and what determines the overall rates of enzyme-catalyzed reactions.

DISCUSSION

I. A. Rose. Dr. Northrop's remarks suggest a possibly important use of isotope effects in determining the rate of mixing of an enzyme-bound substrate with the free pool. Consider the sequence:

$$AH(T) + E \underset{2}{\overset{1}{\rightleftharpoons}} E^{AH(T)} \overset{3}{\rightleftharpoons} E_B{}^{AH(T)} \overset{5}{\longrightarrow} E + AB + H^+(T)$$

and suppose that a primary isotope effect is seen in the reaction with deuterated or tritiated substrate. When studied in the competitive mode (AH vs. AT) the extent of discrimination observed in the initial rate, either because of a primary or secondary kinetic effect, will depend on the concentration of the second substrate, B, being greatest as B → 0 and in the limit at high B for an ordered sequence the discrimination will disappear. At a concentration of B that decreases the isotope effect to half the value obtained at very low B, one has the condition that

$$k_2 E^{AH} = k_5 E_B{}^{AH}$$

In this steady-state condition under which K_m is defined as $K_m = (B)(E^{AH})/E_B{}^{AH}$ the amount of B called for to diminish the isotope effect to half (called $K_{1/2}$) would be $K_m \cdot E_B{}^{AH}/E^{AH} = K_m \cdot k_2/k_5$. From this one calculates that $k_2 = V_{max}/E_t \cdot K_{1/2}/K_m$. Comparison of this value with the V_{max}/E_t for the reverse reaction would indicate if release of the last product, $k_2 E^{AH}$, contributes to the reverse V_{max}. In addition, the same procedure could be applied to measure the k_{off} for a proton (acting as the first substrate) from E_H using the discrimination from tritiated water into product at varying (B). This approach uses the isotope effect to apply to the steady state a variation of the isotope trapping procedure we used to determine the rate of dissociation of ^{14}C-glucose from hexokinase (Rose et al., 1974).

I. A. Rose (communicated). An excellent example that shows the effect of the concentration of a second substrate on an isotope discrimination is seen in the data of Palm (1968) for NADH(T) + Pyruvate → NAD + 2H(T) Lactate, catalyzed by muscle lactate dehydrogenase (see Table 6 in Klinman, this volume). The discrimination factor changes from 1.81 to 1.0 as pyruvate is changed from 0.1 mM to 4.0 mM. It is estimated that at pyruvate ≌ 0.2 mM the isotope effect would be half its value at the low pyruvate limit. The K_m of pyruvate, also ≌ 0.2 mM, indicates that the dissociation rate of NADH is ≌ V_{max} for the forward reaction. Because the reverse V_{max} is about 0.2 x $V_{max}^{forward}$ it would be estimated that the off rate for NADH is 5 x $V_{max}^{reverse}$. This compares well with direct measurements of E^{NADH} → E + NADH by Stinson and Gutfreund (1971) where a factor of 10 was obtained.

D. B. Northrop (edited comments). Dr. Rose's suggestion is elegant and correct. Like the rest of us, however, he has not yet picked up the habit of thinking in terms of the "isotope effect minus one." That is the quantity which is halved at B = $K_{1/2}$, not the isotope effect as such. The concentration of B which will accomplish this in an ordered mechanism can be determined from a plot of data as described for equation 42: the horizontal intercept represents the half-maximal substrate concentration. From equation 42, this concentration is seen to be

$$[B]_{1/2} = \frac{K_{ia} K_b}{K_a}$$

In an ordered mechanism, k_2 may be calculated from the relationship (Cleland, 1963):

$$k_2 = \frac{V_1 K_{ia}}{E_t K_a}$$

Combining these two equations, we obtain

$$k_2 = \frac{V_1 [B]_{1/2}}{E_t K_b}$$

which is the same equation as derived by Dr. Rose.

Dr. Rose's idea is an important link between different types of kinetics. It provides an alternative method for determining k_2 when either the isotope trapping method is impractical (such as with dehydrogenases) or steady-state methods yield imprecise K_{ia} values (which is most of the time). Moreover, it provides an example of something I have been looking for, and that is kinetic methods to corroborate results from isotopic experiments. As it now stands, too much of what I presented cannot be confirmed by independent means.

C. Walsh. This is a comment about some experiments which are relevant to your proposal: Dr. Cheung and I have had occasion, in connection with the stereochemistry of some experiments with chiral pyruvate, to look at the isotope effect in pyruvate carboxylation by pyruvate carboxylase. With deuteropyruvate there is an isotope effect at low substrate; it is thus a V/K and not a V_{max} effect. Spurred on in part by your analysis, we decided to look at V/K effects for tritium. In doing that we were able, in the end, to extract the true isotope effect on the catalytic rate constant for pyruvate carboxylation and to find that it is about 3.1. This is the number which is useful in determining what the chiral split will be with samples of chiral pyruvate. It is, in fact, the best way to check the chiral purity of samples of (^{1}H, ^{2}H, ^{3}H)-pyruvate, so I think it is a useful method, and the experiments bear out the relationships you are talking about (Cheung and Walsh, 1976).

D. B. Northrop. Excellent! I have expressed a concern that the theory behind isotope effects may be running too far ahead of experimental data, and what is most needed now are studies such as yours. Your results point out something else, and that is a new use for the intrinsic isotope effect. We still don't know the extent to which intrinsic isotope effects may be used to answer other questions of enzyme mechanisms, but I suspect it may be very extensive.

ACKNOWLEDGMENTS

This work was supported by USPHS Grant GM-18701, NSF Grant PCM75-16404, and USPHS Career Development Award GM-00254.

LITERATURE CITED

Bardsley, W. G., Crabbe, M. J., and Shindler, J. S. 1973. Kinetics of the diamine oxidase reaction. Biochem. J. 131:459-469.

Bigeleisen, J., and Wolfsberg, M. 1958. Theoretical and experimental aspects of isotope effects in chemical kinetics. Adv. Chem. Phys. 1:15-76.

Bigeleisen, J. 1962. Tritium in the physical and biological sciences. Int. Atomic Energy Agency 1:161-168.

Bright, H. J., and Gibson, Q. H. 1967. The oxidation of 1-deuterated glucose by glucose oxidase. J. Biol. Chem. 242:994-1003.

Bush, K., Shiner, V. J., Jr., and Mahler, H. R. 1973. Deuterium isotope effects on initial rates of the liver alcohol dehydrogenase reaction. Biochemistry 12:4802-4805.

Carter, R. E., and Melander, L. 1973. Experiments on the nature of steric isotope effects. Adv. Phys. Org. Chem. 10:1-27.

Cheung, Y. F., and Walsh, C. 1976. Studies on the intramolecular and intermolecular kinetic isotope effects in pyruvate carboxylase catalysis. Biochemistry 15:3749-3754.

Cleland, W. W. 1963. The kinetics of enzyme-catalyzed reactions with two or more substrates or products. Biochim. Biophys. Acta 67:104-137.

Cleland, W. W. 1975. What limits the rate of an enzyme-catalyzed reaction? Accts. Chem. Res. 8:145-151.

Cleland, W. W. 1977. Determining the chemical mechanisms of enzyme catalyzed reactions by kinetic studies. Adv. Enzymol. 45:273-387.

Cleland, W. W., Gross, M., and Folk, J. E. 1973. Inhibition patterns obtained where an inhibitor is present in constant proportion to variable substrate. J. Biol. Chem. 248:6541-6542.

Lewis, E. S., and Robinson, J. K. 1968. The influence of tunneling on the relation between tritium and deuterium isotope effects. The exchange of 2-nitropropane-2-t. J. Am. Chem. Soc. 90:4337-4344.

More O'Ferrall, R. A., and Kouba, J. 1967. Model calculations of primary hydrogen isotope effects. J. Chem. Soc. B:985-990.

Northrop, D. B. 1975. Steady-state analysis of kinetic isotope effects in enzymic reactions. Biochemistry 14:2644-2651.

O'Leary, M. H. 1977. Heavy-atom isotope effects in enzyme-catalyzed reactions. In: R. L. Schowen and R. Gandour, (eds.), Transition States of Biochemical Processes. Plenum Press.

O'Leary, M. H., and Limburg, J. A. 1977. Isotope effect studies of the role of metal ions in isocitrate dehydrogenase. Biochemistry 16:1129-1135.

Palm, D. 1968. Kinetische Untersuchungen zum Mechanismus der Lactatdehydrogenase mit Tritium-Isotopeneffekten. Eur. J. Biochem. 5:270-275.

Rose, I. A., O'Connell, E. L., Litwin, S., and Bar Tana, J. 1974. Determination of the rate of hexokinase-glucose dissociation by the isotope-trapping method. J. Biol. Chem. 249:5163-5168.

Schimerlik, M. I., Grimshaw, C. E., and Cleland, W. W. 1977. Determination of the rate limiting steps for malic enzyme by the use of isotope effects and other kinetic studies. Biochemistry 16:571-575.

Simon, H., and Palm, D. 1966. Isotope effects in organic chemistry and biochemistry, Angew. Chem. Intl. Edn. 5:920–933.
Stern, M. J., and Vogel, P. C. 1971. Relative tritium-deuterium isotope effects in the absence of large tunneling factors. J. Am. Chem. Soc. 93:4664–4675.
Stern, M. J., and Weston, R. E., Jr. 1974. Phenomenological manifestations of quantum-mechanical tunneling. III. Effect on relative tritium-deuterium kinetic isotope effects. J. Chem. Phys. 60:2815-2821.
Stinson, R. A., and Gutfreund, H. 1971. Transient-kinetic studies of pig muscle lactate dehydrogenase. Biochem. J. 121:235–240.
Swain, C. G., Stivers, E. C., Reuwer, J. F., Jr., and Schaad, L. J. 1958. Use of hydrogen isotope effects to identify the attacking nucleophile in the enolization of ketones catalyzed by acetic acid. J. Am. Chem. Soc. 80:5885-5893.
Vogel, P. C., and Stern, M. J. 1971. Temperature dependence of kinetic isotope effects. J. Chem. Phys. 54:779-796.

Measurement of Isotope Effects by the Equilibrium Perturbation Method

W. W. Cleland

Several methods for determining isotope effects are commonly used. First, the rates for deuterium- and hydrogen-labeled molecules may be directly compared. When reciprocal plots are constructed for each molecule, the ratio of slopes is the V/K isotope effect, and the ratio of intercepts is the V effect. An example of such an experiment is shown in Figure 1, and FORTRAN programs that make least square fits to such data and print out the isotope effects are given at the end of this volume. The limit of detection by this method is generally an isotope effect of 1.10, although Gorenstein (1972) claims to have determined an ^{18}O effect of 1.0204 ± 0.0044 by direct comparison of the spontaneous hydrolysis of [^{18}O] and [^{16}O]-2,4-dinitrophenyl phosphates. It is doubtful that such accuracy can be attained in the study of enzyme-catalyzed reactions!

The second method involves isotope discrimination. One compares the specific activity of a labeled substrate with that of the first portion of labeled product formed, and the ratio of specific activities is the isotope effect. This method is used to measure tritium effects and usually used for heavier isotopes such as ^{13}C, ^{15}N, ^{18}O, etc. For these, the natural abundance of the heavy isotope is often used as the tracer, and the isotope ratio mass spectrometer is used to determine specific activities. This method gives only the isotope effect on V/K, but it is more sensitive than direct comparison. With the isotope ratio mass spectrometer, one can easily determine isotope effects of 1.001 or less, while with tritium (especially if ^{14}C is used as an internal standard and $^{14}C/T$

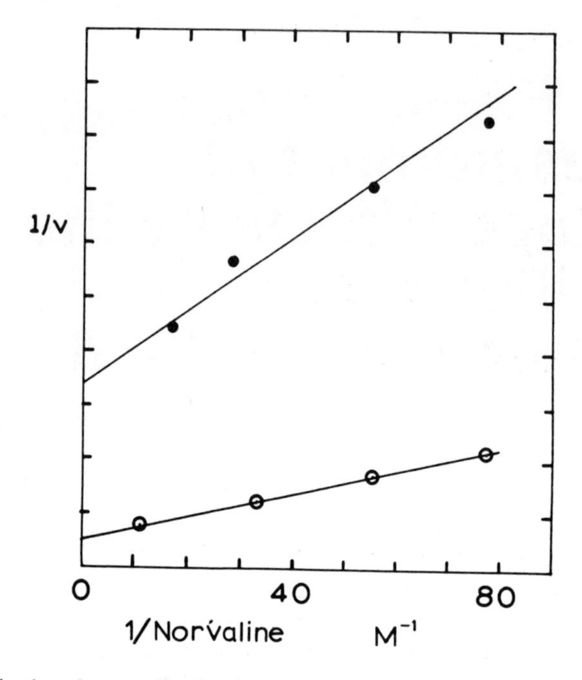

Figure 1. Kinetics of L-norvaline (open circles) and L-norvaline-2-D (closed circles) with glutamate dehydrogenase at pH 9, 1 mM TPN$^+$. The isotope effects on V and V/K are 6.0 ± 0.6 and 3.3 ± 0.5. (See footnote 2 on p. 170.)

ratios are used to measure the discrimination) the limit is probably effects of 1.05.

EQUILIBRIUM PERTURBATION METHOD

While working on malic enzyme in our laboratory, Michael Schimerlik discovered by accident a new method for determining isotope effects. He was trying to determine the effect of deuterium substitution on the equilibrium constant, and had been measuring K_{eq} values by making up reaction mixtures that were close to equilibrium, and seeing which way and how far the reaction went when enzyme was added. When one reactant is varied in such a mixture, a plot of ΔA (the change in absorbance) versus (products/reactants) is a line crossing zero on the ΔA axis at a (products/reactants) ratio equal to K_{eq}. This method worked for normal L-malate, but when Schimerlik used L-malate-2-D (but with normal TPNH!) the absorbance at 340 nm first dropped, and then slowly increased to a final value.

The reactions that occur here are:

$$\text{L-malate-2-D} + \text{TPN}^+ \xrightarrow{k_1} CO_2 + \text{pyruvate} + \text{TPND} \quad (1)$$

$$\text{L-malate-2-H} + \text{TPN}^+ \xleftarrow{k_2} CO_2 + \text{pyruvate} + \text{TPNH} \quad (2)$$

Each reaction initially goes in the direction indicated but gradually comes to equilibrium, at which point the system is at isotopic, as well as chemical, equilibrium. If k_1 is less than k_2, however, reaction 2 initially removes TPNH faster than it is generated by reaction 1, and if one starts with concentrations that will produce final chemical equilibrium, there is an apparent perturbation from equilibrium, followed by a gradual return (Figure 2). What one is actually seeing is a difference spectrum between reactions 1 and 2, and the size of the perturbation relative to the TPNH level is a measure of the isotope effect (a large isotope effect gives a large percent perturbation; a small effect, a low percent perturbation). To a first approximation (we will derive the exact equation later):

$$\text{Fractional perturbation} = \frac{\text{Isotope effect} - 1}{2.72} \quad (3)$$

The sensitivity limit is set by how small a change in the existing absorbance one can measure; thus the limit is about 1.007 where the colored molecule is one of those involved in the perturbation, and about

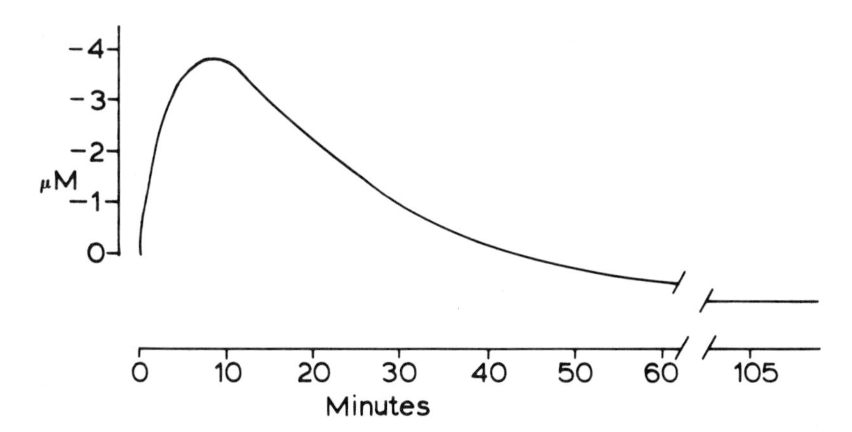

Figure 2. Equilibrium perturbation caused by deuterium isotope effect with malic enzyme. 419 μM malate-2-D, 79 μM TPNH, 20 mM $KHCO_3$ (3.8 mM CO_2), 20 mM $MgSO_4$, 102 μM TPN^+, 3.83 mM pyruvate. Reaction at pH 7.1 and 25° followed at 340 nm. (From Schimerlik, Rife, and Cleland, 1975.)

half of that effect (that is, about 1.003) when the colored molecule is not one of those involved in the perturbation (we will see why shortly). To improve the sensitivity will require improvement in signal-noise ratios in the spectrophotometers we use for such experiments. The limits given here are for a Gilford used with a 10 mV recorder and 0.05 A full scale; custom instrumentation could probably drop these limits by a factor of 2 or 3, and eventually with some effort it may be possible to approach the sensitivity of the isotope ratio mass spectrometer.

The existing sensitivity allows the use of ^{13}C, ^{15}N, ^{18}O, and other similar isotopes, as well as deuterium, but one must have a nearly full label (90% label gives 90% of the full perturbation). Although this requires synthesis of labeled compounds, there is no need for further chemical processing of products, as is generally required for experiments with the mass spectrometer. The necessity for converting the products quantitatively and without isotopic fractionation into a form suitable for the mass spectrometer is the major drawback of the use of this instrument to determine isotope effects.

A further advantage of the equilibrium perturbation method is that rigorous purity is not required for the compounds used, because an inhibitor only slows down the perturbation and does not change its size. It is not necessary to know the exact concentrations of reactants, although approximate values are, of course, needed. In practice, one makes up a premix with all reactants in it, adds enzyme to an aliquot of it, and sees how close one is to equilibrium. One then adds a calculated amount of one reactant to the premix and retests it. This can be done several times. Finally, the premix plus variable amounts of a dilute solution of one reactant are tested until the reaction returns after the perturbation to the starting point. If one misses a little bit, the average of the starting and final absorbances can be used as the baseline to calculate the perturbation with very little loss of accuracy.

There are two problems with this method. First, one must have a reaction with a color change at some wavelength so that the reaction can be followed. Coupled perturbations are possible (see below for an example), but in general the method is most useful for dehydrogenases or enzymes like fumarase where there is a clear action spectrum to the reaction. Second, one must maintain very close temperature control! Most dehydrogenases have ΔH_0 values of about 15 Kcal/mol, and thus the equilibrium position is very temperature sensitive. We saw some beautiful artifacts which resembled perturbations when we tried to use cuvets with thick glass walls and only 1 ml capacity. When such a cuvet is held in the fingers to add enzyme, a surge of heat gradually penetrates the glass and, until final temperature equilibrium is reached, produces in the small liquid volume a very noticeable perturbation from equilibrium. The answer to this problem is to use adder-mixers and never to touch the cuvets once they are at the correct temperature. Control reactions with unlabeled substrates should always be run also to check on the presence of artifacts and prove that the observed perturbations are really caused by the labeled compounds.

EXAMPLES OF THE USE OF THE METHOD

With malic enzyme, Schimerlik, et al. (1975) found that L-malate-2-D gave an isotope effect[1] of 1.45 by the equilibrium perturbation method (Figure 2), compared to an average *V/K* effect of 1.47. In addition to the deuterium effect, a ^{13}C effect of 1.031 was seen with malic enzyme when $^{13}CO_2$ and normal L-malate were the labeled and unlabeled reactants (Figure 3). In this case, the TPNH concentration first increased and then returned to equilibrium, since $^{13}CO_2$ reacted to form malate more slowly than malate formed $^{12}CO_2$. With malic dehydrogenase, L-malate-2-D gave perturbations at pH 9 which gave average isotope effects of 2.16 at 11 μM oxalacetate, and 1.70 at 100 μM oxalacetate (Figure 4). These values can be compared with the *V/K* effect of 1.9.

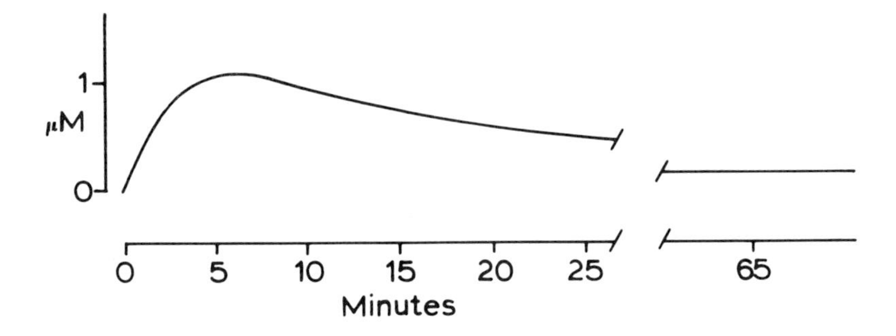

Figure 3. ^{13}C isotope effect with malic enzyme. Reaction run at pH 7.8, 25°, in a stoppered cuvet with 102 μM each of TPN and TPNH, 25 mM $MgSO_4$, 10 mM pyruvate, 1.25 mM malate, and 61.4 mM [^{13}C]-bicarbonate ($^{13}CO_2$ = 2.36 mM; 90% ^{13}C). (From Schimerlik, Rife, and Cleland, 1975.)

[1] Throughout this discussion, isotope effects are given as the rate constant for the light isotope divided by the corresponding rate constant for the heavy isotope; that is, for deuterium, k_H/k_D.

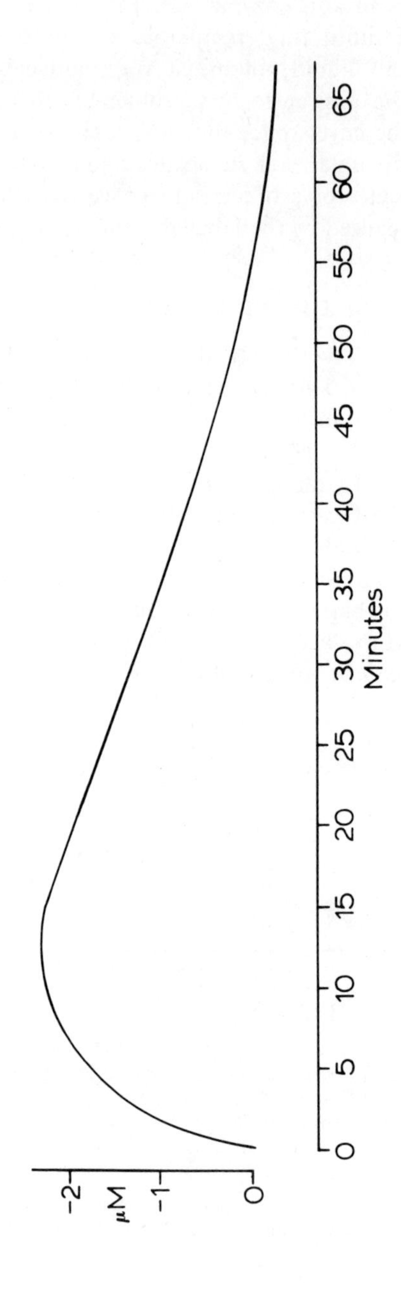

Figure 4. Perturbation with malate-2-D and malate dehydrogenase. Reactants were 1.97 mM DPN, 100 μM oxalacetate, 40 μM DPNH, 1.14 mM malate-2-D at pH 9.3. (From Schimerlik, Rife, and Cleland, 1975.)

With glutamic dehydrogenase, a very large ^{15}N effect of 1.047 was seen by Schimerlik, Rife, and Cleland (1975) when $^{15}NH_4^+$ was used along with α-ketoglutarate and glutamate (Figure 5), but no effect was seen when norvaline and α-ketovalerate were the substrates. In unpublished experiments, Jim Rife in our lab has found that L-glutamate-2-D gives small perturbations, corresponding to an isotope effect of 1.22, which compares with a *V/K* effect of about 1.2. L-norvaline-2-D, on the other hand, gave perturbations that varied with the amount of α-ketovalerate present, and gave an isotope effect as large as 1.7 (Figure 6). Although the *V/K* isotope effect is 3.3, no isotope effect over 1.7 has yet been seen with the equilibrium perturbation method.

John Blanchard in our lab has recently used deuterated malates and fumarates to determine isotope effects for fumarase by the equilibrium perturbation method. Erythro-L-malate-3-D (which is dehydrated to unlabeled fumarate and HDO) gives an inverse isotope effect of 0.92 (Figure 7). There appears to be a rapid catalytic equilibrium on the enzyme between malate and fumarate which is shifted toward fumarate when the malate is deuterated. The proton or deuteron removed here is released to water only after fumarate leaves, and thus deuterium is enriched in the catalytic group (probably a carboxyl) relative to carbon 3 of malate. This is reasonable, since K_{eq} shifts toward fumarate by 7% in D_2O (Thomson, 1960), and there should be little fractionation between water and a protonated carboxyl group.

When fumarate-2,3-D_2 was used with normal L-malate, an inverse isotope effect was again seen, as one would expect from the secondary effects on the catalytic equilibrium on the enzyme. We hope shortly to determine the ^{18}O isotope effect with [^{18}O]-malate synthesized by hydra-

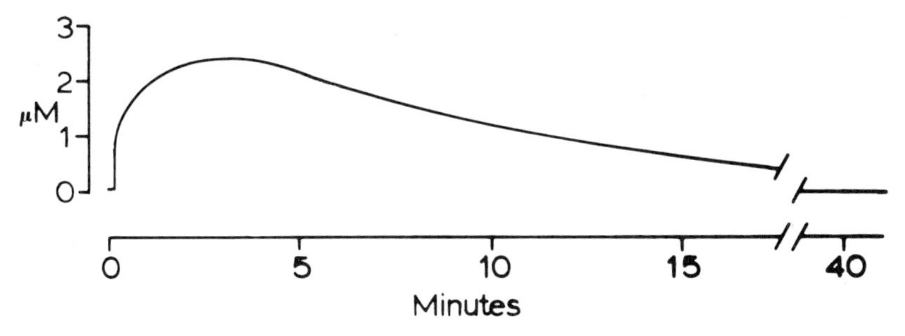

Figure 5. ^{15}N isotope effect with glutamate dehydrogenase. Reactants were 15.3 mM L-glutamate, 1.12 mM α-ketoglutarate, 358 μM $^{15}NH_4^+$ (as sulfate), 1.09 mM TPN^+, 120 μM TPNH at pH 8.1. (From Schimerlik, Rife, and Cleland, 1975.)

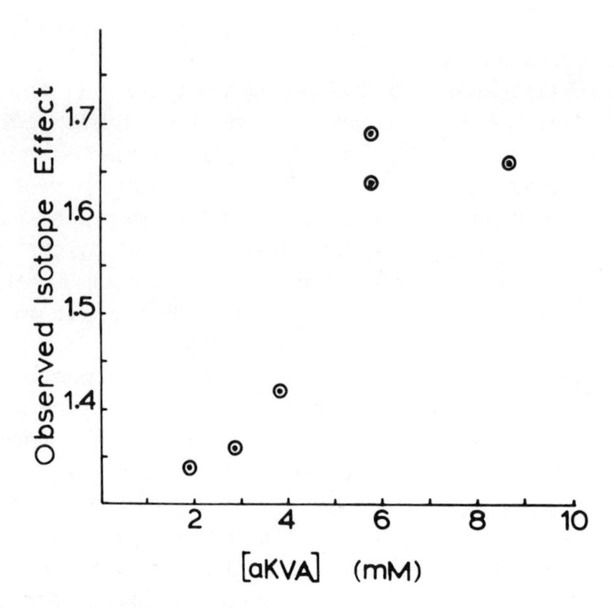

Figure 6. Isotope effects from equilibrium perturbations caused by norvaline-2-D and TPNH with glutamate dehydrogenase. 15.1 mM norvaline-2-D, 150 μm TPNH, 1.14-1.61 mMNH_4^+, 1.181-5.98 mM TPN^+, pH 8.7.

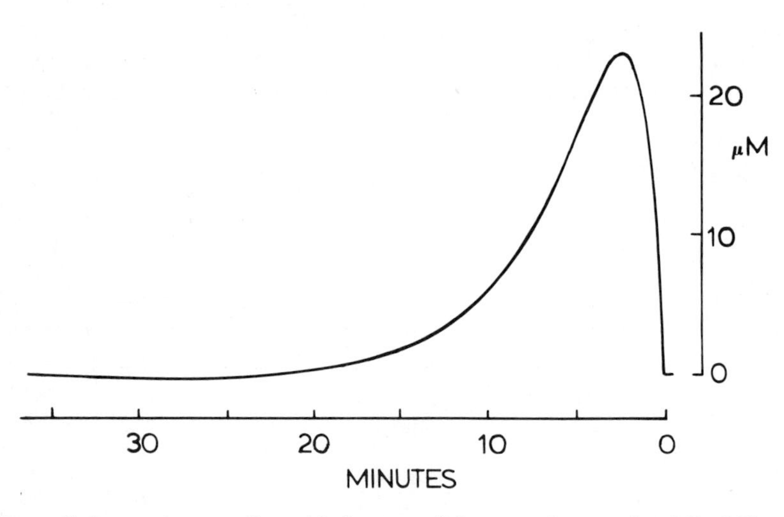

Figure 7. Inverse isotope effect with fumarase. 2.3 mM erythro-L-malate-3-D, 0.58 mM fumarate, pH 7.

tion of fumarate in $H_2{}^{18}O$. This study shows that both primary and secondary isotope effects are readily determined with the equilibrium perturbation method.

The only use of the equilibrium perturbation method reported from other than our lab to date is the isotope effect of 1.20 in forward and 1.02 in reverse directions at pH 7.5 with isocitrate-2-D and TPNH with isocitrate dehydrogenase (O'Leary and Limburg, 1977). However, the observation by Cardinale and Abeles (1968) with proline racemase that the reaction with L-proline in D_2O overshoots the equilibrium point by 20% before returning to final equilibrium is a form of equilibrium perturbation experiment. In essence the L-proline-2-H is racemized to D-proline-2-D without exchange, and L-proline-2-D forms only by the reverse reaction. Thus when the apparent equilibrium point is first reached, the reaction contains largely L-proline-2-H and D-proline-2-D, and the overshoot is an equilibrium perturbation. The equations for this experiment are derived in a later chapter in this volume.

THEORY OF THE EQUILIBRIUM PERTURBATION METHOD

Two Reactant Case

We shall next derive the equations which describe the perturbation and show how one calculates the isotope effect in a given case. In their paper describing the method, Schimerlik, Rife, and Cleland (1975) used the mechanism of malic enzyme as a model. To simplify the algebra, we will assume the simple mechanism:

$$\mathrm{E} \underset{k_2}{\overset{k_1\mathrm{A}}{\rightleftharpoons}} \mathrm{EA} \underset{k_4}{\overset{k_3}{\rightleftharpoons}} \mathrm{EP} \underset{k_6\mathrm{P}}{\overset{k_5}{\rightleftharpoons}} \mathrm{E} \tag{4}$$

in which A contains the heavy isotope and P the light one, and P is the colored molecule (if P contains the heavy isotope instead of A, the direction of the perturbation is simply reversed); k_3 and k_4 are the steps which show the isotope effect. Fumarase with threo-L-malate-2,3-D_2 as A and fumarate as P is an example. Most reactions will include other substrates besides A and P, but we will deal with such situations later.

The rate equation for appearance of labeled P can be derived by the net rate constant method of Cleland (1975):

$$\frac{d\mathrm{P}_D}{dt} = \frac{\dfrac{k_1k_{3D}}{k_2}\mathrm{A}_D\,(\mathrm{E}) - \dfrac{k_{4D}k_6}{k_5}\mathrm{P}_D\,(\mathrm{E})}{1 + k_{3D}/k_2 + k_{4D}/k_5} \tag{5}$$

In equation 5 the subscript D indicates a deuterated molecule, or a rate constant for such a molecule. The comparable rate equation for unlabeled P is:

$$-\frac{dP_H}{dt} = \frac{\frac{k_{4H}k_6}{k_5} P_H (E) - \frac{k_1 k_{3H}}{k_2} A_H (E)}{1 + k_{3H}/k_2 + k_{4H}/k_5} \tag{6}$$

In equation 6 the subscript H refers to unlabeled molecules containing all hydrogen and to rate constants for such molecules. Because initially at $t = 0$ all A is labeled and all P unlabeled, we can write:

$$A_D = A_0 - P_D \tag{7}$$

$$A_H = P_0 - P_H \tag{8}$$

where A_0 and P_0 are the initial concentrations of A and P. We must now remember that deuterium substitution affects K_{eq} (see chapter by Shiner, this volume, and the Note on Fractionation Factors elsewhere in this volume), and thus define:

$$\beta = \frac{K_{eq\ D}}{K_{eq\ H}} = \frac{k_{3D}k_{4H}}{k_{3H}k_{4D}} \tag{9}$$

If we now multiply the numerator and denominator of equation 5 by k_{3H}/k_{3D}, substitute for A_D from equation 7, and remember that $k_{3H}k_{4D}/k_{3D} = k_{4H}/\beta$, we get:

$$\frac{dP_D}{dt} = \frac{\left(\frac{k_1 k_{3H}}{k_2}(A_0 - P_D) - \frac{k_{4H}k_6}{k_5\beta} P_D\right) E}{\alpha\Delta} \tag{10}$$

where:

$$\Delta = 1 + k_{3H}/k_2 + k_{4H}/k_5 \tag{11}$$

$$\alpha = \text{isotope effect} = \frac{k_{3H}/k_{3D} + k_{3H}/k_2 + k_{4H}/k_5\beta}{\Delta} \tag{12}$$

Note that α is the isotope effect for reaction of A to P, and that $\alpha\beta$ is the isotope effect in the reverse direction.

By substitution from equations 8 and 11, equation 6 can now be written:

$$\frac{dP_H}{dt} = \frac{\left(\frac{k_1 k_{3H}}{k_2}(P_0 - P_H) - \frac{k_{4H}k_6 P_H}{k_5}\right) E}{\Delta} \tag{13}$$

We now make the following definitions:

$$K = \frac{P_{H\ eq}}{A_{H\ eq}} = \frac{k_1 k_{3H} k_5}{k_2 k_{4H} k_6} \tag{14}$$

$$k = \frac{k_1 k_{3H}(E)}{k_2\ \Delta} \left(\frac{1 + K}{K}\right) \tag{15}$$

In the present case K is the equilibrium constant, but if there are other substrates besides A and P, their concentrations will occur in K. For any situation, K is the ratio of unlabeled P to A at equilibrium. The constant k is defined assuming that (E) will remain constant during the perturbation. This is a good assumption unless the perturbation is very large, and, as we shall see, any variation in k will only change the time course of the perturbation, and not its size, which is independent of k.

Substituting from equations 14 and 15, equations 10 and 13 now become:

$$\frac{dP_D}{dt} = \left(\frac{K}{1 + K}\right) \frac{k}{\alpha} \left(A_0 - P_D \left(\frac{1 + K\beta}{K\beta}\right)\right) \tag{16}$$

$$\frac{dP_H}{dt} = \left(\frac{K}{1 + K}\right) k \left(P_0 - P_H \left(\frac{1 + K}{K}\right)\right) \tag{17}$$

These equations integrate to:

$$P_D = \frac{A_0 K\beta}{1 + K\beta} \left(1 - e^{-\frac{k(1 + K\beta)t}{\alpha\beta(1 + K)}}\right) \tag{18}$$

$$P_H = \frac{P_0}{1 + K} (K + e^{-kt}) \tag{19}$$

If we add equations 18 and 19 together, we will get an equation for total P as a function of time which describes the time course of the perturbation. Before we do this, however, we will define:

$$\text{app}\ \alpha = \alpha \frac{(\beta + K\beta)}{(1 + K\beta)} = \alpha\beta \frac{(1 + K)}{(1 + K\beta)} \tag{20}$$

and calculate P_∞ by letting t become infinite in equations 18 and 19, and adding them together:

$$P_\infty = \frac{A_0 K\beta}{1 + K\beta} + \frac{P_0 K}{1 + K} = P_0 \tag{21}$$

Because the perturbation returns to the starting point, $P_\infty = P_0$, and thus we can rearrange equation 21 as:

$$\frac{A_0 K\beta}{1 + K\beta} = \frac{P_0}{1 + K} \qquad (22)$$

Substitution of equations 20 and 22 into 18, and addition of equations 18 and 19, give:

$$P = P_0 + \left(\frac{P_0}{1 + K}\right)\left(e^{-kt} - e^{-\frac{kt}{\text{app}\,\alpha}}\right) \qquad (23)$$

Because K is not determined directly during an experiment, we solve equation 22 for K in terms of A_0 and P_0, which are known, and thus rewrite equation 23 as:

$$P = P_0 + P_0{}'\left(e^{-kt} - e^{-\frac{kt}{\text{app}\,\alpha}}\right) \qquad (24)$$

where:

$$P_0{}' = \frac{2}{\frac{1}{A_0} + \frac{1}{P_0} + \sqrt{\left(\frac{1}{A_0} + \frac{1}{P_0}\right)^2 + 4(1/\beta - 1)\left(\frac{1}{A_0}\right)\left(\frac{1}{P_0}\right)}} \qquad (25)$$

While this looks messy, note that if $\beta = 1$, $P_0{}'$ is simply the reciprocal of the sum of reciprocals of A_0 and P_0, and thus is approximately equal to the smaller of A_0 and P_0.

If we differentiate equation 24 and solve for the maximum point of the perturbation, we get:

$$\frac{P_{max} - P_0}{P_0{}'} = (\text{app}\,\alpha)^{-1/(\text{app}\,\alpha - 1)} - (\text{app}\,\alpha)^{-(\text{app}\,\alpha)/(\text{app}\,\alpha - 1)} \qquad (26)$$

which is the equation we use to determine (app α) from the size of the perturbation. If (app α) < 1.1, this is approximated by equation 3, but in practice one should use the table at the end of this volume, which covers all solutions of equation 26 that one will meet in practice.

Several comments should be made on experimental procedure at this point. First, if the starting and final concentrations of P are not exactly identical, the values should be averaged and used as P_0 in equation 26. As long as they differ less than 20%, the resulting error will be small. Second, the time at which the maximum point occurs is given by:

$$t_{max} = \frac{1}{k} \frac{(\text{app } \alpha)}{(\text{app } \alpha - 1)} \ln(\text{app } \alpha) \tag{27}$$

which at values of (app α) near 1 is approximated by:

$$t_{max} = \frac{1}{k} \frac{(\text{app } \alpha + 1)}{2} \tag{28}$$

which makes t_{max} only slightly different from $1/k$. Because a reaction mixture containing only unlabeled reactants that is not quite at equilibrium returns to equilibrium with a time constant k, one can use such mixtures to determine how much enzyme to use for the perturbations. The point where such a mixture has gone 63% of the way to equilibrium corresponds to $1/k$, and thus will approximate the maximum point of a perturbation. This time should be set at 2–5 minutes if possible, because one must wait at least 10 t_{max} to ensure return to equilibrium after the perturbation.

Once (app α) is determined, α and $\alpha\beta$ can be calculated from equation 20. Again, K is not readily available, and, if we substitute its value from a solution of equation 22 into equation 20, we get:

$$\alpha = (\text{app } \alpha)\, z \tag{29}$$

where:

$$z = \frac{1 - A_0/P_0 + \sqrt{(1 - A_0/P_0)^2 + 4A_0/P_0\beta}}{2}$$

$$= \frac{2/\beta}{1 - P_0/A_0 + \sqrt{(1 - P_0/A_0)^2 + 4P_0/A_0\beta}} \tag{30}$$

The first expression is best used when $A_0/P_0 < 1$, and the second when $A_0/P_0 > 1$; if $A_0 = P_0$, $z = 1/\sqrt{\beta}$. Then, $\alpha\beta$ is obtained by multiplying α by the experimental value of β.

Although β can deviate the most from 1.00 when deuterium is the label, equilibrium isotope effects with heavy isotopes such as ^{13}C, ^{15}N, and ^{18}O may in some cases be as large as the kinetic isotope effects seen with these isotopes. A good example is the large ^{15}N effect seen with glutamate dehydrogenase (see the discussion at the end of this chapter). Thus, it may not be true that $\beta = 1$ for such heavy isotopes. When $\beta = 1$ for any isotope, however, $z = 1$, and (app α) $= \alpha$. In such a case

$$P_0' = 1/(1/A_0 + 1/P_0) \tag{31}$$

The above equations all assume 100% labeling with the heavy isotope. If the percent label is less than this, $(1 - \alpha)$ must be divided by the actual fraction of labeling. Thus, for 80% label and an observed α of 1.04, the true α is 1.05.

Other Reactants Besides A and P

When there are reactants other than A and P (as is usually the case), the equations derived above will still hold as long as the concentrations of the other reactants considerably exceed P_0', so that they stay essentially constant during the perturbation. When this is not true (as is again usually the case for at least one reactant, especially if the perturbation is between non-colored reactants), the concentrations of these reactants will change during the perturbation in a manner that tends to decrease the size of the perturbation.

If we expand mechanism 4 to include B and Q as other reactants, K will now include the concentrations of these reactants, and if we let x equal the size of the perturbation at any time, we have (for a normal isotope effect):

$$B = B_0 + x = B_0(1 + x/B_0) \tag{32}$$

$$Q = Q_0 - x = Q_0(1 - x/Q_0) \tag{33}$$

If we now redefine K to include B_0 and Q_0, instead of B and Q (which means K will thus remain a constant), equations 16 and 17 now include the expression: $(1 + x/B_0)/(1 - x/Q_0)$ multiplied by K. Because x^2 terms will be very small, this expression can be rewritten:

$$\frac{(1 + x/B_0)(1 - x/B_0)}{(1 - x/Q_0)(1 - x/B_0)} = \frac{1}{1 - x \sum \frac{1}{C_0}} \tag{34}$$

where:

$$\sum \frac{1}{C_0} = 1/B_0 + 1/Q_0 \tag{35}$$

If there are further reactants, their reciprocal concentrations are also included in $\sum 1/C_0$.

Because $x = P_D + P_H - P_0$, it is clear that equations 16 and 17 must now be solved simultaneously. We have not found an analytical solution to this problem, except in certain special cases, and thus we have resorted to empirical solutions achieved with the aid of a computer program for simultaneous solution of differential equations. The results are that equation 24 is still valid as long as P_0' is replaced by P_0'':

$$P_0'' = \frac{1}{1/P_0' + \frac{1}{y}\sum \frac{1}{C_0}} \tag{36}$$

The value of y must be chosen as shown in Table 1.

It is clear from all of the above that these calculations can become somewhat involved, and as a result at the end of this volume a short FORTRAN program is included which calculates the value of $(P_{max} - P_0)/P_0''$ with which to enter the table and determine app α and the value of z to multiply by app α to get the true value of the isotope effect.

INTERPRETATION OF EQUILIBRIUM PERTURBATION RESULTS

The isotope effect on V/K for an enzymatic reaction can be expressed as:

$$\frac{(V/K)_H}{(V/K)_D} = \frac{k_H/k_D + c_f + c_r/\beta}{1 + c_f + c_r} \tag{37}$$

where β is $K_{eq\,D}/K_{eq\,H}$ and k_H/k_D is the true isotope effect on the bond-breaking step in the forward direction. For the reverse direction, this equation is multiplied by β, so the true isotope effect in the reverse direction is $(\beta\, k_H/k_D)$, the c_f term becomes $\beta\, c_f$ in the numerator, and the c_r term becomes the same in numerator and denominator. The constants c_f and c_r are commitments to catalysis in forward and reverse directions (see the chapter by Northrop).

The isotope effect seen in the equilibrium perturbation method has a similar form (see equation 12), but in place of c_f or c_r one has

Table 1. Proper values of y to be used in equation 36

$\left(1/\sum \frac{1}{C_0}\right)/P_0'$	y
> 0.3	2.2
0.16–0.3	2.3
0.08–0.16	2.4
< 0.08	2.5

expressions that may be the same or include more rate-constant ratios. As an example, consider malic enzyme, which has the following mechanism:

$$E_1 \underset{k_2}{\overset{k_1A}{\rightleftharpoons}} E_2 \underset{k_4}{\overset{k_3B}{\rightleftharpoons}} E_3 \underset{k_6}{\overset{k_5}{\rightleftharpoons}} E_4 \underset{k_8}{\overset{k_7}{\rightleftharpoons}} E_5 \underset{k_{10}}{\overset{k_9}{\rightleftharpoons}} E_6 \underset{k_{12}}{\overset{k_{11}}{\rightleftharpoons}} E_7 \underset{k_{14}P}{\overset{k_{13}}{\rightleftharpoons}} E_8 \underset{k_{16}Q}{\overset{k_{15}}{\rightleftharpoons}} E_9 \underset{k_{18}R}{\overset{k_{17}}{\rightleftharpoons}} E_1 \quad (38)$$

where A, B, P, Q, and R are TPN^+, L-malate, CO_2, pyruvate, and TPNH. k_7 and k_8 are rate constants for the hydride transfer step, and k_9 and k_{10} are those for decarboxylation and carboxylation. The steps corresponding to k_5, k_6, k_{11}, and k_{12} represent conformation changes that precede or follow catalysis. Table 2 shows the expressions for c_f and c_r that correspond to V/K and equilibrium perturbation experiments with either deuterium or ^{13}C-containing substrates. For the ^{13}C effects, the equilibrium perturbation experiment gives the same isotope effect as the V/K one, because the perturbation is between malate and CO_2, the first product released. When deuterium isotope effects are observed, however, the perturbation is between malate and TPNH, the last product to be released, and thus c_r for the equilibrium perturbation experiment includes additional terms not present when V/K is looked at. Because the observed deuterium isotope effect on V/K and that seen in the equilibrium perturbation experiment were nearly identical, however, it is clear that under the conditions of the equilibrium perturbation experiment the additional terms were small.

This will not always be the case, however. With malic dehydrogenase, the isotope effects from perturbations with malate-2-D and DPNH were 2.16 ± 0.05 at 11 μM oxalacetate and 1.70 ± 0.11 at 100 μM oxalacetate (Schimerlik, Rife, and Cleland, 1975). The difference here presumably results from the extra term in c_r, which includes oxalacetate concentration (oxalacetate is released before DPNH). Thus, when mechanisms are ordered, or where the presence of one reactant slows down the release of one involved in the perturbation, it is important to keep the levels of the other reactants low enough to avoid this problem. Because too low a level lowers P_0'', and thus the size of the observed perturbation, however, one may have to try several levels in order to determine the maximum observable isotope effect.

REMAINING PROBLEMS

We should mention one problem that has been encountered for which no explanation has yet been found. This involves the data in Figure 6 for

Table 2. Terms in equation 37 corresponding to mechanism 38

Isotope effect	c_f	c_r
Deuterium (L-Malate-2-D) V/K_{malate}	$\frac{k_7}{k_6}\left(1 + \frac{k_5}{k_4}\right)$	$\frac{k_8}{k_9}\left(1 + \frac{k_{10}}{k_{11}}\left(1 + \frac{k_{12}}{k_{13}}\right)\right)$
Equilibrium Perturbation	$\frac{k_7}{k_6}\left(1 + \frac{k_5}{k_4}\right)$	$\frac{k_8}{k_9}\left(1 + \frac{k_{10}}{k_{11}}\left(1 + \frac{k_{12}}{k_{13}}\left(1 + \frac{k_{14}P}{k_{15}}\left(1 + \frac{k_{16}Q}{k_{17}}\right)\right)\right)\right)$
^{13}C V/K_{malate}	$\frac{k_9}{k_8}\left(1 + \frac{k_7}{k_6}\left(1 + \frac{k_5}{k_4}\right)\right)$	$\frac{k_{10}}{k_{11}}\left(1 + \frac{k_{12}}{k_{13}}\right)$
Equilibrium Perturbation, $^{13}CO_2$	$\frac{k_9}{k_8}\left(1 + \frac{k_7}{k_6}\left(1 + \frac{k_5}{k_4}\right)\right)$	$\frac{k_{10}}{k_{11}}\left(1 + \frac{k_{12}}{k_{13}}\right)$

equilibrium perturbations with norvaline-2-D and TPNH in the presence of glutamate dehydrogenase. Norvaline-2-D gives a large *V/K* isotope effect of 3.3 (Figure 1),[2] but the size of the isotope effects seen in equilibrium perturbations is much smaller and is very sensitive to the level of α-ketovalerate. We have been unable to account for this phenomenon, because the equations predict that low α-ketovalerate should give a higher isotope effect, or at the least the same one, as seen at high levels.

This is the point at which to describe a system which allows us to use a coupled assay to measure an equilibrium perturbation. The system consists of 1-[^{18}O]-sorbitol (prepared by reduction of glucose exchanged with $H_2{}^{18}O$), DPN^+ and excess sorbitol dehydrogenase, which will establish an equilibrium containing small amounts of 1-[^{18}O]-fructose and DPNH. Fructose-1-P, MgADP, and MgATP are added at levels chosen to match the equilibrium constant of fructokinase, and finally fructokinase is added to start the experiment. If there were an ^{18}O isotope effect in the fructokinase reaction, the level of fructose would rise slightly, causing a drop in the DPNH level. In fact, no perturbation is observed, and we estimate that the isotope effect is less than 1.003.

CONCLUDING REMARKS

The equilibrium perturbation method has proven very useful for determining isotope effects on enzymatic reactions and will probably be used widely as soon as knowledge of it becomes disseminated. Its one real drawback is that it can be used only with reversible reactions, but because most enzymatic reactions are reversible, this causes no problem. Only time will tell whether other problems will arise and limit the usefulness of this method, but at the moment it appears to be the method of choice whenever isotope effects are small.

DISCUSSION

S. Benkovic. Mo, do you have to work out a kinetic scheme each time you try to calculate the perturbation? In other words, do you have to know ahead of time whether you have an ordered sequence or a random one, or do you have any general rules yet?

---

[2] More recent experiments have given a lower value for $V/K_{norvaline}$ (about 1.6, in agreement with the values from the equilibrium perturbation experiments) and V (about 2.6), and a higher value for V/K_{TPN} (4.0). We are trying to determine the reasons for these differences.

W. Cleland. No, the observed isotope effect (that is, the α calculated from the perturbation) really does not depend on the mechanism, or at least as far as I can see. Trying to go from there to an intrinsic isotope effect—well, it depends. In certain cases, like fumarase, where we have an equilibrium effect on the rapid catalytic reaction on the enzyme, the equations are somewhat simpler, and then you can go to an intrinsic effect and this is mechanism dependent. But the calculation of α from the size of the perturbation is not mechanism dependent.

W. Jencks. I wonder if you could summarize what you can conclude if you see an equilibrium perturbation isotope effect for an enzyme in which you know relatively little about the mechanism in terms of what step is rate limiting and how that relates to effectively irreversible steps. Presumably there are no effectively irreversible steps when this reaction is at equilibrium.

W. Cleland. You can only use the technique if you can set up an equilibrium situation. Obviously for a completely irreversible enzyme it is not a practical method. What you can conclude, I think, is that the *V/K* isotope effect, or something very close to it, has the value that you see. Then you draw the same sort of conclusions you always would about a *V/K* isotope effect. If it is of reasonable size, you know that this bond breaking step is partly rate limiting, at least for the part of the reaction sequence between the addition of the labeled substrate on the one side and the release of the labeled product on the other. That does not mean there is not a slow rate-limiting step elsewhere in the mechanism, or in other words, it does not prove there would be a V_{max} isotope effect. For example, with malic enzyme, where TPNH release is basically rate limiting at neutral pH, you don't get any V_{max} isotope effect, but you see an equilibrium perturbation effect and a *V/K* isotope effect in the chemical reaction that are both about 1.5. So I think in general you can assume that you would make the same sort of conclusions you would normally make from a *V/K* isotope effect.

R. Matthews. Can you adapt your P_0'' corrections to systems where you are measuring a half reaction equilibrium between the enzyme and a substrate in a ping-pong mechanism?

W. Cleland. (This reply has been altered from that given at the symposium.) The equilibrium perturbation method will work on ping-pong mechanisms as long as all reactants are present. The reactants in the half reaction not involving the isotope effect then serve to maintain the proper ratio of stable enzyme forms, and their levels should be high enough not to be changed during the perturbation. When only the reactants for one half reaction are present, the size of the perturbation is limited by the concentrations of the two stable enzyme forms present, and unless one has a very sensitive method for measuring the concentration of one of these enzyme forms, or unless substrate levels of enzyme are used and the perturbation carried out in a stopped flow machine, no perturbation will be observed.

W. Cleland. (In response to a question about glutamate dehydrogenase.) What we have not done (Harvey Fisher may do it), is to use ^{18}O-containing α-ketoglutarate as a perturbant in the system and look for the ^{18}O effect on cleavage of the carbinolamine to α-iminoglutarate and water. Ketoglutarate will slowly exchange its oxygen with water, but that is rather slow at neutral pH (one can introduce ^{18}O into α-ketoglutarate by exchange at low pH where the degree of hydration of the keto group is greater).

W. Jencks. It is interesting that for the cleavage of an imine near neutrality, the attack of water is normally rate determining (dehydration is rate determining in the reverse direction), so that the observation of the large ^{15}N isotope effect means that the enzyme must be facilitating the attack and loss of water so that the nitrogen attack and loss become slow.

W. Cleland. Yes, that is certainly the case. Apparently the enzyme has found it easier to accelerate the proton transfers and other changes involved with the addition and release of water than those involving ammonia.

J. Kirsch. That 4% ^{15}N isotope effect with glutamic dehydrogenase is extraordinarily large. Do you have any explanation?

W. Cleland. (This reply has been altered from that given at the symposium.) The probable reason why the ^{15}N isotope effect is so large is that it is partly an equilibrium effect. Our current ideas on the mechanism of glutamate dehydrogenase are as follows:

$$\overset{\ominus}{Y}\cdots H{-}\overset{\oplus}{N}H_3\ \ \underset{|}{\overset{COO^{\ominus}\ |}{C}}{=}O\cdots HX^{\oplus}\ \ (\text{TPNH})\ \ \text{I} \leftrightarrow YH\ H_3N\ \ \underset{|}{\overset{COO^{\ominus}\ |}{C}}{=}O\cdots H{-}X^{\oplus}\ \ (\text{TPNH})\ \ \text{II} \leftrightarrow YH\ H_3{-}\overset{\oplus}{N}{-}\underset{|}{\overset{COO^{\ominus}\ |}{C}}{-}O{-}H\cdots X\ \ (\text{TPNH})\ \ \text{III}$$

$$\leftrightarrow YH\ H_2N{-}\underset{|}{\overset{COO^{\ominus}\ |}{C}}{-}O(H)\cdots H{-}X^{\oplus}\ \ (\text{TPNH})\ \ \text{IV} \leftrightarrow YH\ H_2{-}\overset{\oplus}{N}{=}\underset{|}{\overset{COO^{\ominus}\ |}{C}}\ \ O(H)(H\cdots X)\ \ (\text{TPNH})\ \ \text{V} \leftrightarrow YH\ H_2{-}N{-}\underset{|}{\overset{COO^{\ominus}\ |}{C}}{-}H\ \ H{-}O{-}H\cdots X\ \ (\text{TPN}^{\oplus})\ \ \text{VI}$$

$$\leftrightarrow \overset{\ominus}{Y}\cdots H{-}\overset{\oplus}{N}H_2{-}\underset{|}{\overset{COO^{\ominus}\ |}{C}}{-}H\ \ O(H)(H\cdots X)\ \ (\text{TPN}^{\oplus})\ \ \text{VII}$$

In this drawing, TPN^{+} or TPNH is above the plane, and X and the water molecule are below the nucleotide. One further catalytic group would be needed in the interconversion of III and IV, because direct proton transfer from N to O is unlikely. This could be accomplished easily by any convenient OH group:

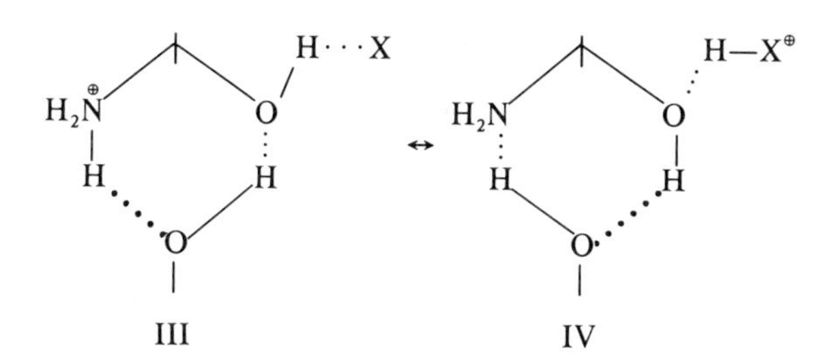

Not shown here are carbons 2-5 of the substrate or the groups which bind the 1 and 5 carboxyls. The slow step here with glutamate is apparently the conversion of II to III, while the low deuterium isotope effect shows that hydride transfer (V to VI) is quite fast.

The effect of ^{15}N substitution on the overall equilibrium constant will be small, possibly 1% in favor of glutamate (the equilibrium ^{13}C effect of replacing H with C on the isotopic atom is 1.012 according to Hartshorn and Shiner (1972), and the ^{15}N effect should be slightly smaller). However, the ^{15}N effect on the equilibrium constants for individual steps will be much larger. From I to II should show an inverse effect similar to that seen on the deprotonation of ammonium ion (1.04 in favor of ammonium ion, according to O'Leary, this volume), while from II to III there will then be a normal effect of about 1.05. There should be no net equilibrium effect in the fast steps converting III to VII, so we will ignore these steps, and model the mechanism as follows:

$$E_1 \underset{k_2}{\overset{k_1 A}{\rightleftharpoons}} (I) \underset{k_4}{\overset{k_3}{\rightleftharpoons}} (II) \underset{k_6}{\overset{k_5}{\rightleftharpoons}} (III) \underset{k_8 P}{\overset{k_7}{\rightleftharpoons}} E_1$$

where I, II, and III are the intermediate enzyme forms shown in the scheme above, and E_1 is the equilibrium mixture of E-TPNH-ketoglutarate and E-TPN$^+$. A and P are ammonium ion and glutamate. We can then write:

$$1.04 = \frac{k_3 k_4^*}{k_4 k_3^*} \qquad\qquad 1.05 = \frac{k_5^* k_6}{k_6^* k_5}$$

where starred rate constants are those for ^{15}N, and unstarred ones are for ^{14}N.

By the procedures of Cleland (this volume), we can derive the apparent isotope effect that will be seen in an equilibrium perturbation experiment as:

$$\alpha = \frac{(1.04)k_5/k_5^* + k_6/(1.01)\,k_7 + (k_5/k_4)(k_3/k_3^* + k_3/k_2)}{1 + k_6/k_7 + (k_5/k_4)(1 + k_3/k_2)}$$

Because the (glutamate)/(ammonium ion) ratio was 40–43 in these experiments, the apparent α seen in the perturbations (1.047) is equal to α, and:

$$\alpha\beta = \frac{k_6/k_6^* + k_6/k_7 + (1.01)(k_5/k_4)(k_3/k_3^* + k_3/k_2)}{1 + k_6/k_7 + (k_5/k_4)(1 + k_3/k_2)}$$

if we take $\beta = 1.01$ for the overall reaction.

It is clear that k_6/k_7 must be small, otherwise no isotope effect (or a small inverse one) would have been seen. Thus C—N bond breaking is considerably slower than the steps involving decomposition of III to V, hydride transfer, and release of glutamate from the enzyme.

There remain two possibilities as to the source of the actual isotope effect. First, if k_5/k_4 is large but k_3/k_2 is small (that is, ammonia attacks ketoglutarate more rapidly than it is protonated to ammonium ion on the enzyme, while ammonium ion is released from the enzyme faster than it is deprotonated to free ammonia), α will equal k_3/k_3*. Because k_4/k_4* should be near unity, k_3/k_3* should be 1.04, which is close to the observed α value of 1.047. In solution, the proton transfers represented by k_3 and k_4 would be rapid and certainly not as slow as required here. It is possible, however, that a conformation change is involved in the conversion of I to II and that Y may not be correctly located for proton transfer except in I. In this case k_3 and k_4 would be limited by the rate of the conformation change, and thus this possibility can not be ruled out.

The second possibility is that k_5/k_4 is small, and that $\alpha = (1.04)k_5/k_5*$, so that $k_5/k_5* = 1.007$, and $\alpha\beta = 1.057 = k_6/k_6*$. In this case C—N bond breaking is the rate limiting step. This possibility seems most likely of the two, because we know that C—N bond cleavage (k_6) is considerably slower than all of the following steps in the mechanism (k_7).

If k_5/k_4 is around unity, the values for k_3/k_3*, k_5/k_5*, and k_6/k_6* calculated in both possibilities must apply. Thus the most likely explanation for the observed α value is a minimum kinetic isotope effect of 1.057 on C—N bond cleavage, and one of 1.007 on C—N bond making, but the possibility that the entire effect is an equilibrium one on the slow deprotonation of ammonium ion can not be ruled out.

ACKNOWLEDGMENT

This research has been supported by grants from NSF (BMS-16134) and NIH (GM 18938).

LITERATURE CITED

Cardinale, G. J., and Abeles, R. H. 1968. Purification and mechanism of action of proline racemase. Biochemistry 7:3970-3978.

Cleland, W. W. 1975. Partition analysis and the concept of net rate constants as tools in enzyme kinetics. Biochemistry 14:3220-3224.

Gorenstein, D. G. 1972. Oxygen-18 isotope effect in the hydrolysis of 2,4-dinitrophenyl phosphate. A monomeric metaphosphate mechanism. J. Am. Chem. Soc. 94:2523-2525.

Hartshorn, S. R., and Shiner, V. J., Jr. 1972. Calculation of H/D, $^{12}C/^{13}C$, and $^{12}C/^{14}C$ fractionation factors from valence force fields derived for a series of simple organic molecules. J. Am. Chem. Soc. 94:9002-9012.

O'Leary, M. H., and Limburg, J. A. 1977. Isotope effect studies of the role of metal ions in isocitrate dehydrogenase. Biochemistry 16:1129-1135.

Schimerlik, M. I., Rife, J. E., and Cleland, W. W. 1975. Equilibrium perturbation by isotope substitution. Biochemistry 14:5347-5354.
Thomson, J. F. 1960. Fumarase activity in D_2O. Arch. Biochem. Biophys. 90: 1-6.

Isotope Effects in Hydride Transfer Reactions

J. P. Klinman

A large number of enzyme-catalyzed oxidation reactions are characterized by the removal of the elements of hydrogen to generate an unsaturated bond:

$$R_1R_2C(H)\text{—}X\text{—}H + \text{Coenz(OX)} \rightleftharpoons R_1R_2C{=}X + \text{Coenz(RED)} + H_3O^+ \quad (1)$$

The coenzyme in equation 1 may be either $NAD(P)^+$ or a flavin, and X represents carbon, oxygen, nitrogen, or sulfur. Kinetic hydrogen isotope effects are powerful tools for investigating the mechanisms of these reactions. Primary kinetic isotope effects can be measured for both C—H and X—H cleavage. In the case of X—H cleavage, when X equals oxygen, nitrogen, or sulfur, replacement of H_2O by D_2O is required in order to study isotope effects, because these heteroatoms exchange their protons rapidly with solvent. As shown in equation 1, the conversion of reduced substrate to oxidized product involves a change in hybridization ($sp^3 \rightarrow sp^2$), indicating that α-secondary kinetic isotope effects can also be measured when either R_1 or R_2 is hydrogen. To date, the most extensive application of isotope effects in these reactions has been the determination of the magnitude of the primary kinetic isotope effect for C—H cleavage. In the case of reactions which are characterized by large primary kinetic isotope effects for C—H cleavage, further exploration of isotope effects can provide considerable insight into mechanism and transition-state structure; for example, the magnitude of solvent isotope effects will indicate the "concertedness" of X—H relative to C—H cleavage, and the magnitude of α-secondary isotope effects will indicate the extent to which the structure of the transition state resembles that of reduced substrate versus oxidized product.

A fundamental mechanistic question in these oxidation-reduction reactions is the mode of hydrogen activation, i.e., is a hydrogen plus two electrons transferred from the carbon of substrate to coenzyme in a concomitant fashion (formally referred to as hydride transfer), or is the transfer of hydrogen and electrons a stepwise process involving the formation of either polar or radical intermediates? In the case of flavin-mediated reactions, it is currently believed that hydride, radical, or proton abstractions may occur, depending on the nature of the substrate (Bruice, 1975). In contrast, $NAD(P)^+$-mediated reactions are normally considered to occur via a formal hydride transfer, although Chan and Bielski (1975) have demonstrated a stereospecific transfer of a hydrogen atom from NADH to a preformed, fumarate-derived radical at the active site of lactate dehydrogenase. To a large extent, the concept of a formal hydride transfer in $NAD(P)^+$-mediated reactions is a consequence of the observation that hydrogen is transferred directly from substrate to coenzyme (Westheimer et al., 1951). The failure to observe exchange of a substrate-derived hydrogen with solvent may result from the inaccessibility of an intermediate to solvent, however, and need not imply the absence of an intermediate in the course of a reaction.[1]

COMPARISON OF KINETIC TO PRODUCT RATIO ISOTOPE EFFECTS IN MODEL DIHYDRONICOTINAMIDE REACTIONS

Steffens and Chipman (1971) and Creighton et al. (1973) have investigated isotope effects in the oxidation of *N*-substituted dihydronicotinamides by trifluoroacetophenone and *N*-methylacridinium, respectively. Both of these reactions are characterized by a *direct transfer of hydrogen* from dihydronicotinamide to substrate. As illustrated in Figure 1, two isotope effects were determined: kinetic isotope effects for the disappearance of NADH versus NADD, $k_R/k_R' = 2k_H/(k_H' + k_D)$, and product isotope effects for the formation of protonated versus deuterated

[1] Xylose isomerase (Rose, O'Connell, and Mortlock, 1969) and Δ^5-ketosteroid isomerase (Wang, Kawahara, and Talalay, 1963), two enzymes that catalyze the activation of hydrogen as a proton, are characterized by a complete transfer of hydrogen from substrate to product.

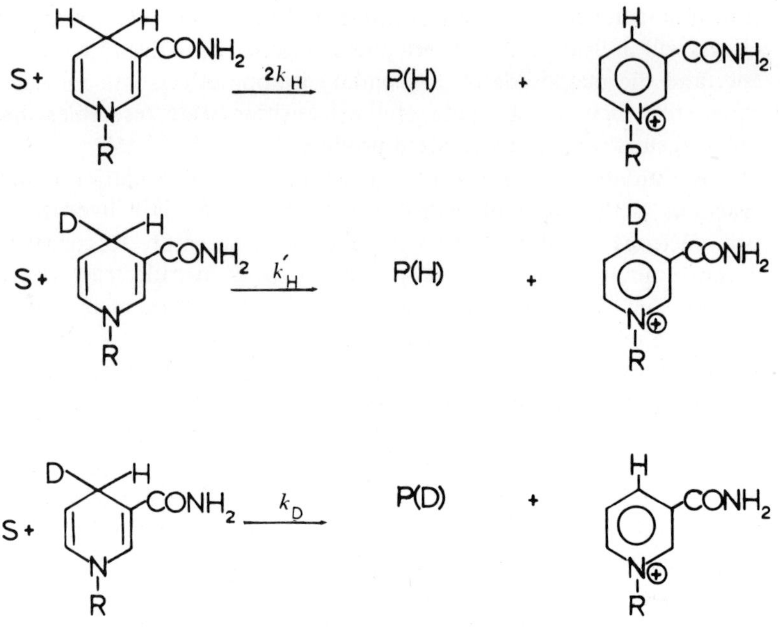

Figure 1. Isotope effect studies in the reduction of activated substrates, S, by *N*-substituted dihydronicotinamide (Creighton et al., 1973; Steffens and Chipman, 1971). Kinetic isotope effects were measured for the disappearance of protonated ($2k_H$) and monodeuterated ($k_H{}' + k_D$) dihydronicotinamide. The ratio of protonated (P(H)) to deuterated (P(D)) product formed from monodeuterated dihydronicotinamide was also determined.

product from monodeuterated dihydronicotinamide, P(H)/P(D).[2] If the reaction shown in Figure 1 proceeds in a single step, bimolecular transfer of hydrogen from dihydronicotinamide to substrate, the relationship between $k_R/k_R{}'$ and P(H)/P(D) is given in equation 2:

$$k_R/k_R{}' = 2k_H/(k_H{}' + k_D) = 2(k_H/k_H{}')/[1 + \mathrm{P(D)/P(H)}] \qquad (2)$$

The ratio $k_H/k_H{}'$ in equation 2 is a secondary isotope effect for the abstraction of hydrogen from a carbon which contains another hydrogen or deuterium. Because the overall reaction involves the conversion of an sp^3 to an sp^2 hybridized carbon at C—4 of the dihydronicotinamide

[2] The ability to compare kinetic and product ratio isotope effects is a unique feature of model systems because of the stereospecific constraints of enzyme reactions.

ring, this isotope effect is expected to fall in the range $k_H/k_H' = 1\text{–}1.3$ (e.g., do Amaral et al., 1973). Using equation 2 and the observed values for k_R/k_R' and P(H)/P(D), Table 1, k_H/k_H' is 0.73 for trifluoroacetophenone and 0.74 for *N*-methylacridinium reduction. These calculated values for k_H/k_H' indicate that equation 2 will not accommodate the available kinetic data; a minimal kinetic scheme involving the steady-state formation of an intermediate which partitions between substrate (k_{-1}) and product ($k_H' + k_D$) has been proposed:

$$\text{S} + \text{[4-H,4-D-dihydronicotinamide (CONH}_2\text{), N-R]} \underset{k_{-1}}{\overset{k_1}{\rightleftharpoons}} \underset{?}{[\text{I}]} \begin{cases} \xrightarrow{k_H'} \text{P(H)} \\ \xrightarrow{k_D} \text{P(D)} \end{cases} \quad (3)$$

The relatively small magnitudes of rate constants for the reduction of trifluoroacetophenone ($k_R = 4.2 \times 10^{-3} \text{M}^{-1}s^{-1}$) and *N*-methylacridinium ($k_R = 2 \times 10^3 \text{M}^{-1}s^{-1}$) compared to constants observed for the formation of stacked dimers from aromatic compounds such as proflavine [$k = 7.9 \times 10^8 \text{M}^{-1}\text{s}^{-1}$ (Turner et al., 1972)] argues that I is not a simple charge transfer complex. It appears that I may be covalent in

Table 1. Isotope effects for the reduction of trifluoroacetophenone and *N*-methylacridinium by *N*-propyl dihydronicotinamide

S	k_R, $\text{M}^{-1}s^{-1}$	k_R/k_R'	P(H)/P(D)	$(k_H/k_H')_{\text{CALC}}$[a]
$CF_3\text{—C(=O)—}C_6H_5$	0.0042[b]	1.16	3.8	0.73
N-methylacridinium ($N^+\text{—}CH_3$)	2040[c]	1.26	5.4	0.74

[a] Secondary isotope effect calculated from k_R/k_R', P(H)/P(D), and equation 2 in text.
[b] Steffens and Chipman, 1971.
[c] Creighton et al., 1973.

nature, possibly a caged radical pair. Although further experimentation will be necessary to elucidate the properties of I, the observed kinetic and product ratio isotope effects require the presence of a kinetically significant intermediate in these two model dihydronicotinamide reactions.

ISOTOPE EFFECTS IN HYDRIDE VERSUS PROTON TRANSFER REACTIONS

A question which frequently arises in a discussion of enzyme-catalyzed oxidation-reduction reactions is whether there are any fundamental differences in the anticipated isotope effects for hydride transfer as opposed to other forms of hydrogen activation. A simple model for the transfer of hydrogen between two atomic centers is shown in Figure 2. Neglecting bending vibrations in the transition state and assuming a linear transfer of H between A and B, the magnitude of the kinetic isotope effect is a function of the temperature in °K, the vibrational frequency in cm^{-1} of the A—H ground state stretch, ω_H, and the transition-state symmetrical stretch, $\omega_H^{\ddagger}$ (Bigeleisen, 1964; Melander, 1960; Westheimer, 1961):

$$\log \frac{k_H}{k_D} = \frac{0.081}{T} (\omega_H - \omega_H^{\ddagger}) \tag{4}$$

Although this formulation of the kinetic isotope effect makes no distinction among the possible modes of hydrogen activation, Swain, Wiles, and Bader (1961) pointed out some important differences be-

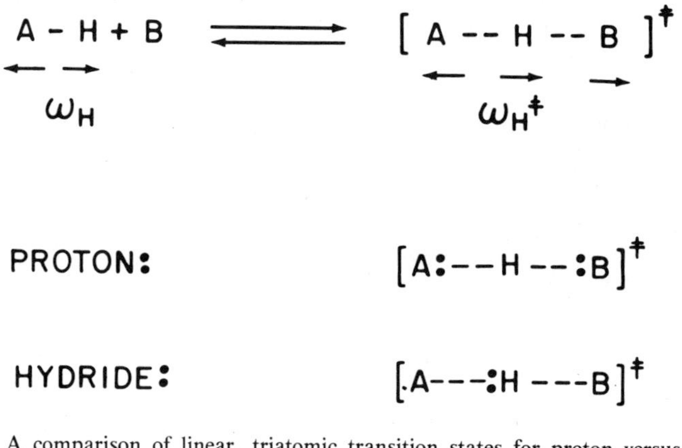

$$A - H + B \rightleftharpoons [A -- H -- B]^{\ddagger}$$

ω_H $\omega_H^{\ddagger}$

PROTON: $[A{:}--H--{:}B]^{\ddagger}$

HYDRIDE: $[A---{:}H---B]^{\ddagger}$

Figure 2. A comparison of linear, triatomic transition states for proton versus hydride transfer reactions.

tween transition states involving proton versus hydride transfer. In the case of hydride abstraction, a hydrogen plus a pair of electrons is transferred between two electrophilic centers. The presence of a single pair of electrons serves to "cement" A, H, and B together, leading to relatively short, strong, and nonpolarizable bonding in the transition state. In contrast, a proton abstraction is characterized by the transfer of hydrogen between two nucleophilic centers; repulsion between electron pairs in the transition state is expected to lead to relatively long, weak, and, importantly, polarizable bonding. As a result of these differences, Swain, Wiles, and Bader (1961) suggested that the magnitude of the observed isotope effect will be less sensitive to changes in substrate structure for a hydride than for a proton transfer reaction.

A study of isotope effects in the oxidation of secondary alcohols by Br_2 was undertaken in an effort to test these theoretical considerations. The reaction can be represented by two kinetically indistinguishable mechanisms, Figure 3, A and B. The distinction between mechanisms A and B concerns whether the hydrogens bound to carbon and oxygen leave as protons or hydride ions. Both C—H and O—H isotope effects were measured for the oxidation of 2-propanol and 1-fluoro-2-propanol, Table 2. The presence of fluorine slows the reaction rate by a factor of 10^3. It was pointed out that the retardation of the rate by fluorine cannot by itself distinguish proton from hydride activation at carbon; mechanisms A and B are both concerted processes and could lead to a development of positive charge at C-2 of the alcohol in the transition state, depending on the precise timing of the C—H and O—H cleavages. Importantly, the C—H isotope effect was found to be essentially unchanged over a 10^3 change in rate, whereas the O—H isotope effect increased from 1.5 to 2.1. On the basis of these observations, the oxidation of propanol by Br_2 was concluded to occur via hydride abstraction from carbon.

In the decade subsequent to the work of Swain, Wiles, and Bader (1961), numerous investigators have measured primary hydrogen isotope effects for simple proton abstraction reactions from carbon (Bell, 1973,

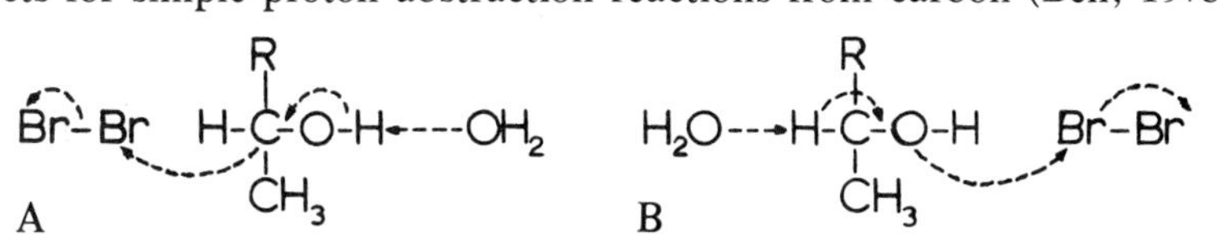

Figure 3. Alternate, kinetically indistinguishable mechanisms for the oxidation of a secondary alcohol by Br_2. In A, C—H cleavage occurs via hydride abstraction whereas in B, C—H cleavage involves proton abstraction (Swain, Wiles, and Bader, 1961).

Table 2. Isotope effects for the oxidation of propanol and fluoropropanol by Br_2

	k_H, $M^{-1}s^{-1}$	k_H/k_D	k_{H_2O}/k_{D_2O}
Propanol	5.9×10^{-2}	2.94	1.49
Fluoropropanol	5.8×10^{-5}	2.83	2.06

From Swain, Wiles, and Bader, 1961.

and references therein). In general, over a wide range in reactivity, the magnitude of the isotope effect is found to be sensitive to changes in the structure of the carbon acid and abstracting base. One example from the extensive literature on this subject is summarized in Table 3: Over a 3×10^4 change in reactivity, the isotope effect for proton abstraction from tricarbomethoxymethane varies from 3.8 to 6.6 (Barnes and Bell, 1970). Unfortunately, very few systematic studies have been conducted which permit a correlation of substrate reactivity with the magnitudes of hydrogen isotope effects in oxidation-reduction reactions.[3] As regards enzyme-catalyzed reactions, alterations in substrate structure are frequently observed to give rise to a wide range of isotope effects. Normally, these observed variations in isotope effect reflect a change in the relative contributions of individual rate constants to the overall kinetic scheme, rather than intrinsic differences in isotope effects.

Table 3. Isotope effects for the base-catalyzed abstraction of a proton from tricarbomethoxymethane

Base	k_H, $M^{-1}s^{-1}$[a]	k_H/k_D[b]
H_2O	$1.48 \times 10^{-2}/55.5$	3.8
$ClCH_2CO_2^-$	0.452	5.0
$CH_3CO_2^-$	6.46	5.5
$(CH_3)_3CCO_2^-$	6.21	6.6

From Barnes and Bell, 1970.
[a] Measured at 0°.
[b] Converted to 25°, assuming k_H/k_D is due entirely to differences in activation energy.

[3] In a preliminary report by Audette, Quail, and Smith (1972), isotope effects were found to vary from 5.3 to 7.9 over an approximately three-fold change in rate for the oxidation of p-CH_3O, -H, and m-NO_2 benzyl alcohols by ferrate ion. On the basis of these results, which contrast with those observed for bromine oxidation of alcohols, and the base-catalyzed properties of the ferrate ion reaction, oxidation was concluded to proceed via a proton abstraction from carbon.

KINETIC COMPLEXITY OF ENZYME SYSTEMS LEADS TO SMALL ISOTOPE EFFECTS

Among the most thoroughly studied enzymes which use $NAD(P)^+$ as coenzyme are the dehydrogenases. Many of these enzymes are characterized by a steady-state kinetic mechanism in which coenzyme (A) and substrate (B) add to the enzyme in a preferred order, equation 5.

$$E + A \underset{k_2}{\overset{k_1}{\rightleftharpoons}} EA + B \underset{k_4}{\overset{k_3}{\rightleftharpoons}} EAB \underset{k_6}{\overset{k_5}{\rightleftharpoons}} ECD \xrightarrow{k_7} ED + C \xrightarrow{k_9} E + D \quad (5)$$

The relevant steady-state kinetic parameters are V_{max}, V_{max}/K_A, and V_{max}/K_B. V_{max} is the rate of conversion of bound substrates to product, whereas V_{max}/K_A and V_{max}/K_B represent rates for the conversion of free A and B to product, respectively. An important feature of V/K terms, which contrasts with V, is their independence of kinetically important processes occurring subsequent to the first irreversible step. V_{max}/K_A is measured at infinite B concentration, which makes the addition of B to EA both irreversible and infinitely fast. Consequently, V_{max}/K_A is equal simply to the rate of addition of coenzyme to enzyme and will not be characterized by an isotope effect (neglecting isotope effects on binding).

The relationship between observed and intrinsic isotope effects on V_{max} under limiting conditions of rate-determining interconversion of ternary complexes (k_5), first product release (k_7), or second product release (k_9) is summarized in Table 4. Clearly, only when the release of both products is fast will the observed isotope effect reflect an intrinsic value. It is of interest that under conditions of rate-determining release of first product, V_{max} reflects an equilibrium constant ($K_{eq} = k_5/k_6$) for the interconversion of ternary complexes. Recent measurements of primary equilibrium isotope effects for the transfer of a hydrogen from NAD(P)H to C-2 of oxaloacetate or to C-1 of acetaldehyde indicate small deviations from unity, $K_{eq_H}/K_{eq_D} = 0.85$ (Cook and Cleland, 1977) and 0.89 (Klinman, unpublished results), in agreement with computed values of Hartshorn and Shiner (1972). The unusual observation of an inverse primary hydrogen isotope effect on V_{max} would be consistent with the presence of kinetically significant steps subsequent to a reversible chemical conversion step.

A comparison of isotope effects on V_{max} and V_{max}/K_B can be quite informative. When V_{max} represents a single rate-determining hydrogen transfer step, the isotope effect on V_{max}/K_B will vary from one to an intrinsic isotope effect, depending on the rate at which substrate dis-

Table 4. The relationship of observed isotope effects on V_{max} and V_{max}/K_B to the intrinsic effect on k_5 for the kinetic mechanism shown in equation 5 in text

Limiting case	V_{max} Term	V_{max} Observed isotope effects	V_{max}/K_B Term	V_{max}/K_B Observed isotope effects
I $k_{5(6)} \ll k_7, k_9$	k_5	Intrinsic	$k_3k_5/(k_4 + k_5)$	$k_4 \gg k_5$: Intrinsic $k_4 \ll k_5$: One
II $k_7 \ll k_{5(6)}, k_9$	$k_7/(1 + k_6/k_5)$	Reflects an equilibrium effect	$k_3k_5k_7/(k_4k_6 + k_5k_7)$	$k_4 \gg k_7$: Equilibrium $k_4 \ll k_7$: One
III $k_9 \ll k_{5(6)}, k_7$	k_9	One	$k_3k_5k_7/(k_4k_7 + k_4k_6 + k_5k_7)$	$k_4, k_7 \gg k_{5(6)}$: Intrinsic $k_4, k_7 \ll k_{5(6)}$: Reflects an equilibrium effect

sociates from ternary complex versus conversion to product (Case I, Table 4); from the relative magnitudes of isotope effects on V_{max}/K_B and V_{max}, the partitioning of EAB between EA (k_4) and products (k_5) can be determined and substrate dissociation constants calculated. In contrast, limitation of V_{max} by a rate-determining dissociation of second product leads to isotope effects on V_{max}/K_B which may be equal to but are frequently *greater than* V_{max} effects (Case III, Table 4).

Horse liver alcohol dehydrogenase is an example of an enzyme characterized by a rate-limiting dissociation of NAD^+ from $ENAD^+$ in the direction of acetaldehyde reduction by NADH(D) (equation 6, p. 186). Consistent with a rate-limiting product release step, there is no isotope effect at infinite concentration of reduced coenzyme and acetaldehyde, $V_H/V_D = 1$. As summarized in Table 5, conversion of the measured kinetic parameter from V to V/K by reduction of the acetaldehyde concentration below its K_m($\leqq$0.3 mM), results in isotope effects significantly greater than one (Bush, Shiner, and Mahler, 1973). The magnitude of the isotope effect at zero acetaldehyde concentration $(V/K)_H/(V/K)_D = 2.7$, indicates that k_5 is a major rate-determining step in the conversion of ENADH and acetaldehyde to $ENAD^+$ and ethanol.

Palm (1968) studied isotope effects in the lactate dehydrogenase reduction of pyruvate by NADH(T) (equation 7, p. 187). In analogy with horse liver alcohol dehydrogenase, the magnitude of the observed isotope effect was found to vary as a function of substrate concentration, Table 6. The decrease in the isotope effect at high pyruvate does not reflect a conversion of V/K to V, however, because tritium isotope effects are always V/K effects (Simon and Palm, 1966). The disappearance of the isotope effect results from a rapid and essentially irreversible conversion of ENADH(T) to ternary complex at high pyruvate concentrations. These data illustrate the disappearance of tritium isotope effects in ordered two-substrate enzyme reactions using isotopically labeled first substrate and high concentrations of second bound sub-

Table 5. Isotope effects for the liver alcohol dehydrogenase-catalyzed reduction of acetaldehyde by NADH(D)

[Acetaldehyde], mM	Isotope effect
0	2.7
0.05	2.21
0.5	1.78
5.0	1.20
∞	1.0

From Bush, Shiner, and Mahler, 1973.

$$E + NADH(D) \underset{k_2}{\overset{k_1}{\rightleftharpoons}} ENADH(D) + CH_3{-}\overset{\overset{\displaystyle O}{\|}}{C}{-}H \underset{k_4}{\overset{k_3}{\rightleftharpoons}} ENADH(D)\cdot CH_3{-}\overset{\overset{\displaystyle O}{\|}}{C}{-}H \underset{k_6}{\overset{k_5}{\rightleftharpoons}} ENAD^+\cdot CH_3CH_2H(D)OH \xrightarrow{k_7} ENAD^+ + CH_3CH_2H(D)OH \xrightarrow{k_9} E + NAD^+ \quad (6)$$

$$\longleftarrow (V/K)_H(V/K)_D \longrightarrow \qquad \longleftarrow V_H/V_D \longrightarrow$$

$$\mathrm{E + NADH(T)} \underset{k_2}{\overset{k_1}{\rightleftharpoons}} \mathrm{ENADH(T) + CH_3{-}\overset{O}{\overset{\|}{C}}{-}CO_2H} \underset{k_4}{\overset{k_3}{\rightleftharpoons}} \mathrm{ENADH(T)\cdot CH_3{-}\overset{O}{\overset{\|}{C}}{-}CO_2H} \underset{k_6}{\overset{k_5}{\rightleftharpoons}} \mathrm{ENAD^+\cdot CH_3{-}CH(T)OHCO_2H} \xrightarrow{k_7} \mathrm{ENAD^+ + CH_3CH(T)OHCO_2H} \xrightarrow{k_9} \mathrm{E + NAD^+} \quad (7)$$

$$\longleftarrow (V/K)_H/(V/K)_T \longrightarrow$$

Table 6. Isotope effects for the lactate dehydrogenase-catalyzed reduction of pyruvate by NADH(T)

[Pyruvate], mM	Isotope effect
0.1	1.81
0.2	1.64
0.42	1.52
0.83	1.25
2.0	1.10
3.3	1.13
4.0	1.0

From Palm, 1968.

strate and point out the importance of isotope effect measurements at more than one substrate concentration. Palm suggested that the low isotope effect at high pyruvate concentration was the result of abortive complexes formed between $ENAD^+$ and pyruvate. Although the formation of such abortive complexes is expected to decrease *V*, *V/K* will be unaffected if pyruvate adds subsequent to the irreversible loss of lactate from the enzyme.

The majority of dehydrogenases show isotope effects on *V* close to one under conditions of optimal substrates, pH, temperature and ionic strength. Although isotope effects on *V/K* are frequently greater than one, consistent with slow dissociation of enzyme coenzyme complexes in the steady state, these effects are most often less than two, Table 7.

Table 7. A comparison of primary hydrogen isotope effects on *V* versus *V/K*

Enzyme	*V*	*V/K*
1) Lactate dehydrogenase: Oxidation of [2-^{2}H] lactate[a]	1	1.5
Reduction of pyruvate by NAD^{3}H[b]	—	1.8
2) Glutamate dehydrogenase: Oxidation of [2-^{2}H] glutamate[c]	1	1.2
3) Liver alcohol dehydrogenase: Reduction of acetaldehyde by NADD[d]	1	2.7
4) Malic enzyme: Oxidation of [2-^{2}H] malate[e]	1	1.5

[a]Cantwell and Dennis, 1970; [b]Palm, 1968; [c]Rife and Cleland, personal communication; [d]Bush, Shiner, and Mahler, 1973; [e]Schimerlik, Grimshaw, and Cleland, 1977.

EXPERIMENTAL ISOLATION OF THE CHEMICAL CONVERSION STEP IN DEHYDROGENASES

Northrop (1975) has discussed the calculation of intrinsic isotope effects from measured deuterium and tritium isotope effects on V/K, assuming an overall equilibrium isotope effect of one. Application of this approach to dehydrogenases is somewhat limited by the presence of small but significant primary equilibrium isotope effects in these reactions (Cook and Cleland, 1977; Klinman, unpublished data). Experimental methods that have resulted in large primary hydrogen isotope effects for a number of dehydrogenases include: 1) the study of enzyme partial reactions by stopped flow kinetics, 2) chemical modification of an enzyme to increase the relative contribution of the C—H cleavage step to the overall steady-state rate, and 3) the study of slowly reacting substrates. Direct isolation of the chemical conversion step in enzyme reactions has the advantage of providing, in addition to intrinsic isotope effects, experimental conditions for investigating other probes of enzyme mechanism and transition-state structure.

Transient kinetic studies of horse liver alcohol dehydrogenase show a biphasic change in reduced coenzyme concentration in the direction of both aldehyde reduction and alcohol oxidation (McFarland and Bernhard, 1972; Shore and Gutfreund, 1970). An isotope effect of 5.2 was observed by Shore and Gutfreund (1970) for the appearance of enzyme-bound NADH(D) from NAD^+ and ethanol or ethanol-d_5; more recently, Hardman et al. (1974) have reported an isotope effect of 4.3 for the oxidation of benzyl alcohol. Pre-steady-state kinetic studies appear ideally suited to a study of the chemical conversion step of dehydrogenases, because the direct observation of changes in enzyme-bound coenzyme concentration circumvents the problem of slow coenzyme dissociation steps, which may limit V, and slow substrate release from ternary complex, which will reduce the magnitude of isotope effects on V/K (cf. Table 4). Although single turnover isotope effects have been measured for a number of dehydrogenases, e.g., glyceraldehyde-3-phosphate dehydrogenase (Trentham, 1971a,b), lactate dehydrogenase (Holbrook et al., 1975), and glutamate dehydrogenase (Fisher, 1973) only in the case of liver alcohol dehydrogenase has a large isotope effect been observed under pre-steady-state conditions.

An alternate approach to isolating the chemical conversion step in the horse liver alcohol dehydrogenase reaction has been explored by Plapp (1970). Chemical modification of a single lysine side chain per enzyme subunit has been found to increase the rate of coenzyme dissociation, leading to large isotope effects on V_{max} under steady-state conditions (Plapp et al., 1973). Recent studies of the oxidation of aromatic alcohols by hydroxybutyrimidylated enzyme indicate an isotope effect of 3.6 for benzyl alcohol oxidation (Dworschack and Plapp, 1976). Such an approach may have general applicability but does not appear to have been attempted with other dehydrogenases to date.

The study of modified substrates has been a particularly useful approach in investigating chemical mechanism in the glutamate dehydrogenase and yeast alcohol dehydrogenase systems. Recent studies on the glutamate dehydrogenase-catalyzed oxidation of norvaline, an analog of glutamate in which the C-5 carboxyl has been replaced by a methyl group, indicate an increase in the V_{max} isotope effect from one to 6 (Figure 1 of Cleland, this volume). Bates, Goldin, and Frieden (1970) observed that glutamate dehydrogenase will catalyze the reduction of dinitrobenzene sulfonates by NADH and NAD(P)H:[4]

SO_3^- NO_2 NO_2 X + H H O C—NH_2 N R → H SO_3^- NO_2 NO_2 X + O C—NH_2 N_+ R

This reaction also occurs readily in the absence of enzyme, and thus represents one of the few dihydronicotinamide-requiring systems that can be studied both in solution and at the enzyme surface. A study of structure-reactivity correlations in the model reaction has been reported by Kurz and Frieden (1975). The observed linear relationship between log k_{cat} and the electron withdrawing properties of para-substituents over a 10^3–10^4 fold change in rate, $\rho = 4.97$, led these authors to suggest a mechanism involving direct hydride transfer; preliminary studies indicate a similar ρ value for the enzyme-catalyzed reaction. In a recent paper Brown and Fisher (1976) report large primary deuterium isotope effects for the reduction of trinitrobenzene sulfonate by NADH either in solution or catalyzed by glutamate dehydrogenase. The observed isotope effects and structure-reactivity correlations suggest a common mechanism for the model and enzyme-catalyzed reactions; the relatively

[4]The stereochemistry of hydrogen transfer from NAD(P)H to trinitrobenzene sulfonate and the observed inhibition by α-ketoglutarate indicate that trinitrobenzene sulfonate reduction occurs at the enzyme active sites (Bates, Goldin, and Frieden, 1970).

small enzyme rate acceleration of 30 M reported by Brown and Fisher (1976) may arise solely from a reduction in the entropy of activation for the chemical conversion step, which is a concomitant of the binding of coenzyme and trinitrobenzene sulfonate to glutamate dehydrogenase (Page and Jencks, 1971).

YEAST ALCOHOL DEHYDROGENASE REACTION

Early studies on the mechanism of yeast alcohol dehydrogenase (YADH) showed a small primary isotope effect on V for acetaldehyde reduction, consistent with a partially rate-determining product release step (Mahler and Douglas, 1957). The observation that YADH will interconvert aromatic substrates at a reduced rate, relative to acetaldehyde, led to an investigation of both structure-reactivity correlations and deuterium isotope effects in this system:

$$X\text{-}C_6H_4\text{-}CHO + \text{NADH(D)} + H^+ \rightleftharpoons X\text{-}C_6H_4\text{-}CH(OH)\text{-}H(D) + \text{NAD}^+ \quad (9)$$

Primary Hydrogen Isotope Effects

A useful feature of the YADH system is its reversibility, making it possible to compare isotope effects and substituent effects in the forward and reverse directions. As shown in Table 8, the deuterium isotope effects are large and comparable for aldehyde reduction, $k_{R,H}/k_{R,D}$ = 3-

Table 8. Deuterium isotope effects for the interconversion of aromatic substrates catalyzed by yeast alcohol dehydrogenase

Substrate	$k_{R,H}/k_{R,D}$ [a]	$k_{O,H}/k_{O,D}$ [b]	$\left(\frac{k_{R,H}/k_{R,D}}{k_{O,H}/k_{O,D}}\right)_{OBS}$	$\left(\frac{k_{R,H}/k_{R,D}}{k_{O,H}/k_{O,D}}\right)_{CALC}$ [c]
p-Br	3.5	4.8	0.73	↑
p-Cl	3.3	4.2	0.78	
p-H	3.0	3.4	0.88	0.69–0.89
p-CH_3	5.4	4.2	1.21	
p-CH_3O	3.4	3.2	1.06	↓

[a] From Klinman, 1972.
[b] From Klinman, 1976.
[c] Calculated from equation 11 in text.

5.4, and alcohol oxidation $k_{O,H}/k_{O,D}$ = 3.2–4.8 (Klinman, 1972, 1976). Despite the relative insensitivity of the isotope effect to substrate structure, rate constants for aldehyde reduction vary over two powers of ten. The linear relation between log k_R and electronic substituent constants, together with the large observed isotope effects, led to the conclusion that C—H cleavage is fully rate determining in the direction of aromatic aldehyde reduction (Klinman, 1972). It is of interest to compare the isotope effect for alcohol oxidation to that observed for aldehyde reduction, because these isotope effects are interrelated by an equilibrium isotope effect for a single rate-determining C—H cleavage step:

$$(k_{R,H}/k_{R,D})/(k_{O,H}/k_{O,D}) = K_{eq,H}/K_{eq,D} \qquad (10)$$

As discussed earlier, the equilibrium isotope effect in the direction of NADH oxidation is $K_{eq,H}/K_{eq,D}$ = 0.89 ± 0.03 (Klinman, unpublished data). The oxidation of dideuterated, rather than monodeuterated, alcohols was studied, and equation 10 must be expanded to include a secondary kinetic deuterium isotope effect:

$$(k_{R,H}/k_{R,D})/(k_{O,H}/k_{O,D}) = (K_{eq,H}/K_{eq,D})/(k_{\alpha,D}/k_{\alpha,H}) \qquad (11)$$

If a range of 1–1.3 is assumed for $k_{\alpha,H}/k_{\alpha,D}$ (do Amaral et al., 1973), the right side of equation 11 is calculated to be 0.69–0.89. With the exception of the p-CH_3 substrate, which appears to be characterized by a partially rate-determining product release step in the direction of alcohol oxidation, the observed ratios of 0.73–1.06 (Table 8) support the conclusion that a single C—H cleavage step is rate determining and indicate that the measured isotope effects reflect *intrinsic* values.

Structure-Reactivity Correlations

Corroborative evidence that the measured rate constants reflect a single step results from an analysis of structure-reactivity correlations in the YADH reaction. Multiple linear regression analysis of the available kinetic constants was carried out in order to determine the contribution of electronic, hydrophobic and steric factors to k_{cat} and substrate binding (Klinman, 1976). Although Michaelis constants rather than binding constants had been measured, binding constants were obtained from a comparison of isotope effects on V/K and V. As summarized in Table 4, under conditions where V_{max} represents a rate-determining chemical conversion step, isotope effects on V_{max}/K_B reflect the partitioning of bound substrate between free substrate and product and permit calculation of substrate dissociation constants. Regression analysis indicated a negligible role for steric factors, while hydrophobic factors were concluded to contribute to aromatic alcohol binding and possibly aromatic aldehyde reduction. The role of electronic factors in substrate binding and turnover is summarized in Figure 4.

The binding of benzaldehydes is seen to increase with electron releasing *para*-substituents, $\rho^+ = -0.92 \pm 0.18$, consistent with a polarization of the bound carbonyl. The observation that $\rho^+ = 2.1$ for aldehyde reduction, together with $\rho^+ = -0.92 \pm 0.18$ for aldehyde binding, indicates a rho value of 1.5 for V/K (where K is a dissociation constant, rather than a Michaelis constant) in the direction of aldehyde reduction. This value is within experimental error of the measured $\rho^+ = 1.5$ for the overall equilibrium interconversion of aldehydes and alcohols, and it is concluded that the distribution of charge at C-1 of the substrate in the transition state is the same as that in the alcohol. By the principle of microscopic reversibility, if we assume a single rate-determining C—H cleavage step in the forward and reverse directions, $\rho^+ = 0$ in the direction of alcohol oxidation. In addition to providing insight into mechanism and transition-state structure in the YADH reaction,[5] structure-reactivity correlations support the conclusion based on deuterium isotope effects that a single C—H cleavage step is rate determining.

pH and Solvent Isotope Effect Studies

An integral part of our understanding of the mode of hydrogen transfer from coenzyme to substrate in the YADH reaction concerns the fate of the proton that must be added to the oxygen of the carbonyl in its conversion to alcohol. Although the uptake of this proton could occur directly from solvent, a mechanism involving acid-base catalysis by an active site residue was considered more likely. A study of the effect of pH on the enzyme rate (Klinman, 1975b) indicated equal but opposite pH titration curves for aldehyde reduction and alcohol oxidation, Figure 5. While it is frequently not possible to obtain unambiguous information concerning acid-base catalysis at enzyme active sites from pH dependences, the observation of opposite pH profiles for the forward

[5]The liver alcohol dehydrogenase-catalyzed oxidation of a series of benzyl alcohols has been studied under pre-steady-state conditions with native enzyme, $\rho^+ \cong -0.3$ (Hardman et al., 1974) and steady-state conditions with modified enzyme, $\rho^+ = -0.2$ (Dworschack and Plapp, 1976). In addition, binding constants for the interaction of substituted benzamides with this enzyme indicate $\rho = -0.8 \pm 0.3$ (Hansch, Kim, and Sarma, 1973). In analogy with the yeast enzyme, liver alcohol dehydrogenase appears to be characterized by the interaction of an active-site electrophile with the oxygen of bound aldehyde and a distribution of charge at C-1 of substrate in the transition state which resembles alcohol.

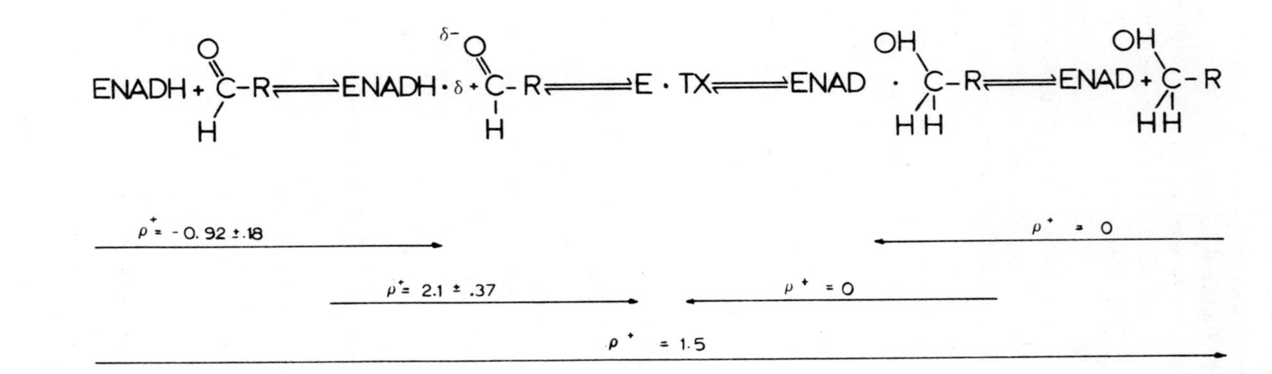

Figure 4. The contribution of electronic factors to aromatic substrate binding and turnover in the yeast alcohol dehydrogenase reaction (Klinman, 1976).

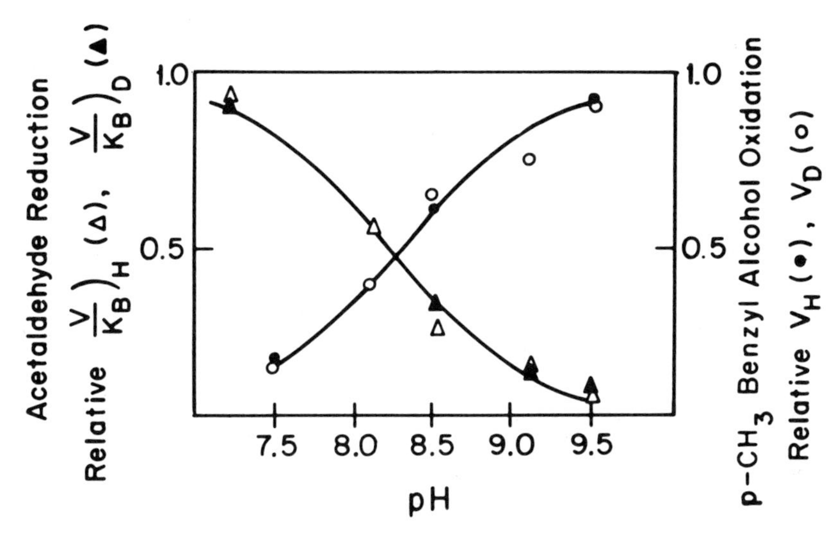

Figure 5. pH dependence of the yeast alcohol dehydrogenase-catalyzed oxidation of *p*-CH_3 benzyl alcohol and reduction of acetaldehyde. The solid lines represent a theoretical titration curve for a group of pK = 8.25 (Klinman, 1975b).

and reverse directions of an enzyme reaction characterized by a pH-dependent equilibrium equation provides strong evidence for such catalysis. From the cross-over point of the curves shown in Figure 5, an active site residue of pK = 8.25 has been implicated in the YADH reaction (Klinman, 1975b). As shown in Figure 6, the simplest mechanism which both incorporates a role for acid-base catalysis and is consistent with the observed lack of charge development at C-1 of the substrate in the transition state involves concerted acid-base catalysis of hydride transfer.[6]

Solvent isotope effects were investigated as a test of the mechanism shown in Figure 6. Catalytic constants were measured in H_2O and D_2O as a function of pH for *p*-CH_3O-benzaldehyde reduction and *p*-CH_3O-benzyl alcohol oxidation. As summarized in Table 9, a pK = 8.21 ± 0.06 was determined for aldehyde reduction and alcohol oxidation in water; this value is within experimental error of a previously determined

[6] Concerted acid-base catalysis by a protonic base of pK = 8.25 would meet the requirements for such catalysis formalized by Jencks (1972), in that the conversion of an aldehyde (pK $\simeq$ −3 to −7, Stewart et al. (1959)) to an alcohol (pK $\simeq$ 15, Ballinger and Long (1959)) involves a large change in pK, and the pK of the catalyst (pK = 8.25) is intermediate between that of substrate and product.

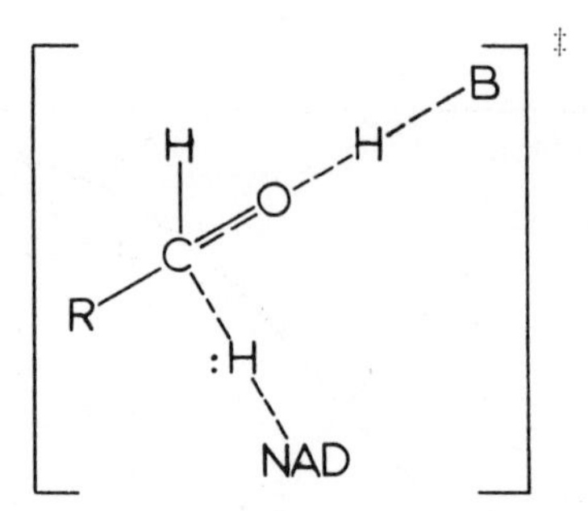

Figure 6. A mechanism involving concerted acid catalysis of hydride transfer for the reduction of carbonyls by NADH.

value of 8.25 (Klinman, 1975b). Surprisingly, D_2O was found to have an extremely small effect on the pK, pK = 8.32 ± 0.03. This result contrasts with a commonly observed $\Delta pK_a = pK_D - pK_H \cong 0.4 - 0.6$. The effect of D_2O on the ionization constant for a monoprotic base is conveniently expressed in terms of fractionation factors, ϕs, for the individual species involved in the equilibrium equation.

$$BH + H_2O \rightleftharpoons B + H_3O^+ \tag{12}$$

$$\Delta pK = \log \frac{H_3O^+}{D_3O^+} \times \frac{B}{B} \times \frac{BD}{BH} \times \frac{D_2O}{H_2O} = \log \frac{\phi BH\ \phi H_2O}{\phi H_3O^+} \tag{13}$$

By definition, $\phi H_2O = 1$; furthermore, for most protonated bases $\phi BH = 1$. The origin of the effect of D_2O on the pK of ionizing bases results primarily from the fractionation factor for ϕH_3O^+ of 0.33. The small ΔpK, 0.11, observed in the YADH system suggests that ϕBH is

Table 9. Solvent isotope effects on pK_a for the reduction of *p*-CH_3O-benzaldehyde and oxidation of *p*-CH_3O benzyl alcohol catalyzed by yeast alcohol dehydrogenase

Substrate	pK	
	H_2O	D_2O
Aldehyde reduction	8.27	8.29
Alcohol oxidation	8.14	8.35
Average	8.21 ± 0.06	8.32 ± 0.03

From Klinman, J. P., Welsh, K., and Creighton, D. J., unpublished data.

similar to ϕH_3O^+; on the assumption that BH is monoprotic or that $\phi B = 1$, the fractionation factor for ϕBH is calculated to be 0.43.

The observed fractionation factors for oxygen and nitrogen bases are close to one (Schowen, 1972), indicating that the active site residue in YADH is probably not a carboxylate, serine, histidine, or lysine. Thiols are characterized by fractionation factors of about 0.5, and two cysteines are known to be present at the active site of this enzyme (Belke, Chin, and Wold, 1975; Klinman, 1975a; Jornvall, Woenckhaus, and Johnscher, 1975). However, the 2.4 Å x-ray structure of horse liver alcohol dehydrogenase indicates a role for two cysteines as ligands for an active site zinc-water (Eklund et al., 1976); the metal content (Klinman and Welsh, 1976) and homologous cysteines (Jornvall, Woenckhaus, and Johnscher, 1975) in YADH are consistent with a similar active site configuration. The fractionation factor for $Zn\text{-}OH_2^{\delta+}$ should not be very different from that for $H{-}OH_2^+$ and provides a likely explanation for the small observed effect of D_2O on the pK of the active site residue in YADH:

$$\overset{\delta+}{ZnOH_2} + H_2O \rightleftharpoons ZnOH + H_3O^+ \tag{14}$$

The effects of D_2O on k_{cat} are summarized in Table 10. For alcohol oxidation there is a small solvent isotope effect, $k_{H_2O}/k_{D_2O} = 1.20$. In contrast, aldehyde reduction is characterized a large *inverse* solvent isotope effect, $k_{H_2O}/k_{D_2O} = 0.50$ and 0.58 for reduction by NADH and NADD, respectively. Because concerted catalysis by a protonic base of hydride transfer is expected to give rise to a primary isotope effect $\geqslant 2$, the observed solvent isotope effects appear to rule out the mechanism illustrated in Figure 6. The observed inverse isotope effect in the direc-

Table 10. Solvent isotope effects on k_{cat} in the YADH-catalyzed reduction of *p*-CH_3O-benzaldehyde by NADH and NADD, and *p*-CH_3O-benzyl alcohol oxidation

Substrate		k_{cat}, s^{-1} H_2O	k_{cat}, s^{-1} D_2O	k_{H_2O}/k_{D_2O}
Aldehyde				
	+ NADH	0.14 ± 0.007	0.28 ± 0.014	0.50 ± 0.05
	+ NADD	0.049 ± 0.002	0.085 ± 0.005	0.58 ± 0.06
Alcohol		0.78 ± 0.02	0.65 ± 0.03	1.20 ± 0.09

From Klinman, J. P., Welsh, K., and Creighton, D. J., unpublished data.

tion of aldehyde reduction is similar to isotope effects observed in model systems characterized by specific acid catalysis (e.g., $k_{H_2O}/k_{D_2O} = 0.54$ for the acid-catalyzed enolization of acetone, Toullec and Dubois (1974)) and suggests that the transfer of a proton from an active site residue to the substrate is uncoupled from carbon-hydrogen activation. A pre-equilibrium transfer of a proton to the carbonyl oxygen to generate a carbonium ion appears unlikely in light of a difference of approximately 13 pK units between the active base (pK = 8.25) and protonated aldehydes (pK = −3 to −7, Stewart et al. (1959)) and the finding that the distribution of charge at C-1 of substrate in the transition state is similar to that of enzyme-bound alcohol ($\rho^+ = 0$ for alcohol oxidation).

Two mechanisms, illustrated in Figure 7, are proposed to be consistent with the available kinetic data for the YADH reaction. According to A, aldehyde reduction occurs by way of a pre-equilibrium transfer of a proton plus one electron to form a neutral, protonated radical intermediate, followed by a rate-determining hydrogen atom transfer (Williams, Shinkai, and Bruice, 1975). A comparison of pKs for the ionization of α-hydroxy radicals ($CH_3—\dot{C}H—OH \rightleftharpoons CH_3—\dot{C}H—O^- + H_3O^+$, pK = 11.51 (Laroff and Fessenden, 1973)) to the putative base indicates that a proton transfer from EBH to the radical anion would be thermodynamically favored, in contrast to a pre-equilibrium formation of a carbonium ion. In an alternate mechanism, B, the inverse isotope effect in the direction of aldehyde reduction results from a displacement of water from zinc by the carbonyl functional group of the aldehyde substrate before a rate-determining hydride transfer. The amount of charge generated at the aldehyde carbonyl upon coordination to zinc will be considerably less than that resulting from a direct proton transfer (for example, pK ($Zn—OH_2^{\delta+}$) = 8.7 (Chaberek, Courtney, and Martell, 1952) versus pK ($H—OH_2^+$) = −1.7), and mechanism B appears to be compatible with the observed charge properties at the reacting bonds of substrate in the transition state of this reaction.

Clearly, further experimental work is necessary to distinguish A from B. Appropriate models for the effect of D_2O on the ionization of zinc-bound water and the displacement of water from zinc by aprotic ligands would be extremely useful in interpreting solvent isotope effects for YADH and other zinc-containing enzymes.[7] A long-standing question

[7] Splinter, Harris, and Tobias (1968) report a rather sizable isotope effect for the ionization of aquopentaamine cobalt (III), ΔpK = 0.48, in contrast to an earlier report of ΔpK = 0.18 (Taube, 1960).

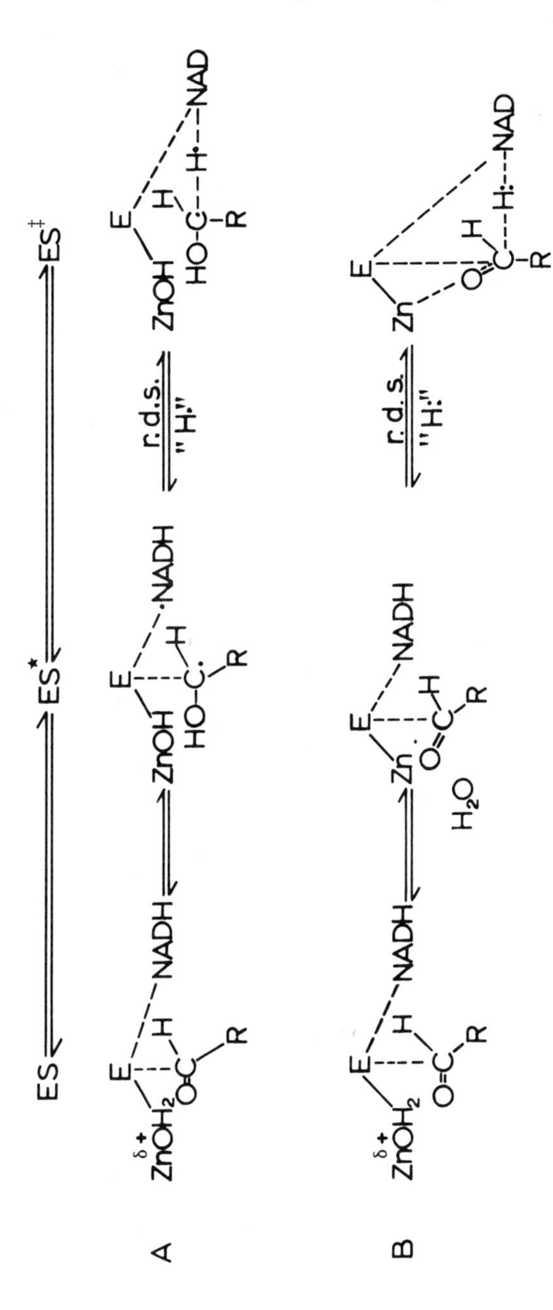

Figure 7. Two proposed mechanisms consistent with the measured isotope effects and structure-reactivity correlations in the yeast alcohol dehydrogenase reaction.

concerning the mechanism of alcohol dehydrogenases is the role of metal versus metal-bound water in the chemical reaction. Numerous investigators have proposed direct coordination to zinc (e.g., Branden et al., 1975; Theorell and Yonetani, 1963). Recent nuclear magnetic resonance studies of cobalt-substituted liver alcohol dehydrogenase indicate *second* sphere complexes between substrate and metal; on the basis of their data Sloan, Young, and Mildvan (1975) proposed catalysis by a metal-bound water. According to mechanism B, Figure 7, an inner sphere coordination complex, ES*, forms subsequent to the ES complexes. Depending on the relative concentrations of the ES and ES* complexes, it may not be possible to detect ES* by static probes such as NMR. It is of interest that isotope effects implicate an intermediate both in the YADH reaction and dihydronicotinamide model reactions (Creighton et al., 1973; Steffens and Chipman, 1971). In contrast to the model systems, intermediate formation is concluded to be rapid relative to C—H cleavage for the reaction catalyzed by YADH.

"INTRINSIC" ISOTOPE EFFECTS IN REACTIONS CATALYZED BY DEHYDROGENASES

The isotope effects that have been observed for liver and yeast alcohol dehydrogenase and glutamate dehydrogenase are summarized in Table 11. In some instances the observed effects are the product of primary and secondary effects, e.g., the oxidation of ethanol-d_5 catalyzed by liver alcohol dehydrogenase (Shore and Gutfreund, 1970) and the oxidation of NADH, dideuterated at C-4, catalyzed by glutamate dehydrogenase (Brown and Fisher, 1976). After correction for secondary effects the magnitude of the primary effect would be reduced. Because of an inverse equilibrium isotope effect in the direction of NAD(P)H oxidation, primary isotope effects are expected to be larger for NAD^+ reduction than for NADH oxidation by a factor of 1.1–1.3.

As discussed earlier, the sensitivity of primary isotope effects to changes in substrate reactivity has been proposed as a means of distinguishing among the possible modes of hydrogen activation in hydrogen transfer reactions. To the extent that the isotope effects summarized in Table 11 represent intrinsic values, the relatively small magnitude and insensitivity of these effects may reflect hydrogen activation as a hydride ion. A persistent problem in evaluating "intrinsic" isotope effects in enzyme reactions is the ability to be certain that a measured effect reflects a single step. As illustrated for the yeast alcohol dehydro-

Table 11. "Intrinsic" isotope effects for dehydrogenases

Enzyme	k_H/k_D
1. Yeast alcohol dehydrogenase, reduction of para-substituted benzaldehydes by NADH:[a]	
p-Br	3.5
p-Cl	3.3
p-H	3.0
p-CH_3	5.4
p-CH_3O	3.4
2. Horse liver alcohol dehydrogenase, oxidation of alcohols by NAD^+:	
Ethanol[b]	5.2
Benzyl alcohol[c,d]	3.6, 4.3
3. Glutamate dehydrogenase, reduction of trinitrobenzene-sulfonate by NADH[e]	4.9
Oxidation of norvaline by $NADP^+$[f]	6.0

[a] From Klinman, 1972.
[b] From Shore and Gutfreund, 1970.
[c] From Dworschak and Plapp, 1976.
[d] From Hardman et al., 1974.
[e] From Brown and Fisher, 1976.
[f] From Figure 1 of Cleland, this volume.

genase reaction, a comparison of the ratio of kinetic isotope effects for the forward and reverse directions of a reversible reaction to an equilibrium effect and the simultaneous investigation of more than one probe of transition-state structure are valuable approaches in evaluating whether C—H cleavage is fully rate determining. Although the presence of equilibrium isotope effects precludes the calculation of unique values for intrinsic from observed isotope effects on *V/K* (Northrop, 1975), it may be possible to estimate intrinsic effects within a narrow range for reactions characterized by large kinetic and small equilibrium hydrogen isotope effects (Schimerlik, Grimshaw, and Cleland, 1977).

CYTOCHROME b_5 REDUCTASE

Microsomal cytochrome b_5 reductase is an amphipathic protein containing a hydrophilic portion, where the flavin prosthetic group is located, and a hydrophobic segment, which serves as an attachment point to the endoplasmic reticulum in vivo. The hydrophilic portion of this protein has been isolated and studied kinetically (Strittmatter, 1966). The enzyme has been found to undergo successive reduction by two electron donors, such as NADH, followed by reoxidation by one electron acceptor, such as cytochrome b_5^{ox}. Stopped flow kinetic studies indicate that the reoxidation of reduced enzyme takes place in two successive steps which are dependent on the nature of the coenzyme, consistent with a release of oxidized coenzyme subsequent to the reoxidation of reduced flavin (equation 15):

$$\text{E—FAD} + \text{NADH} \underset{k_2}{\overset{k_1}{\rightleftharpoons}} \text{E}\begin{matrix}\diagup \text{FAD} \\ \diagdown \text{NADH}\end{matrix} \xrightarrow{k_3} \text{E}\begin{matrix}\diagup \text{FADH}_2 \\ \diagdown \text{NAD}^+\end{matrix} \xrightarrow[k_5]{\text{cyt } b_5^{ox} \ \text{cyt } b_5^{red}}$$

$$\text{E}\begin{matrix}\diagup \text{FADH}\cdot \\ \diagdown \text{NAD}^+\end{matrix} \xrightarrow[k_7]{\text{cyt } b_5^{ox} \ \text{cyt } b_5^{red}} \text{E—FAD} + \text{NAD}^+ \qquad (15)$$

Rate constants and isotope effects for the reduction of enzyme by NADH and its analogs, AcPyrNADH and PyrAlNADH, are summarized in Table 12 under conditions where the reoxidation of reduced enzyme is fast relative to k_3. In contrast to the available data for the reactions catalyzed by dehydrogenases, Table 11, primary isotope effects increase from 3.8-10.4 over a 100-fold decrease in reactivity in the cytochrome b_5 reductase reaction. The isotope effects for AcPyrNADH and PyrAlNADH oxidation are quite large, slightly greater than the maximal values predicted from the loss of a ground-state stretching frequency (cf., Bell, 1973). In attempting to evaluate the mode of hydrogen acti-

Table 12. Primary deuterium isotope effects for the reduction of cytochrome b_5 reductase

Nucleotide	Electron acceptor	k_H, s^{-1}	k_H/k_D
NADH	cyt b_5^{ox}	69.0	3.7
AcPyrNADH	↓	8.3	10.4
PyrAlNADH		0.96	8.7

From Strittmatter, 1966.

vation in the reduction of flavins by NAD(P)H,[8] it would be of value to know whether the isotope effect of 3.4 for NADH oxidation reflects an intrinsic value.

Subsequent to a characterization of the kinetic mechanism of solubilized cytochrome b_5 reductase, the properties of the membrane-bound enzyme were investigated. The observation that added cytochrome b_5 can be completely reduced by reductase led to a model for the microsomal membrane in which bound cytochrome b_5 reductase and cytochrome b_5 undergo diffusion and random collision (Rogers and Strittmatter, 1974; Strittmatter, Rogers, and Spatz, 1972). Hydrogen isotope effects were investigated in an effort to compare the kinetic properties of bound and free reductase (Rogers and Strittmatter, 1974). The steady-state rate equation which is representative of equation 15 when the release of NAD^+ from EFAD·NAD^+ is fast is given below:

$$v = \frac{k_3 k_5 k_7 \,(\text{cyt } b_5)(\text{E}_t)}{k_3 k_5 + k_3 k_7 + k_5 k_7 \,(\text{cyt } b_5)} \tag{16}$$

From equation 16 it can be seen that the magnitude of both the velocity and the observed isotope effect for NADH(D) oxidation (a reflection of k_3) will depend on the cytochrome b_5 concentration. This dependence is shown graphically in Figure 8; the curve in Figure 8 has been generated from values of k_3, k_5, and k_7 measured with *solubilized* reductase at 0° and extrapolated to 25°. The range of isotope effects that can be measured with microsomal enzyme is limited, because an isotope effect of 2.9 is observed at the level of cytochrome b_5 endogenous to microsomes; importantly, a four- and six-fold increase in cytochrome b_5 concentration was found to result in isotope effects of 3.6 and 4.1, respectively (Rogers and Strittmatter, 1974). Although diffusion at a membrane surface is limited to two dimensions and may not occur at the same rate as in solution, both bound and free cytochrome b_5 reductase indicate a similar response of isotope effect to changes in cytochrome b_5 concentration. In addition to supporting the proposed model for the distribution of proteins in the microsomal membrane, these studies illustrate the versatility of isotope effects in addressing questions concerning both structure and mechanism in enzymology.

---

[8] As a result of their observation that cytochrome b_5 reductase catalyzes a direct hydrogen transfer from NADH to $AcPyrNAD^+$, Drysdale, Spiegel, and Strittmatter (1961) suggested a hydride mechanism. Kosower (1966) has proposed radical mechanisms for the reduction of flavins by NAD(P)H.

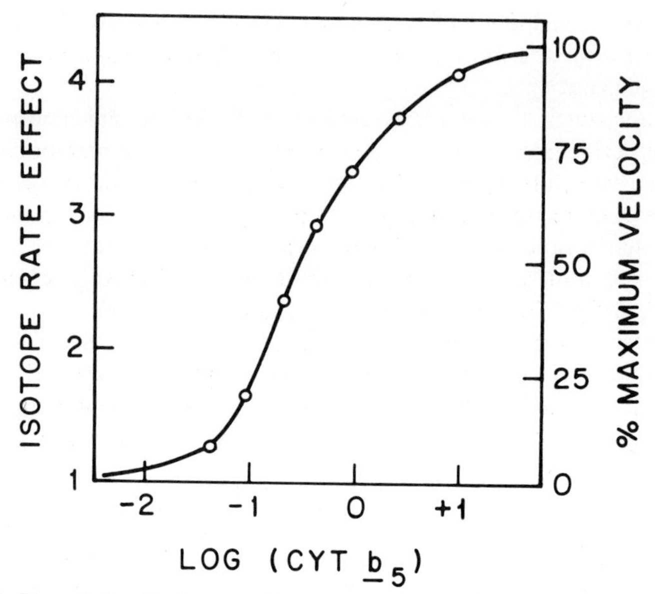

Figure 8. The relationship between isotope effects for NADH(D) oxidation and cytochrome b_5 concentration in the b_5 reductase reaction. The curve was generated from equation 16 in the text, and values for k_3, k_5, and k_7 measured with solubilized enzyme at 0°, extrapolated to 25°. (From Rogers and Strittmatter, 1974, reproduced by permission of The American Society of Biological Chemists, Inc.)

ACKNOWLEDGMENTS

This work was supported by USPHS Grants GM-20627, CA-06927, RR-05539; by NSF Grant BO-40663; and also by an appropriation from the Commonwealth of Pennsylvania.

LITERATURE CITED

Audette, R. J., Quail, J. W., and Smith, P. J. 1972. Oxidation of substituted benzyl alcohols with ferrate (VI) ion. J. Chem. Soc. Chem. Comm. 38–39.

Ballinger, P., and Long, F. A. 1959. Acid ionization constants of alcohols. I. Trifluoroethanol in the solvents H_2O and D_2O. J. Am. Chem. Soc. 81: 1050–1053.

Barnes, D. J., and Bell, R. P. 1970. Kinetic hydrogen isotope effects in the ionization of some carbon acids. Proc. Roy. Soc. Lond. A. 318:421–440.

Bates, D. J., Goldin, B. R., and Frieden, C. 1970. A new reaction of glutamate dehydrogenase: The enzyme-catalyzed formation of trinitrobenzene from TNBS in the presence of reduced coenzyme. Biochem. Biophys. Res. Comm. 39:502–507.

Belke, C. J., Chin, D. C. Q., and Wold, F. 1974. Effect of pH on the reactivity of the active site sulfhydryl groups in yeast alcohol dehydrogenase. Biochemistry 13:3418–3420.

Bell, R. P. 1973. The Proton in Chemistry. Cornell University Press, Ithaca, N.Y.

Bigeleisen, J. 1964. Correlation of kinetic isotope effects with chemical bonding in three-center reactions. Pure Appl. Chem. 8:217–223.

Branden, C. I., Jornvall, H., Eklund, H., and Furugren, B. 1975. Alcohol dehydrogenases. In: P. D. Boyer (ed.), The Enzymes, Vol. XI, pp. 104–190. 3rd Ed. Academic Press, New York.

Brown, A., and Fisher, H. F. 1976. A comparison of the glutamate dehydrogenase-catalyzed oxidation of NADPH by trinitrobenzene sulfonate with the uncatalyzed reaction. J. Am. Chem. Soc. 98:5682.

Bruice, T. C. 1975. Models and flavin catalysis. Prog. Bioorg. Chem. 4:1–87.

Bush, K., Shiner, Jr., V. J., and Mahler, H. R. 1973. Deuterium isotope effects on initial rates of the liver alcohol dehydrogenase reaction. Biochemistry 12: 4802–4805.

Cantwell, A. M., and Dennis, D. 1970. Substrate concentration dependence of deuterium isotope effects on beef heart lactate dehydrogenase. Biochem. Biophys. Res. Comm. 41:1166–1170.

Chaberek, S., Jr., Courtney, R. C., and Martell, A. E. 1952. Stability of metal chelates. II. β-hydroxyethyliminodiacetic acid. J. Am. Chem. Soc. 74: 5057–5060.

Chan, P. C., and Bielski, B. H. J. 1975. Lactate dehydrogenase-catalyzed stereospecific hydrogen atom transfer from reduced nicotinamide adenine dinucleotide to dicarboxylate radicals. J. Biol. Chem. 250:7266–7271.

Cook, P. F., and Cleland, W. W. 1977. Deuterium and tritium isotope effects in liver alcohol dehydrogenase using cyclohexanol. Fed. Proc. 36:665.

Creighton, D. J., Hajdu, J., Mooser, G., and Sigman, D. S. 1973. Model dehydrogenase reactions. Reduction of *N*-methylacridinium ion by reduced nicotinamide adenine dinucleotide and its derivatives. J. Am. Chem. Soc. 95:6855–6857.

do Amaral, L., Bastos, M. P., Bull, H. G., and Cordes, E. H. 1973. Secondary deuterium isotope effects for addition of nitrogen nucleophiles to substituted benzaldehydes. J. Am. Chem. Soc. 95:7369–7374.

Drysdale, G. R., Spiegel, M. J., and Strittmatter, P. 1961. Flavoprotein-catalyzed direct hydrogen transfer between pyridine nucleotides. J. Biol. Chem. 236:2323–2328.

Dworschack, R. T., and Plapp, B. V. 1976. Isotope and substituent effects on the interconversion of aromatic substrates catalyzed by hydroxybutyrimidylated liver alcohol dehydrogenase. Fed. Proc. 35:2003.

Eklund, H., Nordstrom, B., Zeppezauer, E., Soderland, G., Ohlsson, I., Boiwe, T., Soderberg, B.-O., Tapia, O., Branden, C. I., and Akeson, A. 1976. Three-dimensional structure of horse liver alcohol dehydrogenase at 2.4 Å resolution. J. Mol. Biol. 102:27–59.

Fisher, H. F. 1973. Glutamate dehydrogenase-ligand complexes and their relationship to the mechanism of the reaction. Adv. Enzymol. 39:369–417.

Hansch, C., Kim, K. H., and Saram, R. H. 1973. Structure-activity relationship in benzamides inhibiting alcohol dehydrogenase. J. Am. Chem. Soc. 95: 6447–6449.

Hardman, M. J., Blackwell, L. F., Boswell, C. R., and Buckley, P. D. 1974. Substituent effects on the presteady-state kinetics of oxidation of benzyl alcohols by liver alcohol dehydrogenase. Eur. J. Biochem. 50:113–118.

Hartshorn, S. R., and Shiner, V. J. Jr., 1972. Calculation of H/D, $^{12}C/^{13}C$, and $^{12}C/^{14}C$ fractionation factors from valence force fields derived for a series of simple organic molecules. J. Am. Chem. Soc. 94:9002–9012.

Holbrook, J. J., Liljas, A., Steindel, S. J., and Rossmann, M. G. 1975. Lactate dehydrogenase. In: P. D. Boyer (ed.), The Enzymes, Vol. XI, pp. 191–292. 3rd. Ed. Academic Press, New York.

Jencks, W. P. 1972. General acid-base catalysis of complex reactions in water. Chem. Rev. 72:705–718.

Jornvall, A., Woenckhaus, C., and Johnscher, G. 1975. Modification of alcohol dehydrogenases with a reactive coenzyme analog. Identification of labelled residues in the horse liver and yeast enzymes after treatment with nicotinamide-5-bromacetyl-4-methyl-imidazole dinucleotide. Eur. J. Biochem 53:71–81.

Klinman, J. P. 1972. The mechanism of enzyme-catalyzed reduced nicotinamide adenine dinucleotide-dependent reduction. Substituent and isotope effects in the yeast alcohol dehydrogenase reaction. J. Biol. Chem. 247:7977–7987.

Klinman, J. P. 1975a. The interaction of an epoxide with yeast alcohol dehydrogenase: Evidence for binding and the modification of two active site cysteines by styrene oxide. Biochemistry 14:2568–2574.

Klinman, J. P. 1975b. Acid-base catalysis in the yeast alcohol dehydrogenase reaction. J. Biol. Chem. 250:2569–2573.

Klinman, J. P. 1976. Isotope effects and structure-reactivity correlations in the yeast alcohol dehydrogenase reaction. A study of the enzyme-catalyzed oxidation of aromatic alcohols. Biochemistry 15:2018–2026.

Klinman, J. P., and Welsh, K. 1976. The zinc content of yeast alcohol dehydrogenase. Biochem. Biophys. Res. Comm. 70:878–884.

Kosower, E. M. 1966. Charge transfer complexes in flavin chemistry and biochemistry. In: E. Slater (ed.), Flavins and Flavoproteins, pp. 1–14. Vol. 8. Elsevier Publishing Co., New York.

Kurz, L. C., and Frieden, C. 1975. A model dehydrogenase reaction. Charge distribution in the transition state. J. Am. Chem. Soc. 97:677–679.

Laroff, G. P., and Fessenden, R. W. 1973. Equilibrium and kinetics of the acid dissociation of several hydroxyalkyl radicals. J. Phys. Chem. 77:1283–1288.

Mahler, H. R., and Douglas, J. 1957. Mechanism of enzyme-catalyzed oxidation-reduction reactions. I. An investigation of the yeast alcohol dehydrogenase reaction by means of the isotope rate effects. J. Am. Chem. Soc. 79:1159–1166.

McFarland, J. T., and Bernhard, S. A. 1972. Catalytic steps during the single-turnover reduction of aldehydes by alcohol dehydrogenase. Biochemistry 11:1486–1493.

Melander, L. 1960. Isotope Effects on Reaction Rates. Ronald Press, New York.

Northrop, D. B. 1975. Steady-state analysis of kinetic isotope effects in enzymic reactions. Biochemistry 14:2644–2651.

Page, M. I., and Jencks, W. P. 1971. Entropic contribution to rate accelerations in enzymic and intramolecular reactions and the chelate effect. Proc. Nat. Acad. Sci. USA 68:1678–1683.

Palm, D. 1968. Kinetische untersuchungen zum mechanismus der lactatdehydrogenase mit tritium-isotopeneffekten. Eur. J. Biochem. 5:270–275.

Plapp, B. V. 1970. Enhancement of the activity of horse liver alcohol dehydrogenase by modification of amino groups at the active sites. J. Biol. Chem. 245:1725–1733.

Plapp, B. V., Books, R. L., and Shore, J. D. 1973. Horse liver alcohol dehydrogenase. Amino groups and rate limiting steps in catalysis. J. Biol. Chem. 248:3470–3475.

Rogers, M. J., and Strittmatter, P. 1974. Evidence for random distribution and translational movement of cytochrome b_5 in endoplasmic reticulum. J. Biol. Chem. 249:895–900.

Rose, I. A., O'Connell, E. L., and Mortlock, R. P. 1969. Stereochemical evidence for a *cis*-enediol intermediate in Mn-dependent aldose isomerases. Biochim. Biophys. Acta 178:376–379.

Schimerlik, M. I., Grimshaw, C. E., and Cleland, W. W. 1977. Determination of the rate limiting steps for malic enzyme by the use of isotope effects and other kinetic studies. Biochemistry 16:571–575.

Schowen, R. L. 1972. Mechanistic deductions from solvent isotope effects. Prog. Phys. Org. Chem. 9:275–332.

Shore, J. D., and Gutfreund, H. 1970. Transients in the reactions of liver alcohol dehydrogenase. Biochemistry 9:4655–4659.

Simon, H., and Palm, D. 1966. Isotope effects in organic chemistry and biochemistry. Angew. Chem. Intl. Edn. 5:920–933.

Sloan, D. L., Young, J. M., and Mildvan, A. S. 1975. Nuclear magnetic resonance studies of substrate interaction with cobalt substituted alcohol dehydrogenase from liver. Biochemistry 14:1998–2008.

Splinter, R. C., Harris, S. J., and Tobias, R. S. 1968. The solvent isotope effect on the dissociation of the aquopentamminecobalt (III) ion. Inorg. Chem. 7:897–902.

Steffens, J. J., and Chipman, D. M. 1971. Reactions of dihydronicotinamides. I. Evidence for an intermediate in the reduction of trifluoroacetophenone by 1-substituted dihydronicotinamides. J. Am. Chem. Soc. 93:6694–6696.

Stewart, R., Gatzke, A. L., Mocek, M., and Yates, K. 1959. Deuterium isotope effects in organic cations. Chem. Ind. 331–332.

Strittmatter, P. 1966. NADH-cytochrome b_5 reductase. In: E. Slater (ed.), Flavins and Flavoproteins, pp. 325–340. Vol. 8. Elsevier Publishing Co., New York.

Strittmatter, P., Rogers, M. J., and Spatz, L. 1972. The binding of cytochrome b_5 to liver microsomes. J. Biol. Chem. 247:7188–7194.

Swain, C. G., Wiles, R. A., and Bader, R. F. W. 1961. Use of substituent effects on isotope effects to distinguish between proton and hydride transfers. Part I. Mechanism of oxidation of alcohols by bromine in water. J. Am. Chem. Soc. 83:1945–1950.

Taube, H. 1960. The D_2O—H_2O solvent effect on a complex ion equilibrium. J. Am. Chem. Soc. 82:524–526.

Theorell, H., and Yonetani, T. 1963. Liver alcohol dehydrogenase-diphosphopyridine nucleotide (DPN)-pyrazole complex: A model of a ternary intermediate in the enzyme reaction. Biochem. Z. 338:537–553.

Toullec, J., and Dubois, J. E. 1974. CH—CD and H_2O—D_2O isotope effects on the forward and reverse rates of keto-enol tautomerization of acetone in acidic media. J. Am. Chem. Soc. 96:3524–3532.

Trentham, D. R. 1971a. Reactions of D-glyceraldehyde 3-phosphate dehydrogenase facilitated by oxidized nicotinamide-adenine dinucleotide. Biochem. J. 122:59–69.

Trentham, D. R. 1971b. Rate determining processes and the number of simultaneously active sites of D-glyceraldehyde-3-phosphate dehydrogenase. Biochem. J. 122:71–77.

Turner, D. H., Flynn, G. W., Lundberg, S. K., Faller, L. D., and Sutin, N. 1972. Dimerization of proflavin by the laser raman temperature-jump method. Nature (London) 239:215–217.

Wang, S. F., Kawahara, F. S., Talalay, P. 1963. The mechanism of the Δ^5-3-ketosteroid isomerase reaction: Absorption and fluorescence spectra of enzyme-steroid complexes. J. Biol. Chem. 238:576–585.

Westheimer, F. H., Fisher, H. F., Conn, E. E., and Vennesland, B. 1951. The enzymatic transfer of hydrogen from alcohol to DPN. J. Am. Chem. Soc. 73:2403.

Westheimer, F. H. 1961. The magnitude of the primary kinetic isotope effect for compounds of hydrogen and deuterium. Chem. Rev. 61:265–273.

Williams, R. F., Shinkai, S., and Bruice, T. C. 1975. Radical mechanisms for 1,5-dihydroflavin reduction of carbonyl compounds. Proc. Nat. Acad. Sci. USA. 72:1763–1767.

Hydrogen Isotope Effects in Proton Transfer to and from Carbon

I. A. Rose

Certainly proton abstraction from a tetrahedral carbon is one of the fundamental elementary steps in enzyme catalysis. It is well established by the many examples of intramolecular proton transfer combined with partial loss of the migrating proton that proton abstraction from carbon acids is catalyzed by particular amino acid residues of enzymes in their basic forms (Rose, 1970). The enzymes for which direct evidence exists catalyze proton transfer between neighboring carbons (sugar isomerases), allylic carbons (transaminases), and different positions of the same carbon (epimerases). In addition, it is established for α,β elimination processes (dehydratases) and most recently for biotin-dependent carboxylases. However, it is likely for *all* stereospecific enzyme reactions that the enzyme and not the solvent will prove the source of the base in transfers from carbon acids.

The examination of the nature of proton transfer reactions between substrate and protein amino acid for all of these enzymes is still in various stages of finding out *what* is happening in the structural sense: what is the mechanism, what is the rate profile, what is the nature of the base, and what are the changes in bond order in going to the transition state. The *how,* which grows from a late stage of the *what*, asks what are the mechanisms for lowering the activation energy: increasing the basicity of the base, increasing the acidity of the substrate, the contribution made by entropic factors, etc.

The tool of primary and secondary hydrogen isotope effects makes a unique contribution to all of these problems.[1] The necessity to isolate the proton transfer step kinetically is often a major factor if the effects of temperature, pH, substrate changes, and anything else that influences rate are to be attributed to changes at this step. Therefore, many isotope effect studies have had as their initial goal a determination of the degree to which the chemical step is kinetically isolated. In general, the usual kinetic expressions for V_{max} and V_{max}/K_m apply in comparing rates with two isotopic forms using modified rate constants, k_{2H} or k'_{2D} or $k''_{2D\alpha}$ or $k''_{2D\beta}$, depending on the isotope and its relation to the bond being broken.[2]

By varying experimental conditions it is usually possible to vary the observed isotope effect and thereby to choose the best condition for isolating the chemical step. Such variables as temperature, a poorer or slower substrate or cofactor may show whether a maximum isotope effect has been reached. In these cases the same general rate equations with different variables are compared.

This chapter shows some ways that isotope effect logic combined with other tools of mechanism study can be used to inquire about proton transfer in enzyme reaction catalysis.

PRIMARY ISOTOPE EFFECTS AND THE LOCATION OF PROTON EXCHANGE SITES: PYRUVATE KINASE

The simplest case one can hope for in attempting to characterize a proton abstraction step is if it can be studied in the semi-isolation of an equilibrium isotope exchange and in the absence of the overall reaction:

$$E + SH \overset{1}{\rightleftharpoons} E\cdot S{-}H \overset{2}{\rightleftharpoons} E^{S}_{H} \overset{x}{\rightleftharpoons} ES^{-} + H^{+}$$

The equation for the rate of proton exchange at $(SH) < K_1$ is:

$$v = (k_1 k_2 k_x/(k_{-1}k_x + k_{-1}k_{-2} + k_x k_2))(E_T)\cdot(SH) \tag{1}$$

The extent to which step 2 is rate determining may be judged by comparing the observed isotope effect with that expected for this step from model system studies. This is necessarily an equilibrium study because

---

[1] The use of isotope effects to answer questions in systems as complex as the whole rat should not be overlooked (Katz and Crespi, 1970).

[2] The single prime ($'$) is for a rate constant for the bond broken (primary effect) and the double prime ($''$) for the rate constant influenced by substitution of D at the carbon (α effect) or neighboring (β effect). Here $k_{2H}/k''_{2D\beta}$ represents the secondary β-deutero effect on k_2.

enzyme concentration is very low compared with substrate and reversal is the only route for regeneration of E. However, for an isotope exchange at equilibrium, the exchange step x is actually irreversible resulting from the size of the water pool. A mixture of SH and ST is incubated in D_2O, and the initial rates of appearance of D in reisolated substrate, or as viewed by NMR, and of T in the water give a measured effect v_H/v_T that can be compared with the isotope effect in model reactions.[3] The equation that relates the observed isotope effect to the intrinsic effect, k_{2H}/k'_{2T}, is:[4]

$$\left(\frac{v_H}{v_T}\right)_{exch} = \frac{k_{2H}}{k'_{2T}} \cdot \frac{k_{-1}k_x + k_{-1}k'_{-2T} + k_xk'_{2T}}{k_{-1}k_x + k_{-1}k_{-2H} + k_xk_{2H}} \tag{2}$$

The full isotope effect, k_{2H}/k'_{2T}, will be seen only if $k_x > k_{-2H}$ *and* $k_{-1} > k_{2H}$. Thus, a dissipation of the discrimination between SH and ST that is inherent in step 2 can be caused by a slow exchange step, allowing partial equilibration of E^{SH} and E^{S}_{H}, or, which is also important, by the incomplete return of enriched E^{ST}/E^{SH}, arising from the discrimination step 2, to the substrate pool and therefore allowing its passage through step 2.

An example of such a study is given by Robinson and Rose (1972) with pyruvate kinase. As the name of the enzyme suggests, primary attention has been given to the phosphoryl transfer step of pyruvate kinase:

---

[3]The discrimination between SH and ST in the exchange for D_2O has advantages of using a single incubation and of giving the largest rate difference possible. The use of the combination SD + ST in H_2O is the least desirable because $(k_H/k_T)_{obs} = (k_D/k_T)_{obs}^{3.26}$ and an experimental error is more than cubed in estimating the exchange rate of SH. If precision is excellent, however, the preconditions $k_x > k_{-2D}$ and $k_{-1} > k_{2D}$ have greater chance of being met, and therefore the observed effect will be closer to the inherent isotope effect of step 2 if SD and ST are compared. It should be recalled that the Swain equations cannot be used to relate the three isotope rates unless the observed isotope effect is intrinsic, i.e., k_2/k'_2. The method of Northrop (1975) relating the two kinetic isotope effects can be used to calculate the intrinsic effect if sufficiently accurate data are available.

[4](Editor's note): If there is no effect of tritium substitution on k_2/k_{-2}, this equation can be written in a form similar to that used by Northrop and Cleland in this volume:

$$\left(\frac{v_H}{v_T}\right)_{exch} = \frac{k_{2H}/k'_{2T} + k_{-2H}/k_x + k_{2H}/k_{-1}}{1 + k_{-2H}/k_x + k_{2H}/k_{-1}}$$

where k_{-2H}/k_x and k_{2H}/k_{-1} are commitments to catalysis.

$$ATP + H_3C{-}CO{-}CO_2^- \rightleftarrows ADP + H_2C{=}C(CO_2^-)(O{-}\textcircled{P})$$

That the enolization of pyruvate-3*T* could be divorced from the phosphoryl transfer step was shown by the ability of several dianions such as P_i to replace the requirement for ATP for detritiation (Rose, 1960). Note from Table 1 that the primary isotope effects with analogs of ATP were quite high and showed variation from v_H/v_T = 14–26.[5] With the highest of these, methylphosphonate at pH 9.1, the proton abstraction step may be truly rate limiting. On the other hand, the possibility should be considered that in all cases the intrinsic isotope effect rather

Table 1. Isotope effect in pyruvate kinase catalyzed enolization of pyruvate*

$$O{=}C(CO_2^-){-}CH_3(T) \xrightarrow[D_2O]{Mg^{2+},\ K^+,\ RPO_3^{=}} O{=}C(CO_2^-){-}CH_2D + T^+,\ H^+$$

Activator	pD	μmol/min/mg enzyme v_{x_H}[a]	v_{x_T}[a]	$\left(\frac{v_H}{v_T}\right)_{exch}$
$HOPO_3^{=}$	9.0	23.6	1.57	15.0
$H_3CPO_3^{=}$		25.1	0.97	26.0
$ADPOPO_3^{=}$		96.0	38.0	2.5
$HOPO_3^{=}$	7.6	25.0	1.5	17.0
$H_3CPO_3^{=}$		12.0	0.7	17.0
$ADPOPO_3^{=}$		8.6	2.1	4.0

From Robinson and Rose (1972).

$$^{a}v_x = \frac{-3(\text{pyruvate})\text{-}\ln(1 - \text{fraction exchanged})}{\text{time}\cdot(\text{mg of enzyme})}.$$

[5] When enolpyruvate-P (PEP) is treated with *Escherichia coli* Pase at pH 7.5 in TOH, the pyruvate (isolated as lactate) has 5.3% of the specific activity of the water protons. Therefore, the ketonization of enolpyruvate in unbuffered solution, a single irreversible step, shows k_H/k_T = 18.9.

than the rate profile alone may vary when different analogs are used if, as originally proposed (Rose, 1960), the analog serves as the electrophile in the enolization. A study of the secondary isotope effect in the exchange process is being undertaken and is clearly most desirable both to answer this question and to evaluate the degree of sp^2 character in the transition state of the enolization process. This study should confirm that enolpyruvate is the intermediate in the proton exchange reaction.

After the partial reaction is examined in this way, one would like to integrate the information into the overall reaction catalyzed by the enzyme. The presence of the substrate ATP may make profound changes in our ability to study the proton abstraction step for several reasons. An example is seen in the enolpyruvate kinase (a better name than pyruvate kinase) study when ATP is present. Free enzyme can be generated now by the formation of ADP + PEP as well as by return to pyruvate. Figure 1 shows the current view of the enolpyruvate kinase mechanism without specifying the path between the proton and the medium. The appearance of tritium in the water from pyruvate-T will be the result of the contribution of the net reaction and the exchange. Unless the pathway for the exchange reaction is different than when the analog is used, the same rate equation will obtain in this case because, as shown by Northrop (1975) and by Cleland (1975), the ratio of rates of two competing isotopic forms of substrate includes steps up to the first irreversible step. Of course, the rate constants for steps 1, 2, and *x* may be different if the second substrate combines with one or all of the intermediates in the pyruvate enolization process. The concentration of ES^- certainly will be lowered by the additional steps of the reaction. In order to separate the two routes for regenerating free enzyme (the exchange and the net reaction in the presence of the second substrate), the best course is to use equilibrium conditions and eliminate the extra enolization resulting from net reaction. This can be done easily in the pyruvate kinase case because the equilibrium for the net reaction greatly favors pyruvate and ATP. As can be seen from Table 1, the exchange rate that depends on ATP is much more pH dependent. It is possible that when ATP is used, the rate equation for isotope exchange may

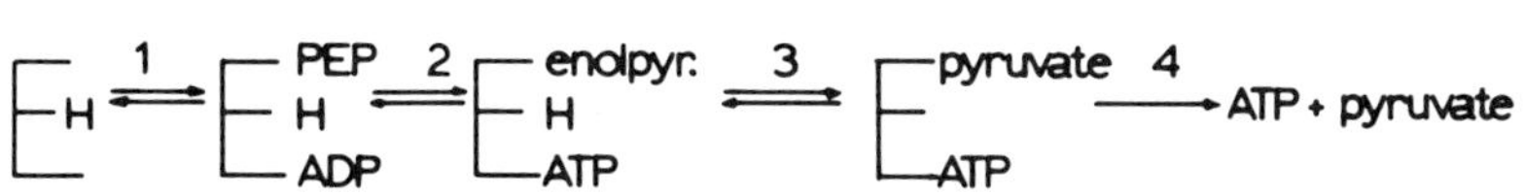

E–H ⇌(1) E(PEP, H, ADP) ⇌(2) E(enolpyr., H, ATP) ⇌(3) E(pyruvate, ATP) →(4) ATP + pyruvate

Figure 1. Pyruvate kinase mechanism.

include the phosphoryl transfer. No such transfer occurs with the analogs (Rose, 1960). The low value of the isotope effect for exchange with ATP present suggests that either the dissociation of the proton or of the pyruvate from the enzyme might be limiting.

In approaching this question of the origin of the low primary isotope effect with ATP, a more versatile approach was found in the use of either PEP-3*T* or TOH in studying the reaction from PEP to pyruvate. With PEP-T the partitioning of tritium into water or pyruvate was studied (Figure 2). With TOH the partitioning into PEP and pyruvate was observed. From a comparison of the two partitionings one can obtain the additional ratio: partitioning of pyruvate-T to PEP-T and TOH. The comparison is valid because in both cases pyruvate-T must randomize the three positions by internal rotation and enolize. The notable difference in the two paths is that an isotope effect in k_{-3} will impose a potential obstacle to the formation of TOH relative to pyruvate-T.

Table 2 shows that starting with PEP-T at alkaline pH with cobalt, about 2.2 times as much tritium is found in the water as in the pyruvate trapped as lactate. This corrects to about 110% exchange for each turnover and from equation 2 shows that enolization plus proton release exceeds product release by ~ seven-fold. This supports the argument for rapid equilibration of the 3 H's of enzyme-bound pyruvate and indicates a slow release of products which, of course, would have explained

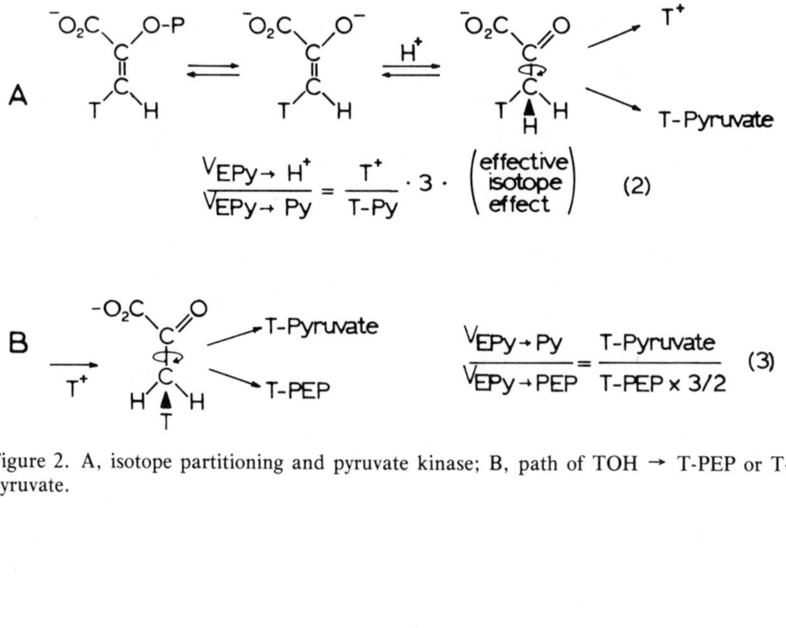

$$\frac{V_{EPy \rightarrow H^+}}{V_{EPy \rightarrow Py}} = \frac{T^+}{T\text{-}Py} \cdot 3 \cdot \left(\begin{array}{c}\text{effective}\\ \text{isotope}\\ \text{effect}\end{array}\right) \quad (2)$$

$$\frac{V_{EPy \rightarrow Py}}{V_{EPy \rightarrow PEP}} = \frac{\text{T-Pyruvate}}{\text{T-PEP} \times 3/2} \quad (3)$$

Figure 2. A, isotope partitioning and pyruvate kinase; B, path of TOH → T-PEP or T-pyruvate.

Table 2. Proton dissociation relative to substrate dissociation

	Mg^{2+}, pH 7.3	Co^{2+}, pH 8.8
$E \cdot PEP\text{-}T \rightarrow E \cdot Pyr\text{-}T \begin{cases} \nearrow T^{+} \\ \searrow Pyr\text{-}T \end{cases}$	$\frac{0.053 \times 3 \times 6.5}{1} = 1.04$ [a]	$\frac{2.24 \times 3 \times (1)}{1} \leqq 7$ [a]
$PEP \xrightarrow{T^{+}} E \cdot Pyr\text{-}T \begin{cases} \nearrow Pyr\text{-}T \\ \searrow PEP\text{-}T \end{cases}$	$\frac{1}{0.154 \times 3/2} = 4.3$ [b]	$\frac{1}{0.15 \times 3/2} = 4.4$ [b]
s.a. TOH/s.a. pyruvate-T	6.5 [c]	274,000/347,000 $\leqq$ 1 [c]
$E \cdot Pyr\text{-}T \begin{cases} \nearrow T^{+} \\ \searrow PEP\text{-}T \end{cases}$	4.5 [d]	31 [d]

[a] Applying equation 2.
[b] Applying equation 3.
[c] Effective isotope effect used in applying equation 2 [a].
[d] Product of the two partitions above.

a low primary isotope that might have been observed in the opposite direction. Clearly, the release of products is slow compared to enolization and exchange of the two original protons of the substrate for T of the medium. The slow formation of PEP-T from TOH under the same conditions shows that the path from E·pyruvate to free PEP is > 31-fold slower than the enolization plus proton release path. The formation of PEP could be limited by phosphoryl transfer (ATP → ADP) or dissociation of PEP or ADP.

When the same experiment is done at pH 7.3 with Mg^{2+} (Table 2), discrimination against TOH in formation of pyruvate appears, $v_H/v_T = 6.5$, so that the product dissociation step is no longer fully rate limiting. The return to free PEP is still slower under these conditions than proton release.

Reasons for considering PEP release and not phosphoryl transfer to be the slow step in generating free PEP are: 1) A D_2O effect of ~ 3 on V_{max} was found for pyruvate formation at pH < 7.5, showing that any slow step before making the C—H bond is eliminated at substrate saturation and therefore could not be phosphoryl transfer; 2) In spite of the D_2O effect, no discrimination against PEP-3T (a secondary isotope effect) was observed (Robinson and Rose, 1972), indicating that k_{off} of PEP is slower than steps up to the ketonization step; and 3) When phosphoryl transfer is made slow by using dGDP the D_2O effect disappears (Table 3). This result is consistent with the step-wise mechanism in which phosphoryl transfer and ketonization are not concerted; otherwise, the D_2O effect would have increased.

A final question of interest about the pattern of rates is whether the substrate proton is introduced before or after phosphoryl transfer or perhaps randomly (Figure 3). The route through reaction $-1h$ is ruled out by the slow rate of PEP release relative to proton exchange. If proton exchange occurs at the substrate's ternary complex step h_1, then with a poor nucleoside diphosphate such as dGDP, when the rate-

Table 3. D_2O effect on PEP + XDP → Pyruvate + XTP

Medium[a]	pH or pD[b]	ADP	2dGDP
H_2O	7.24	100[c]	5.2[c]
D_2O	7.16	36	5.2

[a] TES (0.1 M), XDP (3 mM), $MgCl_2$ (6 mM), PEP (1 mM), KCl (50 mM).
[b] 0.40 added to measured pH in D_2O.
[c] Relative rates.

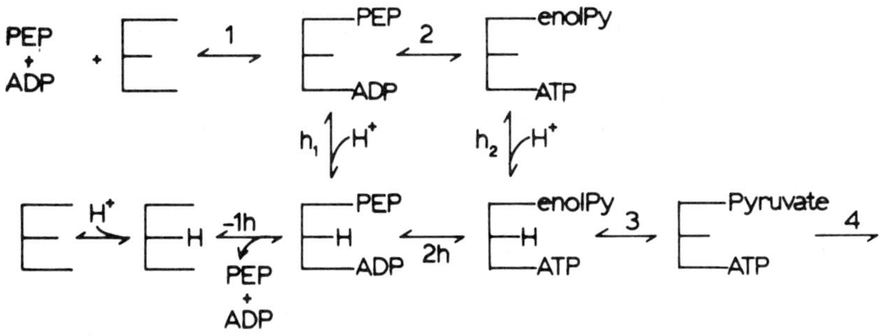

Figure 3. Alternate routes for interaction with the proton substrate.

limiting step becomes phosphoryl transfer instead of —C—H bond formation, then 1) the transfer to water from PEP-3T should be decreased and 2) the discrimination against TOH should disappear just as the D_2O effect did. Table 4 shows results of these experiments: the exchange into water of T from the PEP-3T decreased significantly but the discrimination against TOH did not decrease, as required if entry of the proton preceded the rate-limiting step, but increased significantly. Combined with the loss of the D_2O effect and the decrease in overall rate with dGDP, a simple explanation might be that the poorer nucleoside triphosphate, in addition to its weaker phosphoryl transfer ability, is less firmly bound in the ternary products complex and thereby increases the release of pyruvate relative to its enolization. These results are consistent with the conclusion that proton exchange occurs largely at the E·enolpyruvate·ATP stage, h_2. When ATP analogs are used it is clear that proton exchange must be occurring at the analogous complex containing BH···enolpyruvate··PO_4^{3-}. Therefore, the primary isotope effects measured for enolization of pyruvate-T involve the same steps with ATP and the analogs.

The different effects of pH on the exchanges with ATP and analogs (Table 1) have been noted earlier (Robinson and Rose, 1972; Rose,

Table 4. Effect of a phosphoryl acceptor on proton exchanges[a]

Process	ADP	dGDP
PEP-T → TOH/Pyr-T	0.05	0.017
PEP $\xrightarrow[\text{XDP}]{\text{TOH}}$ Pyr-T, $(\text{TOH/Pyr-T})_{s.a.}$	8.6	11.6
V_{max}	100.0	5.2

[a]Conditions as in Table 3, pH 7.24.

1960). The analog must be a dianion in order for the exchange to occur. Here, pH seems to alter the amount of enzyme participating without changing the rate profile as revealed by the isotope effect study (Table 1). With ATP the pH effect is entirely different: the exchange rate falls sharply below pH 9 and increasing isotope discrimination is observed in the exchange reaction. A simple explanation is seen by comparing equations 1 and 2. The exchange rate is decreased by decreasing k_2 or increasing any term in the denominator in equation 1, but the increase in isotope effect requires that $k_{-1}k_x$ be increased relative to the other terms. This would occur from a decrease in step 2. The rates of product liberation or proton release or both were increased preferentially with acidity. Evidence has been presented (Robinson and Rose, 1972) that at pH 8 with ADP as phosphoryl acceptor the release of product is rate limiting for the net forward reaction. Because V_{max} in this direction increases as the pH is lowered from 9 to 7.5 (Plowman and Krall, 1965), it is likely that release of pyruvate is rate determining for proton exchange also with ATP as activator. With P_i in the complex, proton abstraction is always rate limiting.

ROLE OF ENZYME-PROTON DISSOCIATION IN DETERMINING RATES OF ENZYMES: ACONITASE

Little attention has been given to the role of protonation and deprotonation of enzymes in determining catalytic rates. It is probably the fact that the form of the enzyme generated as a product in Δ^5-3-ketosteroid isomerase is the same as the starting form which allows the high turnover rate of the *Pseudomonas* enzyme, $\backsim 10^5$ sec^{-1} (Talaley and Boyer, 1965). The intramolecular proton transfer occurs with almost no exchange with the medium. Because of the high rate it is likely that not only are the ionization states identical, but also the conformational states of the free enzyme are the same in the beginning and end of the reaction process. In the case of carbonic anhydrase, $k_{cat} \leqq 10^5$, a proton transfer from the medium is known to be rate limiting and respond to buffer catalysis (Silverman and Tu, 1975). In the case of aconitase, the dehydration of citrate leaves the enzyme in the protonated state. It is likely that a conformational change along with deprotonation of the free enzyme are important in determining the overall catalytic rate. This is surprising because aconitase is a slow enzyme, $k_{cat} \leqq 10^2$ sec^{-1}.

Our present view of the proton transfers in aconitase is based on the complete retention of the transferred proton in going from citrate

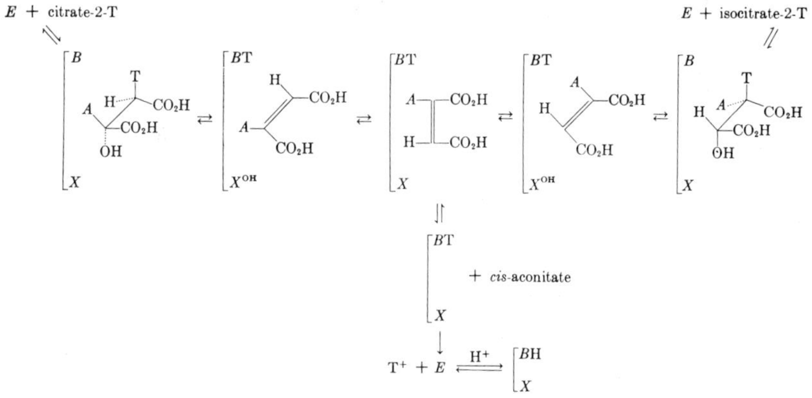

Figure 4. Aconitase reaction scheme.

to isocitrate and the finding that the dehydration alternate product, *cis*-aconitate, leaves a protonated enzyme that can be rescued by another molecule of *cis*-aconitate (Rose and O'Connell, 1967) before proton dissociation (Figure 4). It has been found that at 5 mM *cis*-aconitate about 50% of the label derived from 2-methyl citrate-2T can be captured and found in aconitase products. From this we can calculate the rate of dissociation of the proton to be 300 times the V_{max} for loss of *cis*-aconitate, A, using the formula (Rose et al., 1974):

$$k_x/k_{(A \rightarrow C + iC)} = K_{1/2}/K_M$$

When *cis*-aconitate was partially converted in tritiated water to citrate and isocitrate, the two products were shown to have specific activities below and above that of the water, respectively, (Figure 5). The experiment was repeated beginning with 2 mM citrate with the same

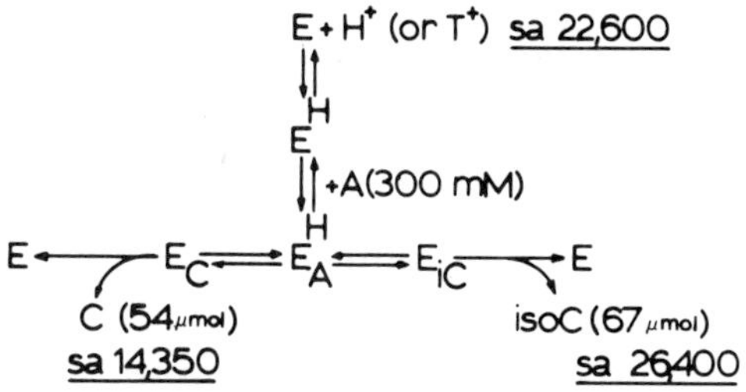

Figure 5. Conversion of *cis*-aconitate to citrate and isocitrate by aconitase in TOH.

results. The occurrence of the 15% inverse tritium effect for the hydration of *cis*-aconitate to form isocitrate is difficult to explain either as a primary or secondary effect but has an easy explanation based on Figure 4: When T is rejected in the path $E_A^H \rightarrow$ citrate the specific activity of E_A^H rises and this is reflected in the higher specific activity of the isocitrate, the formation of which apparently has a lower effective tritium isotope effect by 1.8-fold. The average specific activity of the two products is about 8% below that of the water, which indicates that almost none of the T of E_A^H returned to the medium and all of that which combined with enzyme initially was caught up in product formation.[6]

Thomson et al. (1966) reported an extensive study of beef liver aconitase acting with deuterated citrate or isocitrate or with D_2O. Although no rate differences were found with deuterated substrates, the effect of D_2O with *cis*-aconitate as substrate was to decrease the rate of formation of citrate to 26% of the rate in H_2O, compared with a fall to 68% for isocitrate production. In general, D_2O had the effect of lowering *all* rates by 70–75%, even the conversion of protonated citrate to protonated isocitrate, indicating some kind of medium effect. An explanation for the increase by 2.6-fold in the production of isocitrate compared with citrate in D_2O may be the same as the explanation for the difference between the partitioning of H^+ and T^+ when *cis*-aconitate is the substrate, namely a larger primary isotope effect in the citrate branch.

However, it is clear that Figure 4 is not sufficient to explain all of Thomson's data. In particular, the ratio of products formed from isocitrate should change in favor of *cis*-aconitate by at least 2.6-fold when isocitrate-3D was used as substrate because the path from $E_A^{H(D)}$ to A + H^+ (D^+) should have no isotope effect and the path to citrate was already shown to have one. To explain this, the figure should be expanded to include two conformations of E_A^H, conformations that are particular for citrate and isocitrate formation (Figure 6), as required by T transfer and stereochemistry (Rose and O'Connell, 1967). These two conformations, which have the opposite faces of *cis*-aconitate turned toward the proton-donating site on the enzyme, are interconverted slowly. *cis*-Aconitate must act preferentially with E_I^H to explain the iso-

[6] When, in a situation such as this, the proton dissociation from free enzyme is prevented by high substrate there may be a discrimination against forming tritiated product if the base is $-NH_2$. In this case the conjugate acid formed in TOH, $-NH_2T^+$, will offer a choice in the proton transfer step which, if it is rate limiting, will favor H over T.

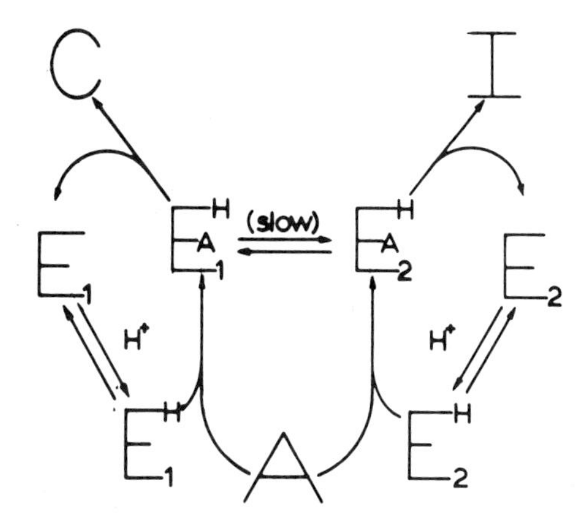

Figure 6. A, *cis*-aconitate; C, citrate; I, isocitrate.

tope effect in the citrate/isocitrate ratio, while isocitrate must give *cis*-aconitate from E_{2A}^{H} to explain the lack of effect of deuterated isocitrate on the citrate/*cis*-aconitate ratio.

Figure 6 does not explain the failure to see a primary deuterium isotope effect on the V_{max} of citrate utilization. It will be noted that four new forms of enzymes are proposed in the scheme: E_1, E_2, E^H, and E_2^H. Villafranca (1974) has recently provided evidence that the forms EH and E are not rapidly interconvertible relative to the time scale of the reactions catalyzed. This conclusion follows from the result that the enzyme produced from citrate, the so-called E^H form, gives inactive complexes with analogs of *cis*-aconitate. Thus, *trans*-aconitate is competitive when *cis*-aconitate is substrate, but it is noncompetitive toward citrate and isocitrate. From Villafranca's data it is possible to calculate the rate of conversion of E^H to E, the form that reacts with citrate. From the replot of the intercepts as a function of *trans*-aconitate [Figure 3A from (Villafranca, 1974)], $k_{EH \rightarrow E} = 46$ sec^{-1}. This compares with a steady state V_{max} for isocitrate of 59 sec^{-1} (Villafranca, 1974). Because the proton-capture data cited earlier implied a much faster ionization rate, this suggests that a *conformational* change following proton dissociation from E^H is completely rate determining for the dehydration reactions. This can explain why Thomson et al. (1966) found no primary isotope effects with either deutero citrate or deutero isocitrate.

When the conversion of *cis*-aconitate to isocitrate was measured at saturation with different concentrations of tricarballylate as inhibitor, Villafranca (1974) found tricarballylate to be noncompetitive, although it was competitive with isocitrate as substrate. We calculate $k_{E \rightarrow EH}$ = 240 sec^{-1}. This is slow for protonation of a base, consistent with the idea that these forms differ in conformation as well as charge. Because citrate and isocitrate formation from *cis*-aconitate together would be about 150 sec^{-1}, the flux in this direction seems to be limited by regeneration of the starting form of the enzyme in this case as well.

The questions of interest to the chemical mechanism of hydration of *cis*-aconitate have not yet been approached experimentally. Clearly, the fact that reactions subsequent to product formation are the rate-determining steps limits greatly the kind of initial rate study that can be applied. The fact that a primary isotope effect determines the partitioning of *cis*-aconitate between citrate and isocitrate implies that a secondary isotope effect study with *cis*-aconitate-2*T* may be informative, although the addition of H^+ to C-2 of *cis*-aconitate-2*T* may be expected to give an inverse isotope effect whether it precedes or follows the addition of OH to C-3. Earlier studies dealing with addition of water to a double bond by fumarase (Berman, Dinovo, and Boyer, 1971; Hansen, Dinovo, and Boyer, 1969; Schmidt et al., 1969) and by enolase (Dinovo and Boyer, 1971; Shen and Westhead, 1973) should be consulted as examples of the uses of primary and secondary effects to determine the sequence of addition of H^+ and OH^-. The enolase case (Shen and Westhead, 1973) is particularly useful because a primary deuterium effect, $V_H/V_D \leqq 3$, was found without difference in K_m. Thus, the on/off rate of glycerate-2-P is fast relative to subsequent steps. The C—H bond-breaking step may not be fully isolated for initial rate studies, however, because the primary deuterium effect is quite pH-dependent. The other slow step is probably subsequent to proton abstraction because a very significant exchange between TOH and unreacted glycerate-2-P was found under conditions where the free product of the enolase reaction, enolpyruvate-P, was trapped (Dinovo and Boyer, 1971).

SECONDARY ISOTOPE EFFECTS AND TRANSITION-STATE STRUCTURE: FRUCTOSE-1,6-P$_2$ ALDOLASE

It is of utmost importance in working with secondary isotope effects to distinguish between kinetic and equilibrium effects. Only when the effect is truly kinetic can further study be expected to give information about the mechanism. The standard approach is to compare the effect measured by initial rates with experimental or calculated values for equilibrium effects. If it can be established that an effect is kinetic, the isotope effect can give unique insight into the nature of the transition state. In an effort to learn more about the reaction of muscle aldolase, Biellmann, O'Connell, and Rose (1969) examined the effects of T substitution at the C_3 and C_4 positions on the cleavage of fructose-1,6-P_2. Current views of the muscle aldolase reaction sequence suppose that glyceraldehyde-P departs before protonation of the enzyme-dihydroxyacetone-P enamine, as in Figure 7, where DG represents fructose-1,6-P_2. This sequence is preferred because slowing the protonation step by carboxypeptidase treatment of the aldolase (Rose, O'Connell, and Mehler, 1965) does not alter isotope exchange between glyceraldehyde-P and fructose-1,6-P_2 (G $\rightleftharpoons$ DG). Because the C—C bond cleavage is made irreversible by trapping the triose-Ps formed with coupled enzymes, the isotope effects measured with 3-T or 4-T fructose-1,6-P_2 reflect the rate pattern up to and including formation of the first product, glyceraldehyde-3-P, and do not contain contributions from later rate constants (Biellmann, O'Connell, and Rose, 1969). The regeneration of free enzyme for catalysis requires a proton transfer to form the second product, dihydroxyacetone-P (DH). However, once the fructose-1,6-P_2-3T has been committed to reaction by passing the first irreversible step, the liberation of tritium reflects only rates up to some step which must be at or before release of glyceraldehyde-P. The specific activity of early product was 22 and 25% lower than that of reaction product when all the substrate was consumed (Biellmann, O'Connell, and Rose, 1969), corresponding to v_H/v_T = 1.33 and 1.37 for 3-*T* and 4-*T*, respectively. The equilibrium isotope effect at C-3 may be approximated by the value 1.26 which we derive from two of the fractionation factors listed in Table 5, those for oxalacetate-3T and PEP-3T, as shown in Figure 8.

The isotope effect for fructose-1,6-P_2-4T can be compared with the value for addition of HCN to 4-methoxybenzaldehyde, K_D/K_H = 1.276 (do Amaral et al., 1973) which, when taken to the 1.44 power and viewed in the cleavage direction, gives K_H/K_T = 1.42. The close agreement between the observed isotope effects and these values for equilibria of model reactions suggests that release of glyceraldehyde-3-P may be the rate-limiting step for the first half of the aldolase reaction and that the whole process from free β-fructose-1,6-P_2 through carbinolamine formation with the ϵ-NH_2 of the lysine residue, the Schiff's base step, and the aldol cleavage to form E_G^{D-} are all in rapid equilibrium.

We have reexamined this conclusion recently by establishing the concentration of E_G^{D-} in the steady state. The incubation conditions were arranged so that fructose-1,6-P_2-5T was generated continuously

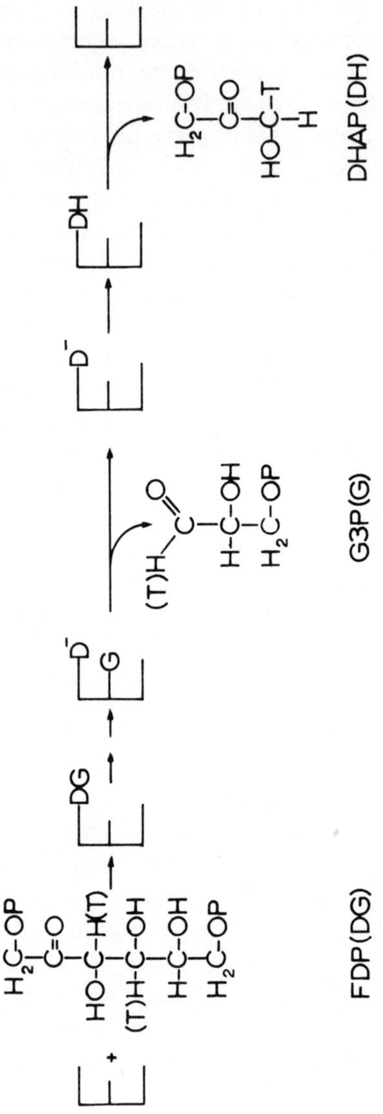

Figure 7. Aldolase reaction scheme.

Table 5. Tritium fractionation factors

$\Phi = K_{eq}$ for $TOH + C{-}H \rightleftharpoons H_2O + C{-}T$		
T-Solute	Φ	Components of calculation
TOH	1.00	By definition
3T-Oxaloacetate	0.856	Transcarboxylase[a] + KDPG aldolase[b]
3T-Pyruvate	0.78	KDPG aldolase[b]
3T-Enolpyruvate-2-P	0.68	Pyruvate kinase[a] + KDPG aldolase[b]

[a] Rose, I. A., unpublished, 1976.

[b] Meloche, H. P., personal communication, 1976.

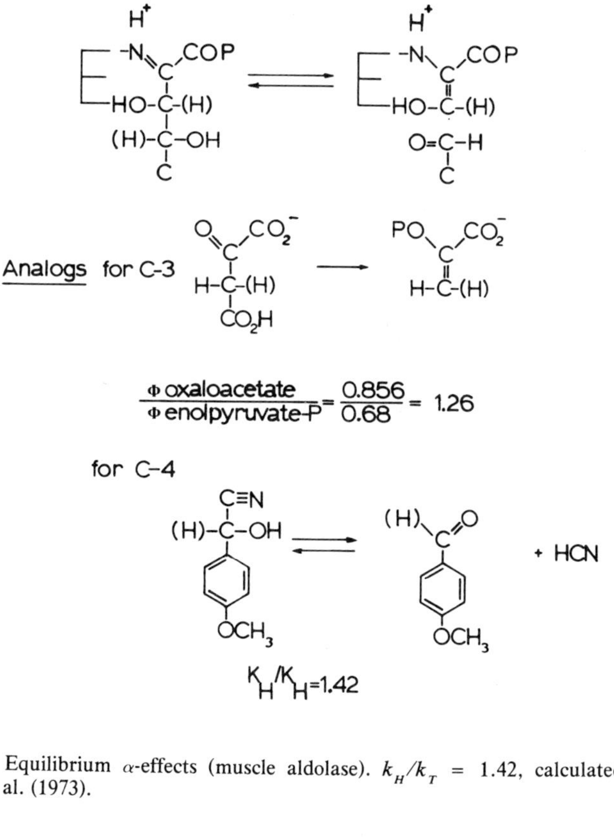

Analogs for C-3

$$\frac{\Phi \text{ oxaloacetate}}{\Phi \text{ enolpyruvate-P}} = \frac{0.856}{0.68} = 1.26$$

for C-4

+ HCN

$K_H/K_H = 1.42$

Figure 8. Equilibrium α-effects (muscle aldolase). $k_H/k_T = 1.42$, calculated from do Amaral et al. (1973).

by phosphofructokinase in the presence of such a large excess of muscle aldolase that all of the generated substrate would be bound. Triose-P isomerase was present to detritiate any glyceraldehyde-3-P-2T that was liberated by the aldolase. When the reaction was stopped in the steady state, the presence of tritium in the recovered fructose-1,6-P_2 and in glyceraldehyde-3-P gave a measure of the bound forms of each. When compared with the counts in the water, this provided the turnover rate of all those pools of bound intermediates that assay as fructose-1,6-P_2 and glyceraldehyde-3-P, respectively. Similar experiments were repeated with fructose-6-P-3T to determine the concentration and turnover of the bound forms that analyze as dihydroxyacetone-P. The combined results and the calculated group rate constants are shown in Table 6 for a study done at 4°C. The important point is that E_G^{D-} was present at very low concentration relative to the other forms, which indicates a high rate constant for the departure of glyceraldehyde-3-P, 22 sec^{-1}, compared with ~ 0.5 sec^{-1} for the turnover of the enzyme. If 22 sec^{-1} is rate limiting, the return of E_G^{D-} to DG must be > 22 sec^{-1}. The rate of enzyme generation in the reverse direction is only about 0.2 sec^{-1}. Therefore, it is most likely that glyceraldehyde-P departure is not rate limiting but follows the rate-limiting step. Although there could be an additional slow step between E_G^{D-} and E^{D-} + G, which would result in a large equilibrium isotope effect and give no insight into the mechanism, a simpler possibility is that the aldol cleavage could be rate limiting and the transition state would have a large amount of sp^2 character at both C_3 and C_4. Further isotope effect studies need to be undertaken to determine whether C_4 is a carbonium ion

$$(\mathrm{H{-}\overset{+}{\underset{|}{C}}{-}OH}) \atop \mathrm{C}$$

or a carbonyl in the transition state and whether in the condensation direction with glyceraldehyde-3-P-1T there is little isotope effect, as expected if the transition state is largely sp^2 at C_4 of the FDP being formed.

If muscle aldolase is treated with carboxypeptidase, the carboxy terminal tyrosines are removed from the four subunits and there is a 50-fold decrease in V_{max} of FDP cleavage. The alteration is confined to the reactions after step 2 because the exchange rate at equilibrium of glyceraldehyde-3-P with FDP is not altered, whereas the exchange of DHAP with FDP is diminished 10–100-fold, depending on the pH of the experiment (Rose, O'Connell, and Mehler, 1965). When the treated

Table 6. Partitioning of intermediates in the steady state

$$5T\text{-}F6P \xrightarrow[PFK]{ATP} \text{Aldolase-FDP forms} \underset{-1}{\overset{1}{\rightleftharpoons}} E^{*D-}_{G} \xrightarrow{2} 2T\text{-}G3P \xrightarrow{TIM} T^{+}$$

$$3T\text{-}F6P \longrightarrow \text{Aldolase-FDP} \rightleftharpoons E^{*D-}_{G} \longrightarrow$$

$$E^{*D-} \overset{3}{\rightleftharpoons} E^{*DH} \xrightarrow{4} T\text{-}DHAP \xrightarrow{TIM} T^{+}$$

at 2.5 sec, 4°C ΣE-FDP forms = 3200 cpm, a
E^{D-}_{G} = 170 cpm, b
ΣE-DHAP forms = 4300 cpm, c
TOH = 9400 cpm, d

$$\frac{\vec{k}_1 \cdot k_2}{\overleftarrow{k}_{-1} + k_2} = \frac{b + d}{a \times 2.5 \text{ sec}} = 1.2 \text{ sec}^{-1}$$

$$k_2 = \frac{d}{b \times 2.5 \text{ sec}} = 22 \text{ sec}^{-1}$$

$$\vec{k}_{3,4} = \frac{d}{c \times 2.5 \text{ sec}} = 0.9 \text{ sec}^{-1}$$

$$k_{cat} = \frac{d}{(a + c) \times 2.5 \text{ sec}} = 0.5 \text{ sec}^{-1}$$

at steady state $V_{max}/E_T = 0.5 \text{ sec}^{-1}$ at 4°C.

enzyme was studied in the direction of condensation, primary isotope effects with DHAP-D and DHAP-T were $v_H/v_D = 6.8$ and $v_H/v_T = 18.5$. Therefore, the conversion of the DHAP-enzyme Schiff's base to enamine seems to be well isolated kinetically. When the secondary isotope effect was studied with 3R-DHAP-3T and glyceraldehyde-3-P with the carboxypeptidase-treated enzyme, $v_H/v_T = 1.14$ was found. If one can use the pyruvate/PEP equilibrium in tritiated water as a reference (Table 5), a value of 1.15 is expected.[7] This result indicates that —C—H cleavage and formation of enamine are greatly advanced in the transition state. However, the pyruvate kinase equilibrium is probably not the best model for

[7] An experimental value of 1.28 was reported (Biellmann, O'Connell, and Rose, 1969) with yeast aldolase, suggesting the need for additional study.

$$\begin{array}{c} C \\ | \\ H_2C{-}OH \end{array} \rightarrow \begin{array}{c} C \\ \| \\ H{-}C{-}OH \end{array}$$

because the secondary isotope effect with aldolase and fructose-1,6-P_2-3T was much greater, 1.33. Correcting this value to represent replacement of one carbon with H requires its multiplication by 0.91 as seen from the fractionation factors of pyruvate and oxaloacetate (Table 5). Thus, a minimal value for the isotope effect with dihydroxyacetone-P-3T should be $1.33 \times 0.91 = 1.21$. In this view the transition state is less than 0.14/0.21, or less than 67% of the way toward sp^2 character. We believe this conclusion can be transferred to the reaction with the native enzyme, where only a small isotope effect ($v_H/v_T = 1.04$) was seen (Biellmann, O'Connell, and Rose, 1969), probably because here the proton abstraction step is not isolated. This assumes that, although the rate of proton abstraction has been changed ∽ 500-fold by carboxypeptidase treatment of the enzyme, the mechanism has remained the same.

We have not spoken here of the use of hydrogen isotope effects to define a reaction mechanism chemically. An unsolved problem is the bacterial enzyme L-ribulose-5-P 4-epimerase. The interconversion of ribulose-5-P to D-xylulose-5-P occurs without exchange of H or O from water (McDonough and Wood, 1961), similar to the UDP-glucose-4′-epimerase, except that the enzyme lacks a pyridine nucleotide. No isotope effect was seen when D-xylulose-4T and D-xylulose-5-P-1^{14}C were compared by Salo et al. (1972) with the *Aerobacter aerogenes* enzyme, but variation in the data would have obscured a secondary effect. A reversible aldol cleavage at C_3—C_4 has been proposed (Deupree and Wood, 1972) and could be tested by carefully looking for small secondary isotope effects like those found with FDP aldolase.

Activation of a —C—H bond neighboring a carbonyl carbon is generally known to proceed by charge delocalization toward an enol transition state so that secondary isotope effect studies are usually important at the more detailed level of inquiry. However, the possibility of bimolecular electrophilic substitution as proposed for C-metal substitutions has recently been considered for biotin-mediated substitution of —C—H by CO_2 based on stereochemical (Retey and Lynen, 1965) and kinetic (Mildvan, Scrutton, and Utter, 1966) considerations. An experiment with transcarboxylase (Figure 9) showing tritium transfer between reactants has been interpreted in support of a mechanism that uses biotin as the common base for activation of both pyruvate and

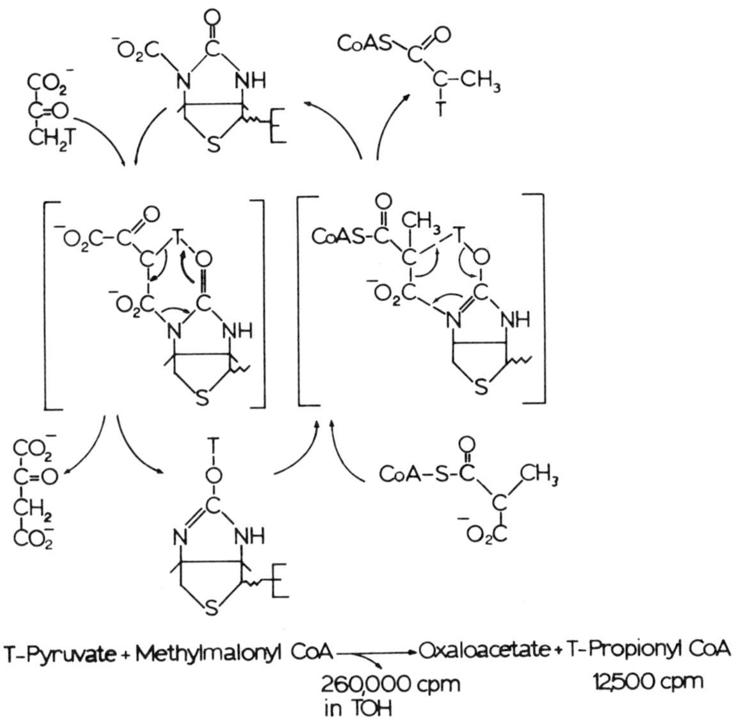

T-Pyruvate + Methylmalonyl CoA → Oxaloacetate + T-Propionyl CoA
260,000 cpm in TOH · 12,500 cpm

Figure 9. Concerted electrophilic substitution mechanism for oxaloacetate-propionyl CoA transcarboxylase.

methylmalonyl CoA (Rose, O'Connell, and Solomon, 1976). Experiments are in progress to determine the secondary tritium isotope effect of this and other biotin-dependent carboxylations and for suitable model reactions.

SUMMARY

The results presented in this chapter, parts of larger studies with pyruvate kinase, aconitase, and aldolase, demonstrate the following. First, primary isotope effects on proton exchange partial reactions can be determined because there are three H isotopes, allowing the exchange rates of two isotopes to be determined in a medium of the third. In the pyruvate kinase reaction the large primary isotope effect seen when P_i is used to activate the enolization of pyruvate is lost when ATP is used.

Second, this change in H-exchange isotope effect for enolization of pyruvate is not due to a change in mechanism. The same three steps are necessary for the isotope exchange. The phosphoryl transfer from ATP to enolpyruvate is independent of enolization, although the presence of ATP changes the relative rates of the steps responsible for proton exchange. Third, proton abstraction in the aconitase reaction fails to show a primary isotope effect because the subsequent regeneration of enzyme by release of that proton is rate limiting for the overall reaction. The enolase reaction appears to show the most promise of having a —C—H bond-breaking step rate limiting for a β-elimination reaction type. Fourth, secondary isotope effects may be due to equilibria or kinetic factors. The former, independent of mechanism, must be excluded before the isotope effect can be used to illuminate the mechanism. Such a dichotomy is examined in the cleavage of 3T and 4T fructose-1,6-P_2 by aldolase. It is concluded that the aldol cleavage reaction is rate limiting for production of glyceraldehyde-3-P from fructose-1,6-P_2 and that bond breaking has progressed greatly in the transition state. Fifth, based on large primary isotope effects, two cases of kinetically well isolated proton abstraction steps from carbon acids are described: pyruvate enolization with pyruvate kinase activated by P_i and enamine formation of the dihydroxyacetone-P as a Schiff's base on carboxypeptidase-treated aldolase of skeletal muscle. These enzymes therefore will lend themselves to further study of the mechanism of the proton abstraction step in enol- and enamine-related reactions.

LITERATURE CITED

Berman, K., Dinovo, E. C., and Boyer, P. D. 1971. Relationships of pH to exchange rates and deuterium isotope effects in the fumarase reaction. Bioorg. Chem. 1:234–242.

Biellmann, J. F., O'Connell, E. L., and Rose, I. A. 1969. Secondary isotope effects in reactions catalyzed by yeast and muscle aldolase. J. Am. Chem. Soc. 91:6484–6488.

Cleland, W. W. 1975. Partition analysis and the concept of net rate constants as tools in enzyme kinetics. Biochemistry 14:3220–3224.

Deupree, J. D., and Wood, W. A. 1972. L-Ribulose 5-phosphate 4-epimerase from *Aerobacter aerogenes.* Evidence for a role of divalent metal ions in the epimerization reaction. J. Biol. Chem. 247:3093–3097.

Dinovo, E. C., and Boyer, P. D. 1971. Isotopic probes of the enolase reaction mechanism. Initial and equilibrium isotope exchange rates: primary and secondary isotope effects. J. Biol. Chem. 246:4586–4593.

do Amaral, L., Bastos, M. P., Bull, H. D., and Cordes, E. H. 1973. Secondary deuterium isotope effects for addition of nitrogen nucleophiles to substituted benzaldehydes, Table III. J. Am. Chem. Soc. 95:7369–7374.

Hansen, J. N., Dinovo, E. C., and Boyer, P. D. 1969. Initial and equilibrium ^{18}O, ^{14}C, ^{3}H exchange rates as probes of the fumarase reaction mechanism. J. Biol. Chem. 244:6270–6279.

Katz, J. J., and Crespi, H. L. 1970. Isotope effects in biological systems. In: C. J. Collins and N. S. Bowman (eds.), Isotope Effects in Chemical Reactions, pp. 286–353. Van Nostrand-Reinhold, New York.

McDonough, M. W., and Wood, W. A. 1961. The mechanism of pentose phosphate isomerization and epimerization studied with T_2O and $H_2{}^{18}O$. J. Biol. Chem. 236:1220–1224.

Mildvan, A. S., Scrutton, M. C., and Utter, M. F. 1966. Pyruvate carboxylase VII. A possible role for tightly bound manganese. J. Biol. Chem. 241:3488–3498.

Northrop, D. B. 1975. Steady-state analysis of kinetic isotope effects in enzymic reactions. Biochemistry 14:2644–2651.

Plowman, E. M., and Krall, A. R. 1965. A kinetic study of nucleotide interactions with pyruvate kinase. Biochemistry 4:2809–2814.

Rétey, J., and Lynen, F. 1965. Biochemical function of biotin. IX. Steric process of the carboxylation of propionyl CoA. Biochem. Z. 342:256–271.

Robinson, J. L., and Rose, I. A. 1972. The proton transfer reactions of muscle pyruvate kinase. J. Biol. Chem. 247:1096–1105.

Rose, I. A. 1960. Studies on the enolization of pyruvate by pyruvate kinase. J. Biol. Chem. 235:1170–1177.

Rose, I. A. 1970. Enzymology of proton abstraction and transfer reactions. In: P. D. Boyer (ed.), The Enzymes. 3rd Ed. Vol. 2, pp. 281–320.

Rose, I. A., and O'Connell, E. L. 1967. Mechanism of aconitase reaction. I. The hydrogen transfer reaction. J. Biol. Chem. 242:1870–1879.

Rose, I. A., O'Connell, E. L., Litwin, S., and Bar-Tana, J. 1974. Determination of the rate of hexokinase-glucose dissociation by the isotope-trapping method, ftnt. 2, p. 5166. J. Biol. Chem. 249:5163–5168.

Rose, I. A., O'Connell, E. L., and Mehler, A. H. 1965. Mechanism of the aldolase reaction. J. Biol. Chem. 240:1758–1765.

Rose, I. A., O'Connell, E. L., and Solomon, F. 1976. Intermolecular tritium transfer in the transcarboxylase reaction. J. Biol. Chem. 251:902–904.

Salo, W. L., Fossitt, D. D., Berill, R. D., Kirkwood, S., and Wood, W. A. 1972. L-Ribulose 5-phosphate 4-epimerase from *Aerobacter aerogenes.* Kinetic isotope effect with tritiated substrate. J. Biol. Chem. 247:3098–3100.

Schmidt, D. E., Jr., Nigh, W. G., Tanzer, C., and Richards, J. H. 1969. Secondary isotope effects in the dehydration of malic acid by fumarate hydratase. J. Am. Chem. Soc. 91:5849–5854.

Shen, T. Y. S., and Westhead, E. W. 1973. Divalent cation and pH-dependent primary isotope effects in the enolase reaction. Biochemistry 12:3333–3337.

Silverman, D. N., and Tu, C. K. 1975. Buffer dependence of carbonic anhydrase catalyzed oxygen-18 exchange at equilibrium. J. Am. Chem. Soc. 97:2263–2269.

Talaley, P., and Boyer, J. 1965. Preparation of crystalline Δ^5-3-ketosteroid isomerase from *Pseudomonas testosteroni.* Biochem. Biophys. Acta 105:389–392.

Thomson, J. F., Nance, S. L., Bush, K. J., and Szczepanik, P. A. 1966. Isotope and solvent effects of deuterium on aconitase. Arch. Biochem. Biophys. 117:65–74.

Villafranca, J. J. 1974. The mechanism of aconitase action. Evidence for an enzyme isomerization by studies of inhibition by the tricarboxylic acids. J. Biol. Chem. 249:6149–6155.

Studies of Enzyme Reaction Mechanisms by Means of Heavy-atom Isotope Effects

M. H. O'Leary

Kinetic isotope effects for isotopes of hydrogen have been used as mechanistic tools in enzymology for a number of years (Kirsch, this volume; Klinman, this volume; Richards, 1970; Simon and Palm, 1966) but interest in kinetic isotope effects for other elements, most notably carbon, nitrogen, oxygen, and sulfur, has developed more slowly. Heavy-atom isotope effects recently have come under closer scrutiny of enzymologists and it is appropriate at this point to review the methodology of heavy-atom isotope effect studies and the interpretation of these effects.

One of the most severe limitations on progress in this field has been the magnitudes of the effects that are obtained. The rate difference resulting from isotopic substitution for carbon, nitrogen, oxygen, or sulfur is never more than a few percent (Fry, 1970; MacColl, 1974). Standard kinetic techniques ordinarily are not adequate for making such measurements, and a whole new set of techniques has been developed. Interpretation, too, has been a serious problem, but recent advances in both theory and experiment promise to improve this situation.

MEASUREMENT OF HEAVY-ATOM ISOTOPE EFFECTS

Because of the small sizes of heavy-atom isotope effects, direct measurement of rate constants for normal and isotopically substituted substrates usually does not provide data of the accuracy required for heavy-atom isotope effect studies. However, the advent of high-precision computer-linked spectrophotometers has led to limited use of this direct technique (Hegazi, Borchardt, and Schowen, unpublished results).

The standard method for measurement of heavy-atom isotope effects is the competitive method. A mixture of normal and isotopically labeled substrates is reacted and the isotopic composition of the substrate or product is measured as a function of the extent of reaction. If the labeling isotope is radioactive, the specific activity of the starting material or product is measured. If the labeling isotope is stable, isotope-ratio mass spectrometry is used for the measurement of isotopic composition.

Calculation of isotope effects from such isotope ratio data has been described in detail elsewhere (Bigeleisen and Wolfsberg, 1958). In essence, for reaction of a mixture of labeled and unlabeled substrates (S* and S, respectively) to form the corresponding products (P* and P) according to equation 1:

$$S + S^* \xrightarrow{\text{enzyme}} P + P^* \tag{1}$$

if the rate constants for reaction of the normal and labeled substrates are k and k^*, respectively, the kinetic isotope effect can be obtained by measurement of the product composition (P*/P) for a sample of substrate that has been converted completely into product with that of a sample obtained after only a few percent reaction by use of equation 2:

$$\frac{k}{k^*} = \frac{(P^*/P)\ 100\%\ \text{reaction}}{(P^*/P)\ 5\%\ \text{reaction}} \tag{2}$$

A small correction must be made for the extrapolation to 0% reaction (Bigeleisen and Wolfsberg, 1958). Alternatively, it is possible to use the change in composition of the substrate during the reaction to calculate the isotope effect (Bigeleisen and Wolfsberg, 1958).

Isotopic composition measurements of the required precision are difficult to make. For radioactive isotopes, specific activities must be measured to approximately ±0.1%; therefore, extreme care in sample preparation and measurement is necessary. For stable isotopes, the natural abundance of carbon-13, nitrogen-15, or oxygen-18 can be used, but the precision in the isotope ratio (P*/P) must then be about ±0.000002. Such precision is attainable only on mass spectrometers specifically designed for the measurement of isotope ratios. The limited accessibility of such machines has been one of the principal impediments to progress in studies of heavy-atom isotope effects. Because of the necessity for rapid sample changing, isotope ratio mass spectrometers are used only for studies with highly volatile samples such as CO, CO_2, and N_2. Thus, many substrates and products of interest in enzymatic systems require chemical conversion before isotopic analysis. Obviously, such conversion must be rigorously free of isotope effects.

The use of isotopically enriched samples is sometimes advantageous. If enriched samples are used, the precision required in the isotope ratio measurements is not as great as for natural abundance samples. However, isotopic enrichment creates a whole new set of problems. The desired materials must be accessible by synthesis specifically designed to incorporate the isotope label. Sample handling and measuring procedures are much more subject to errors because of contamination when enriched samples are used than when natural abundance samples are used.

INTERPRETATION OF HEAVY-ATOM ISOTOPE EFFECTS

At least two complementary approaches to the interpretation of heavy-atom isotope effects in enzymatic reactions are possible: The interpretation can be based on the comparison of observed with calculated isotope effects, or it can be based on the comparison of isotope effects observed in enzymatic reactions with those for corresponding model reactions. The latter technique has generally proven more successful in studies of enzymatic reactions, but it is wise to be mindful of calculated isotope effects as well.

Enzymatic reactions differ from the corresponding model reactions in the numbers and rates of various reaction steps. Because of the existence of substrate binding, product dissociation, and perhaps conformation change steps, enzymatic reactions usually have more steps than their nonenzymatic counterparts. The magnitudes of isotope effects in enzymatic reactions reflect not only transition-state structures, but also relative rate constants for various reaction steps. The balance between these two factors for various simple enzyme mechanisms will be explored in succeeding paragraphs.

General Theory

A qualitative approach to isotope effects can provide useful guidelines for predicting approximate magnitudes of heavy-atom isotope effects. Within the framework of the transition-state theory, isotope effects can be understood in terms of the difference in structure between reactants and transition state. For most heavy-atom isotope effects, the more general expression for isotope effects (Buddenbaum and Shiner, this volume; MacColl, 1974; Van Hook, 1970) can be approximated as a product of two factors (equation 3).

$$\frac{k}{k^*} = \left[\begin{array}{c}\text{zero-point}\\ \text{energy}\\ \text{factor}\end{array}\right]\left[\begin{array}{c}\text{imaginary}\\ \text{frequency}\\ \text{factor}\end{array}\right] \tag{3}$$

The zero-point energy factor is related to the difference in vibration frequencies, and thus to the difference in bonding, for the two isotopic species of reactants in ground state and transition state. The imaginary frequency factor is the ratio of (imaginary) vibration frequencies for passage along the reaction coordinate. This latter factor is always equal to or greater than unity. The magnitude of the zero-point energy factor depends on the degree of change of bonding to the isotopic atom in going from ground state to transition state. If no change in bonding occurs, this factor is unity. If bonding to the isotopic atom decreases, this factor is greater than unity, and thus the expected isotope effect is greater than unity (that is, the lighter isotope reacts more rapidly than the heavier isotope). Conversely, if the isotopic atom is more strongly bonded at the transition state than in the ground state, the zero-point energy factor will be smaller than unity. In that case, the predicted isotope effect may be either greater than unity, exactly unity, or less than unity, depending on the magnitudes of the two factors. For this reason, heavy-atom isotope effects in bond-making reactions may be more difficult to interpret than those in bond-breaking reactions. It is possible, for example, that in a bond-making reaction the two factors might exactly cancel, giving rise to no isotope effect (i.e., $k/k^* = 1.0$), even though the bond-forming step is the rate-determining step.

Enzymatic Reactions

Because heavy-atom isotope effects are usually measured by competitive techniques, the effects observed are effects on $V_{\max}/K_m$. For this reason, heavy-atom isotope effects do not always give information about the "rate-determining step" in the overall reaction. Instead, these isotope effects provide information about steps up to and including the first irreversible step in the reaction.

Enzymatic reactions do not usually proceed in a single step, and the existence of a number of steps must be taken into account in interpreting the magnitudes of heavy-atom isotope effects. It is often possible to assume that only a single reaction step will show a significant heavy-atom isotope effect. Furthermore, it is often possible to arrange reaction conditions so that this step is irreversible under the conditions of the experiment. Under those circumstances, for the simple mechanism of equation 4,

$$\mathrm{E + S} \underset{k_2}{\overset{k_1}{\rightleftharpoons}} \mathrm{ES} \xrightarrow{k_3} \mathrm{E + P} \tag{4}$$

assuming that there are no isotope effects on k_1 and k_2, the relationship between the individual rate constants, the actual isotope effect on k_3, and the observed isotope effect is given in equations 5 and 6.

$$k/k^* \text{ (observed)} = \frac{k_3/k_3^* + \mathrm{R}}{1 + \mathrm{R}} \tag{5}$$

$$\mathrm{R} = k_3/k_2 \tag{6}$$

For the more general mechanism of equation 7,

$$\mathrm{E + S} \underset{k_2}{\overset{k_1}{\rightleftharpoons}} \cdots \underset{k_{i+1}}{\overset{k_i}{\rightleftharpoons}} \cdots \ \mathrm{E + P} \tag{7}$$

in which the *ith* step is subject to an isotope effect and this step is reversible, the observed isotope effect is given in equations 8 to 10. K and K^* are overall equilibrium constants (products/reactants) for reaction of unlabeled and labeled substrates.

$$k/k^* \text{ (observed)} = \frac{k_i/k_i^* + \mathrm{R} + \mathrm{R}'K/K^*}{1 + \mathrm{R} + \mathrm{R}'} \tag{8}$$

$$\mathrm{R} = \frac{k_i}{k_{i-1}}\left\{1 + \frac{k_{i-2}}{k_{i-3}}\left[1 + \frac{k_{i-4}}{k_{i-5}}\left(\cdots\right)\right]\right\} \tag{9}$$

$$\mathrm{R}' = \frac{k_{i+1}}{k_{i+2}}\left\{1 + \frac{k_{i+3}}{k_{i+4}}\left[1 + \frac{k_{i+5}}{k_{i+6}}\left(\cdots\right)\right]\right\} \tag{10}$$

The continued product in equation 9 is taken from the *ith* step all the way back to k_2. The continued product in equation 10 is taken from step $i + 1$ forward to the first irreversible step, or for a reversible reaction, to the step in which the isotopic product is released.

Factors R and R′ are often called partitioning factors. The former factor indicates the fate of the intermediate immediately preceding the step in which the change in bonding to the isotopic atom occurs. When this factor is much less than unity, the step is almost entirely rate limiting and the observed isotope effect very nearly equals k_i/k_i^*. When this factor is much greater than unity, steps before that step are rate limiting and the observed isotope effect is much smaller than k_i/k_i^*. A similar situation obtains with regard to factor R′.

DECARBOXYLATIONS

Carbon isotope effects on decarboxylations are among the most studied heavy-atom isotope effects, both in nonenzymatic reactions and in enzymatic reactions (O'Leary, 1977). Two features of these reactions make them especially attractive for study. In the first place, the product of the decarboxylation, carbon dioxide, can easily be isolated and analyzed for its carbon isotopic composition. In the second place, the principal change at the isotopic carbon atom in going from ground state to transition state is the stretching of a carbon-carbon single bond (equation 11).

$$\mathrm{R{-}C(\overset{\cdot\cdot}{=}O)(\overset{\cdot\cdot}{-}O)^{-}} \rightleftharpoons \left[\mathrm{R{\cdots}C(\overset{\cdot\cdot}{=}O)(\overset{\cdot\cdot}{=}O)}\right]^{\ddagger} \rightarrow \mathrm{R} + \mathrm{CO_2} \tag{11}$$

As a result, the transition states for a variety of decarboxylations appear to be rather similar and give carbon isotope effects k^{12}/k^{13} in the range 1.03–1.06 near room temperature when decarboxylation is entirely rate determining. This experimental result has been obtained for a large number of nonenzymatic decarboxylations and has been shown to be theoretically reasonable by means of a variety of calculations (Huang et al., 1968; Stern and Vogel, 1971; Vogel and Stern, 1971). We may assume that a similar isotope effect would be observed in an enzymatic decarboxylation if the decarboxylation step were entirely rate determining. Differences between predicted and observed isotope effects can then be interpreted in terms of partitioning of various intermediates (equations 7 to 10).

Acetoacetate Decarboxylase

The decarboxylation of acetoacetic acid is catalyzed by primary amines and by an enzyme from *Cl. acetobutylicum* (Fridovich, 1972). The mechanism of the amine-catalyzed reaction and the enzyme-catalyzed reaction appear to be remarkably similar (Figure 1); both involve an intermediate Schiff base. Carbon isotope effects for both the enzyme-catalyzed reaction and the aminoacetonitrile-catalyzed reaction are shown in Table 1.

The carbon isotope effect on the primary amine-catalyzed reaction is pH dependent because the rate-determining step in the reaction changes with pH (Guthrie and Jordan, 1972). At low pH, nucleophilic attack on the carbonyl carbon atom is rate determining and a small carbon isotope effect is observed. At high pH, the decarboxylation step is rate determining and the carbon isotope effect increases. Presumably, if the measurements in Table 1 could be extended to even higher pH, a larger carbon isotope effect could be obtained.

The carbon isotope effect on the enzyme-catalyzed decarboxylation of acetoacetic acid is significantly different from unity but significantly smaller than that which would be expected if decarboxylation were entirely rate determining (Table 1). Furthermore, the isotope effect does not vary with pH. This result can be understood in terms of the mechanism of Figure 1 if return of the Schiff base intermediate to starting materials occurs at approximately the same rate as decarboxylation of the intermediate. That is, the partitioning factor R (equation 6) is approximately

$$R\text{-}NH_2 + CH_3\overset{\overset{\displaystyle O}{\|}}{C}CH_2CO_2^- \underset{k_2}{\overset{k_1}{\rightleftharpoons}} CH_3\overset{\overset{\displaystyle R\text{-}NH^+}{\|}}{C}CH_2CO_2^-$$

$$\downarrow k_3$$

$$R\text{-}NH_2 + CH_3\overset{\overset{\displaystyle O}{\|}}{C}CH_3 + CO_2 \longleftarrow CH_3\overset{\overset{\displaystyle R\text{-}NH}{|}}{C}{=}CH_2 + CO_2$$

Figure 1. Mechanism of the decarboxylation of acetoacetic acid. For the model reaction, the catalyst is a primary amine ($N{\equiv}CCH_2NH_2$ in the case of the data in Table 1). For the enzymatic reaction, the catalyst is acetoacetate decarboxylase and the amino group is the ε-amino group of a lysine residue of the enzyme.

Table 1. Carbon isotope effects on the decarboxylation of acetoacetic acid at 30°

Catalyst	pH	k^{12}/k^{13}
Aminoacetonitrile	3.6	1.031
Aminoacetonitrile	4.1	1.032
Aminoacetonitrile	5.0	1.036
Acetoacetate decarboxylase	5.3	1.019
Acetoacetate decarboxylase	6.0	1.017
Acetoacetate decarboxylase	7.2	1.019

Data from O'Leary and Baughn, 1972.

unity. This suggestion is confirmed by results of Hamilton (1959), who studied exchange between the ketone carbonyl oxygen of acetoacetate and water catalyzed by acetoacetate decarboxylase. The rate of oxygen exchange indicates that the partitioning factor is approximately unity.

Thus, the mechanism of action of acetoacetate decarboxylase involves not a single rate-determining step, but at least two steps that have very similar activation energies. This lack of a single well defined rate-determining step appears to be a common phenomenon in enzyme mechanisms.

Pyruvate Decarboxylase

The thiamine pyrophosphate-dependent pyruvate decarboxylase from yeast functions by the mechanism shown in Figure 2 (Schellenberger, 1967; Ullrich et al., 1970). The carboxyl carbon isotope effect on the decarboxylation is $k^{12}/k^{13} = 1.008$ at pH 6.8, 25° (O'Leary, 1976). The small magnitude of this isotope effect indicates that the partitioning factor (equation 6) is large. That is, decarboxylation is not rate determining. Instead, the slow step is probably attack of the coenzyme on the carbonyl carbon of the substrate (k_1). Note, however, that k_1 does not appear in the equation for the isotope effect (equation 5). The important issue is the ratio k_3/k_2. The small isotope effect obtained in this case indicates that this ratio is large; that is, the covalent thiamine pyrophosphate—pyruvate adduct almost always decarboxylates and infrequently returns to enzyme + substrate. Thus, k_2 is a slow step (i.e., has a high-energy transition state), and k_1 must also be slow. In the present case, k_1 probably represents a unimolecular reaction of the protein—thiamine pyrophosphate—pyruvate complex, rather than a bimolecular reaction of pyruvate with the protein—thiamine pyrophosphate complex.

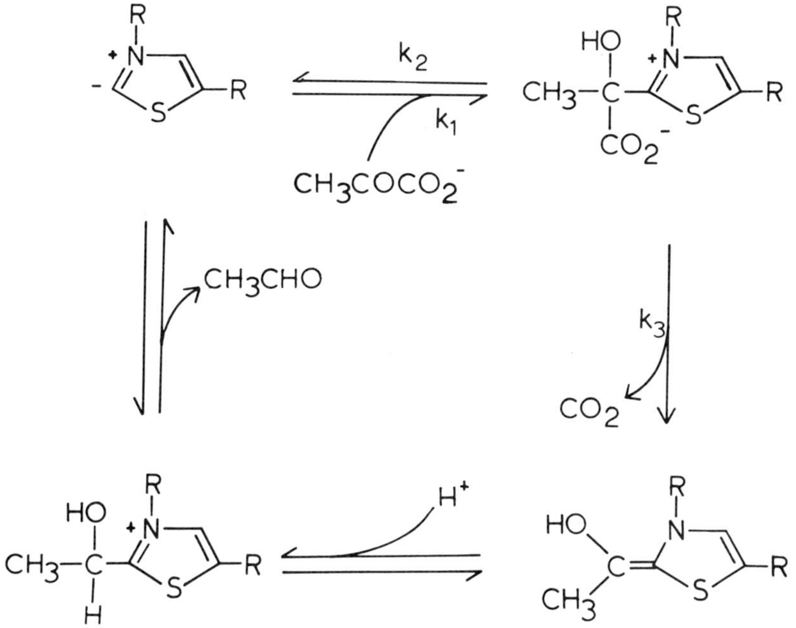

Figure 2. Mechanism of action of the thiamine pyrophosphate-dependent pyruvate decarboxylase from yeast.

Glutamate Decarboxylase

This enzyme catalyzes the decarboxylation of glutamic acid by means of a bound pyridoxal 5′-phosphate cofactor, as shown in Figure 3 (Boeker and Snell, 1972). In the absence of substrate, the coenzyme is bound to the enzyme by means of a Schiff base linkage between the aldehyde carbon of the coenzyme and an ε-amino group of a lysine residue of the enzyme. When substrate combines with the enzyme, Schiff base interchange occurs, forming an enzyme-bound Schiff base between glutamate and pyridoxal 5′-phosphate. Carbon isotope effects on the enzymatic decarboxylation have been obtained at a variety of pH values (Table 2). The carbon isotope effect is significantly different from unity under all conditions studied, but it is always significantly smaller than that expected for rate-determining decarboxylation. Again, a partitioning factor near unity would explain the magnitude of the effect nicely.

The pH dependence of this isotope effect appears to arise in an interesting way: Decarboxylation of the glutamate—pyridoxal 5′-phosphate Schiff base can occur only if the pyridine nitrogen of the

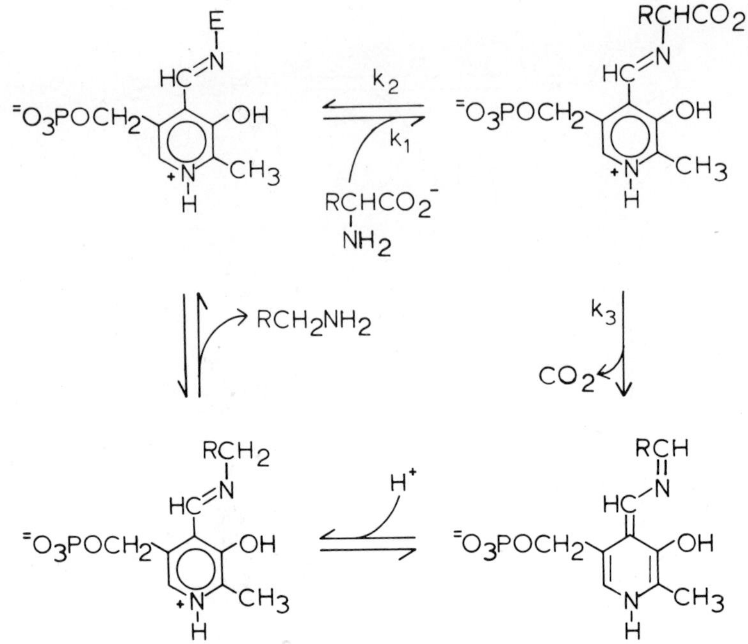

E
HC=N
$^{=}O_3POCH_2$
OH
CH_3
+N
H
k_2
k_1
$RCHCO_2^-$
NH_2
$RCHCO_2^-$
HC=N
$^{=}O_3POCH_2$
OH
CH_3
+N
H
RCH_2NH_2
k_3
CO_2
RCH_2
HC=N
$^{=}O_3POCH_2$
OH
CH_3
+N
H
H^+
RCH
N
HC
$^{=}O_3POCH_2$
OH
CH_3
N
H

Figure 3. Mechanism of the enzymatic decarboxylation of glutamic acid by the pyridoxal 5′-phosphate-dependent glutamate decarboxylase.

Schiff base is protonated. The positively charged nitrogen serves as an electron sink and promotes the decarboxylation. The Schiff base interchange, on the other hand, can presumably proceed (albeit at somewhat different rates) whether or not this nitrogen is protonated.

Table 2. Carbon isotope effects on the enzymatic decarboxylation of glutamic acid at 37°

pH	k^{12}/k^{13}
3.6	1.016
4.0	1.014
4.4	1.015
4.7	1.017
5.2	1.019
5.5	1.022

Data from O'Leary, Richards, and Hendrickson, 1970.

Thus, the partitioning ratio k_3/k_2 varies with pH because of the different pH dependences of the two rate constants. For this reason, the observed isotope effect increases with increasing pH (R in equation 5 decreases with increasing pH, whereas k_3^{12}/k_3^{13} is probably independent of pH; thus, the observed effect increases with increasing pH). It is interesting to note that this interpretation can only hold if R is not too different from unity. This is consistent with our assumption that k_3^{12}/k_3^{13} is in the range 1.03–1.06.

COMPLICATIONS

In favorable cases heavy-atom isotope effects on enzymatic reactions can be measured and interpreted with relative ease, provided proper facilities are available. However, a number of factors should be noted which may make such data more difficult to obtain or to interpret.

Experimental Problems

The necessity for a high-precision isotope ratio mass spectrometer has already been mentioned. If the substrate or product being measured is not highly volatile, chemical conversion of the sample to a form suitable for mass spectrometry must be undertaken. This conversion must be shown to be free of isotope fractionations, so that variations in isotopic composition during sample handling do not obscure the real effects.

Relatively large amounts of enzyme are often necessary for measurement of isotope effects. Typically, each isotope effect measurement requires about 100 ml of a solution roughly 0.01 M in substrate and sufficient enzyme to convert that material completely to products in a few hours. After a procedure for measuring an isotope effect has been developed, a half dozen or more measurements under each set of reaction conditions are desirable to demonstrate the reproducibility of the procedure. Development of such a procedure may require investments of time and reagents that are at least as great as those involved in the actual measurements. Thus, the cost of enzymes and of substrates may become prohibitive.

The development and debugging of an isotope effect procedure may be sufficiently complicated to require a year or more of work. The entire isolation and analysis procedure must be demonstrated to be free of isotope effects. All reagents and solutions must be shown to be free of contaminants that could affect the isotopic composition measurements. This is a particular problem with decarboxylations because of the ubiquity of carbon dioxide in the environment. The enzyme must be sufficiently stable so that it does not decompose before the 100% conversion sample is finished. An error of only a few percent in that conversion would cause a large error in the observed isotope effect. The substrate must be sufficiently stable so that nonenzymatic decomposition does not compete with enzymatic reaction.

Problems of Interpretation

The interpretations of the carbon isotope effects on decarboxylations given above are particularly clear and are not necessarily typical of isotope effects that will be obtained in other systems. A large number of model decarboxylations are known, and we have a reasonably clear understanding of the magnitude of the carbon isotope effect expected in the carbon-carbon bond-breaking step. In addition, carbon isotope effects on steps other than the carbon-carbon bond-breaking step are expected to be negligible and the decarboxylation step is usually irreversible. However, these advantages are not shared by many systems that would be desirable to study.

Isotope effects in appropriate model reactions are available for a variety of enzyme-catalyzed reactions (Fry, 1970), but many reactions of interest in enzymology have not been so investigated. In a number of cases, the assumption made above that only a single step in the reaction mechanism will be subject to a significant isotope effect is probably not correct. When that is so, treatment of isotope effects becomes much more complicated.

In studies of decarboxylations we have largely depended upon our ability to predict the isotope effect on the reaction step involving bond changes to the isotopic atom. Decarboxylations are particularly advantageous in that regard because of the theoretical and nonenzymatic isotope effect data available. In cases where such reliable prediction is not possible, it is often useful to measure the isotope effect for the same enzyme under a variety of conditions and to interpret the change in isotope effect with various parameters, rather than interpreting the magnitude of the isotope effect itself. Among the parameters that may be used in this way are temperature, substrate structure, coenzyme structure, metal ion, and pH.

To date, the most successful use of this method has involved variations of pH. For example, the enzymatic decarboxylation of glutamic acid discussed previously shows a pH-dependent isotope effect that gives information regarding the role of the coenzyme. Similar pH variations for chymotrypsin-catalyzed reactions are described in the next section.

ESTERASES

Chymotrypsin

This enzyme catalyzes the hydrolysis of a variety of esters and amides. Its preferred substrates are esters and amides of amino acids with hydrophobic side chains (Hess, 1971). Nitrogen isotope effects on the chymotrypsin-catalyzed hydrolysis of *N*-acetyl-L-tryptophanamide are shown in Table 3. The nitrogen isotope effect is significantly different from unity, indicating that carbon-nitrogen bond breaking plays some role in determining the overall rate. However, it is not certain what magnitude of nitrogen isotope effect should be expected in such a reaction. The only nitrogen isotope effects that have been measured in nonenzymatic amide hydrolysis are those measured by Harbison (O'Leary and Kluetz, 1972), who found isotope effects in the range k^{14}/k^{15} = 1.00–1.01 for several amide hydrolyses. Nitrogen isotope effects in other organic reactions are generally in the range 1.00–1.025 (Fry, 1970).

The reaction of an amide with the serine hydroxyl group at the active site of chymotrypsin probably occurs by way of a tetrahedral intermediate (Figure 4), which decomposes to form an acyl enzyme.

Table 3. Nitrogen and oxygen isotope effects on the chymotrypsin-catalyzed hydrolysis of esters and amides at 25°

Substrate	Isotope	pH	k/k^*	Reference
N-acetyl-L-tryptophanamide	Amide nitrogen	6.7	1.006	a
N-acetyl-L-tryptophanamide	Amide nitrogen	8.0	1.010	a
N-acetyl-L-tryptophanamide	Amide nitrogen	9.4	1.006	a
N-acetyl-L-tryptophan methyl ester	Ether oxygen	8.0	1.007	b
N-acetyl-L-tryptophan ethyl ester	Ether oxygen	6.8	1.018	c
N-carbomethoxy-L-tryptophan ethyl ester	Ether oxygen	6.8	1.012	c

a) O'Leary and Kluetz, 1972; b) O'Leary and Marlier, unpublished results; c) Sawyer and Kirsch, 1975.

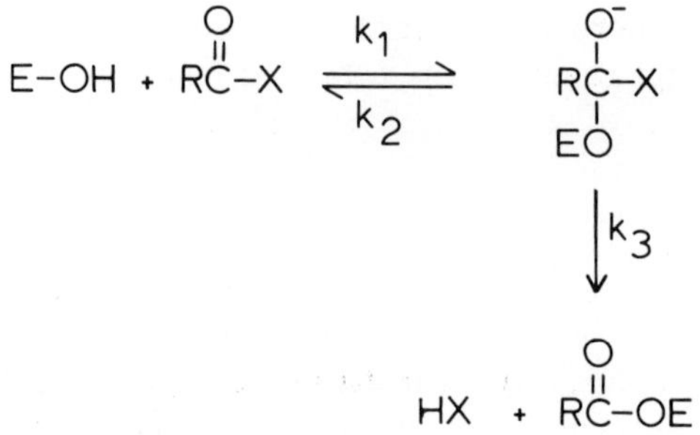

$$E\text{-}OH + RC(=O)\text{-}X \underset{k_2}{\overset{k_1}{\rightleftharpoons}} RC(O^-)(OE)\text{-}X \xrightarrow{k_3} HX + RC(=O)\text{-}OE$$

Figure 4. Mechanism of acylation of chymotrypsin by esters (X = OR) and amides (X = NH_2). The enzyme hydroxyl group shown is that of serine-195.

Steps subsequent to formation of the acyl enzyme will not affect the observed isotope effect.

The second step in Figure 4 is the carbon-nitrogen bond-breaking step, and there clearly should be a substantial nitrogen isotope effect on that step. However, there also might be a substantial nitrogen isotope effect on the first step because resonance in the starting amide makes the carbon-nitrogen bond appreciably stronger than an ordinary carbon-nitrogen single bond. This extra strength is destroyed on going to the tetrahedral intermediate. Thus, we expect that there will be both a kinetic isotope effect and an equilibrium isotope effect on the formation of the tetrahedral intermediate. The magnitudes of these isotope effects are unknown. The kinetic effect might be responsible wholly or in part for the observed nitrogen isotope effect.

A second difficulty which obscures the interpretation of these nitrogen isotope effects results from the structure of the transition state for the carbon-nitrogen bond-breaking step. The product of that step must be NH_3, not NH_2^-. Thus, a proton must be transferred to the leaving nitrogen at some point before carbon-nitrogen bond cleavage is complete. This proton presumably comes from histidine 57, and the transition state can be represented as

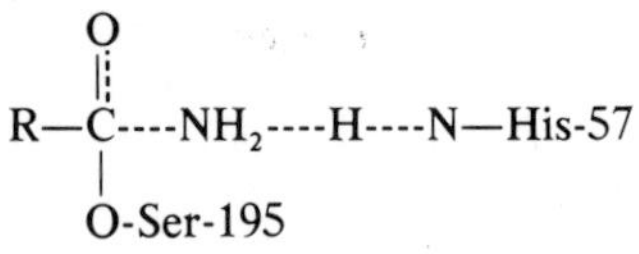

$$R\text{—}\overset{\overset{\textstyle O}{\vdots\|}}{\underset{\underset{\textstyle O\text{-}Ser\text{-}195}{|}}{C}}\text{----}NH_2\text{----}H\text{----}N\text{—}His\text{-}57$$

The magnitude of the nitrogen isotope effect will be governed by two different factors: the degree of breaking of the carbon-nitrogen bond, and the degree of formation of the nitrogen-hydrogen bond. These two effects work in opposite directions. The more broken the carbon-nitrogen bond is, the larger the isotope effect will be. The more formed the nitrogen-hydrogen bond is, the smaller the isotope effect will be.

Neither of these factors can be estimated with certainty. Nitrogen isotope effects on simple carbon-nitrogen bond-breaking reactions are often in the range $k^{14}/k^{15} = 1.010–1.025$. The nitrogen isotope effect resulting from proton transfer may be even larger than this. The equilibrium constant for equation 12 is 1.0393 ± 0.0004 at 25° (O'Leary and Young, unpublished results).

$$^{15}NH_3 + {}^{14}NH_4^+ \rightleftharpoons {}^{14}NH_3 + {}^{15}NH_4^+ \qquad (12)$$

Thus, the magnitudes of the nitrogen isotope effects in the chymotrypsin-catalyzed hydrolysis of *N*-acetyl-L-tryptophanamide fail to indicate the relative rates of formation and decomposition of the tetrahedral intermediate that presumably occurs during the hydrolysis. However, the pH dependence of the isotope effect provides a valuable clue to the interpretation of the effects. If either formation or decomposition of the tetrahedral intermediate were solely responsible for the observed isotope effect, the effect would not be expected to vary with pH. The structure of the transition state should not vary with pH (even though the rate of passage through that transition state might vary with pH), and thus the observed isotope effect should not vary with pH. On the other hand, partitioning factors of the type used in equations 5 and 8 might easily vary with pH, and this variation in partitioning might give rise to the observed variation in isotope effect with pH. Thus, it seems likely that the transition states for formation and decomposition of the tetrahedral intermediate are of very similar energy and neither step is entirely rate determining.

Oxygen isotope effects on the chymotrypsin-catalyzed hydrolysis of esters are also given in Table 3. Unlike the nitrogen isotope effects, these isotope effects appear to be pH independent (Sawyer and Kirsch, personal communication). The oxygen isotope effect on the chymotrypsin-catalyzed hydrolysis of *N*-acetyl-L-tryptophan methyl ester is similar in magnitude to the oxygen isotope effect observed in the alkaline hydrolysis of methyl benzoate in water ($k^{16}/k^{18} = 1.006$ (O'Leary and Marlier, unpublished results) under conditions where formation of the tetrahedral intermediate is rate determining (Shain and Kirsch, 1968). Thus, the oxygen isotope effects in chymotrypsin-catalyzed reactions may reflect the isotope effect on formation of the tetrahedral intermediate.

Papain

Nitrogen isotope effects for the papain-catalyzed hydrolysis of *N*-benzoyl-L-argininamide are given in Table 4. These effects are more than a factor of two larger than the corresponding isotope effects in chymotrypsin-catalyzed reactions (Table 3). An isotope effect of this magnitude must indicate that the carbon-nitrogen bond-breaking step is essentially entirely rate determining; that is, R (equation 3) is much smaller than unity. Although papain and chymotrypsin presumably operate by similar mechanisms (Glazer and Smith, 1971), the difference in isotope effects and the lack of a pH dependence in the papain effects indicate that there are subtle kinetic differences between the two enzymes.

Table 4. Nitrogen isotope effects on the papain-catalyzed hydrolysis of *N*-benzoyl-L-argininamide at 25°

pH	k^{14}/k^{15}
4.0	1.023
6.0	1.024
8.0	1.021

Data from O'Leary, Urberg, and Young, 1974.

CONCLUSIONS

For certain types of enzyme-catalyzed reactions, heavy-atom isotope effects can provide useful information about the rates of the various steps in the mechanism. It is implicit in this discussion that considerable other information about the enzyme mechanism is available. Kinetic isotope effects can only be interpreted in terms of existing mechanisms; they can seldom be used to gain insight into enzymatic reactions of unknown mechanism.

Theoretical and experimental limitations have thus far largely limited studies of heavy-atom isotope effects in enzymatic reactions to cases in which the principal change in bonding to the isotopic atom is the breaking of a bond. Reactions in which bond formation plays an important part promise to be more difficult to interpret, but progress in this direction is being made.

DISCUSSION

R. Cornelius. You estimated the lower limit of detection of a carbon isotope effect to be approximately 0.0002. Would it be possible to extend this lower limit by using enriched samples?

M. H. O'Leary. Yes and no. The precision of the isotope ratio measurements is increased when you use enriched samples, but in terms of measuring isotope effects, a new set of experimental difficulties appear. Consider, for example, a carbon isotope effect for a reaction in which carbon dioxide is being formed. What happens if you pick up a trace of carbon dioxide, say, a percent or so, from an extraneous source? If you are working with natural abundance carbon, the contaminant will make only a very small difference in the observed carbon ratio. The difference will be 1% of the difference between the carbon ratio of the contaminating carbon dioxide and the sample carbon dioxide. Thus, this 1% contaminant will not be a serious problem. However, when you are working with enriched samples, the same contamination will produce a much larger error. Again, the error will be 1% of the difference between contaminant and sample, but that is now a very large difference and thus produces quite a large error. At least in the case of carbon, the present precision is sufficiently good that we are unlikely to improve it by enrichment.

I. A. Rose. I don't understand why a heavy-atom isotope effect larger than unity can be observed in a reaction in which the net change is the formation of a new bond to the isotopic atom.

M. H. O'Leary. The answer lies in the imaginary frequency factor. Recall that the predicted isotope effect depends on the zero-point energy change on going from reactants to transition state and on the imaginary frequency factor. The zero-point energy term will be either greater or less than unity, depending on whether bonds are being broken or made at the isotopic atom. The imaginary frequency factor, however, is always greater than unity, independent of whether bonds are being made or broken. Thus, a bond-making reaction may have an isotope effect larger than unity if the predominant term in the theoretical expression is the imaginary frequency factor.

Because of this difficulty, I feel much less confident about our ability to understand heavy-atom isotope effects in bond-making reactions than those in bond-breaking reactions.

W. W. Cleland. Isn't it true, though, that if you measure an isotope effect for the reaction in one direction and Shiner can calculate the equilibrium isotope effect, then you automatically have the isotope effect in the other direction? Thus, rather than having to interpret isotope effects in bond-making reactions, you can actually study and interpret the isotope effect for bond breaking.

M. H. O'Leary. Yes, for reversible reactions, that is true. A better comparison could probably be made by measuring, rather than calculating, the equilibrium isotope effect.

ACKNOWLEDGMENT

Work in the author's laboratory was sponsored by the National Science Foundation.

LITERATURE CITED

Bigeleisen, J., and Wolfsberg, M. 1958. Theoretical and experimental aspects of isotope effects in chemical kinetics. Adv. Chem. Phys. 1:15–76.

Boeker, E. A., and Snell, E. E. 1972. Amino acid decarboxylases. The Enzymes, 3rd Ed. 6:217–253.

Fridovich, I. 1972. Acetoacetate decarboxylase. In: P. D. Boyer (ed.), The Enzymes, 3rd Ed. 6:255–270. Academic Press, New York.

Fry, A. 1970. Heavy-atom isotope effects in organic reaction mechanism studies. In: C. J. Collins and N. S. Bowman (eds.), Isotope Effects in Chemical Reactions, pp. 364–414. Van Nostrand-Reinhold, New York.

Glazer, A. N., and Smith, E. L. 1971. Papain and other plant sulfhydryl proteolytic enzymes. In: P. D. Boyer (ed.), The Enzymes, 3rd Ed. 3:502–546. Academic Press, New York.

Guthrie, J. P., and Jordan, F. 1972. Amine-catalyzed decarboxylation of acetoacetic acid; the rate constant for decarboxylation of a β-imino acid. J. Am. Chem. Soc. 94:9136–9141.

Hamilton, G. A. 1959. Unpublished doctoral dissertation, Harvard University, Cambridge, Mass.

Hess, G. P. 1971. Chymotrypsin—chemical properties and catalysis. In: P. D. Boyer (ed.), The Enzymes, 3rd Ed. 3:213–248. Academic Press, New York.

Huang, T. T.-S., Kass, W. J., Buddenbaum, W. E., and Yankwich, P. E. 1968. Anomalous temperature dependence of kinetic carbon isotope effects and the phenomenon of crossover. J. Phys. Chem. 72:4431–4446.

MacColl, A. 1974. Heavy-atom kinetic isotope effects. Annu. Repts. Chem. Soc. 71B:77–101.

O'Leary, M. H. 1976. Carbon isotope effect on the enzymatic decarboxylation of pyruvic acid. Biochem. Biophys. Res. Comm. 73:614–618.

O'Leary, M. H. 1977. Enzymic catalysis of decarboxylation. Bioorg. Chem. In press.

O'Leary, M. H., and Baughn, R. L. 1972. Acetoacetate decarboxylase: Identification of the rate-determining step in the primary amine catalyzed reaction and in the enzymic reaction. J. Am. Chem. Soc. 94:626–630.

O'Leary, M. H., and Kluetz, M. D. 1972. Nitrogen isotope effects on the chymotrypsin-catalyzed hydrolysis of *N*-acetyl-L-tryptophanamide. J. Am. Chem. Soc. 94:3585–3589.

O'Leary, M. H., Richards, D. T., and Hendrickson, D. W. 1970. Carbon isotope effects on the enzymatic decarboxylation of glutamic acid. J. Am. Chem. Soc. 92:4435–4440.

O'Leary, M. H., Urberg, M., and Young, A. P. 1974. Nitrogen isotope effects on the papain-catalyzed hydrolysis of *N*-benzoyl-L- argininamide. Biochemistry 13:2077–2081.

Richards, J. H. 1970. Kinetic isotope effects in enzymic reactions. In: P. D. Boyer (ed.), The Enzymes, 3rd Ed. 2:321–333. Academic Press, New York.
Sawyer, C. B., and Kirsch, J. F. 1975. Kinetic isotope effects for the chymotrypsin catalysed hydrolysis of ethoxyl-^{18}O labeled specific ester substrates. J. Am. Chem. Soc. 97:1963–1964.
Schellenberger, A. 1967. Structure and mechanism of action of the active center of yeast pyruvate decarboxylase. Angew. Chem. Intl. Edn. 6:1024–1035.
Shain, S. A., and Kirsch, J. F. 1968. Absence of carbonyl oxygen exchange concurrent with the alkaline hydrolysis of substituted methyl benzoates. J. Am. Chem. Soc. 90:5848–5854.
Simon, H., and Palm, D. 1966. Isotope effects in organic chemistry and biochemistry. Angew. Chem. Intl. Edn. 5:920–933.
Stern, M. J., and Vogel, P. C. 1971. Relative ^{14}C-^{13}C kinetic isotope effects. J. Chem. Phys. 55:2007-2013.
Ullrich, J., Ostrovsky, Y. M., Eyzaguirre, J., and Holzer, H. 1970. Thiamine pyrophosphate-catalyzed enzymatic decarboxylation of α-oxo acids. Vitamins and Hormones 28:365–398.
Van Hook, W. A. 1970. Kinetic isotope effects: Introduction and discussion of theory. In: C. J. Collins and N. S. Bowman (eds.), Isotope Effects in Chemical Reactions, pp. 1-89. Van Nostrand-Reinhold, New York.
Vogel, P. C., and Stern, M. J. 1971. Temperature dependences of kinetic isotope effects. J. Chem. Phys. 54:779–796.

Derivation of an Isotope Effect from the Proline Racemase Overshoot in D_2O

W. W. Cleland

In their study of proline racemase, which catalyzes the reaction

$$\text{L-proline} \rightleftharpoons \text{D-proline} \quad (1)$$

Cardinale and Abeles (1968) reported that the time course of the reaction, which was followed by the optical rotation of the solution, was normal in H_2O, but not in D_2O. In D_2O the optical rotation dropped initially as expected, but, instead of approaching zero asymptotically, it went through zero and became of opposite sign before finally returning to zero. The overshoot amounted to 20.7% of the initial optical rotation and was attributed by the authors to a primary deuterium isotope effect and the inability of protons to exchange while proline was present on the enzyme. Thus, they showed that in the early stages of the reaction (up to 13% conversion to product) the remaining substrate contained no deuterium, while the product contained one deuterium atom per molecule.

Cardinale and Abeles used an analog computer to model this system and to show that it did indeed produce an overshoot, but they did not solve the equations for this system analytically. Because this is essentially an equilibrium perturbation experiment (when the optical rotation reaches zero the first time, one has largely L-proline-2-H and D-proline-2-D present), the equations that describe it are of some interest and will be derived here.

The reaction can be written:

$$A \xrightarrow{k_1} B \underset{k_2}{\overset{k_2}{\rightleftharpoons}} C \quad (2)$$

where A is L-proline-2-H, B is D-proline-2-D, and C is L-proline-2-D. The interconversion of B and C occurs with equal rates, but k_1 is greater than k_2 because of the isotope effect, and the conversion of A to B is irreversible because the 2-hydrogen is removed to form HDO during the reaction, and the reaction is run in D_2O. Because proline was used in the experiment at a level about 20 times its K_m, we can assume saturation of the enzyme, and we will also assume equality of the K_m values, which is essentially correct (the reported values of 2.3 and 3.8 mM for L- and D-prolines may not be significantly different, because the lowest level of substrate that could be used was 1.5 K_m).

If A_0 is the initial concentration of A, then:

$$A + B + C = A_0 \tag{3}$$

At final equilibrium we will have:

$$t = \infty, \quad A = 0, \quad B = C = A_0/2 \tag{4}$$

Because the optical rotation at any time is proportional to $A + C - B$, or substituting from equation 3, to $2(A + C) - A_0$, the optical rotation relative to that at the beginning of the experiment is:

$$R = 2(A + C)/A_0 - 1 \tag{5}$$

with a negative value indicating overshoot.

Then, because A/A_0 is the fraction of enzyme occupied by A,

$$-\frac{dA}{dt} = k_1 E_t (A/A_0) \tag{6}$$

which, because

$$V = k_1 E_t \tag{7}$$

integrates to:

$$A = A_0 e^{-Vt/A_0} \tag{8}$$

Now:

$$B = A_0 - A - C = A_0 (1 - e^{-Vt/A_0}) - C \tag{9}$$

and because the proportion of enzyme containing B or C is B/A_0 or C/A_0,

$$\frac{dC}{dt} = k_2 E_t (B/A_0) - k_2 E_t (C/A_0) = \frac{k_2 E_t}{A_0} [A_0 (1 - e^{-Vt/A_0}) - 2C] \tag{10}$$

Because the isotope effect on V is:

$$\alpha = V_H/V_D = k_1/k_2 \tag{11}$$

we can replace k_2E_t by V/α and rearrange equation 10 to give:

$$\frac{dC}{dt} + \left(\frac{2\,V}{\alpha\,A_0}\right)C = \frac{V}{\alpha}(1 - e^{-Vt/A_0}) \tag{12}$$

The solution of this equation upon integration is:

$$C = \frac{A_0}{2}\left[1 + \left(\frac{\alpha}{2-\alpha}\right)e^{-2Vt/\alpha A_0}\right] - \left(\frac{A_0}{2-\alpha}\right)e^{-Vt/A_0} \tag{13}$$

except where $\alpha = 2$, in which case it is:

$$C = \frac{A_0}{2}(1 - e^{-Vt/A_0}) - \frac{V}{2}\,te^{-Vt/A_0} \tag{14}$$

Because we now have A from equation 8 and C from equation 13 or 14 as functions of time, we can evaluate R in equation 5 as:

$$R = \left(\frac{\alpha}{2-\alpha}\right)e^{-2Vt/\alpha A_0} - 2\frac{(\alpha - 1)}{(2 - \alpha)}e^{-Vt/A_0} \tag{15}$$

or, when $\alpha = 2$:

$$R = e^{-Vt/A_0}(1 - Vt/A_0) \tag{16}$$

We now solve for the minimum point of the curve by differentiating equation 15 or 16 and setting the derivatives equal to zero. For equation 16 this gives:

$$t_{min} = 2\,A_0/V \tag{17}$$

which is the lowest t_{min} possible (t_{min}, as given by equation 19 below, will increase as α gets either larger or smaller than 2). Substitution of t_{min} into equation 16 gives:

$$R_{min} = -e^{-2} = -0.135 \tag{18}$$

as the value when $\alpha = 2$.

For the general case, we differentiate equation 15 to obtain:

$$t_{min} = \frac{A_0\,\alpha}{V(\alpha - 2)}\ln(\alpha - 1) \tag{19}$$

which, when substituted into equation 15, gives:

$$R_{min} = \left(\frac{\alpha}{2 - \alpha}\right) (\alpha - 1)^{-2/(\alpha - 2)} - \frac{2(\alpha - 1)}{(2 - \alpha)} (\alpha - 1)^{-\alpha/(\alpha - 2)} \quad (20)$$

A curve of R_{min} versus α (Figure 1) is a smooth curve showing no break at the unique point where $\alpha = 2$ (when α is between 1 and 2, the first term in equation 20 is smaller than the second, but when α is above 2 this is reversed). Note the similarity of form of equation 20 to the comparable equation for an equilibrium perturbation experiment (equation 26 of Cleland's chapter).

As a nice confirmation of this analysis, the experimental value of R_{min} of -0.207 from Cardinale and Abeles (1968) gives an isotope effect of 2.6, in excellent agreement with the value of 2.5 for V_H/V_D determined directly by these authors.

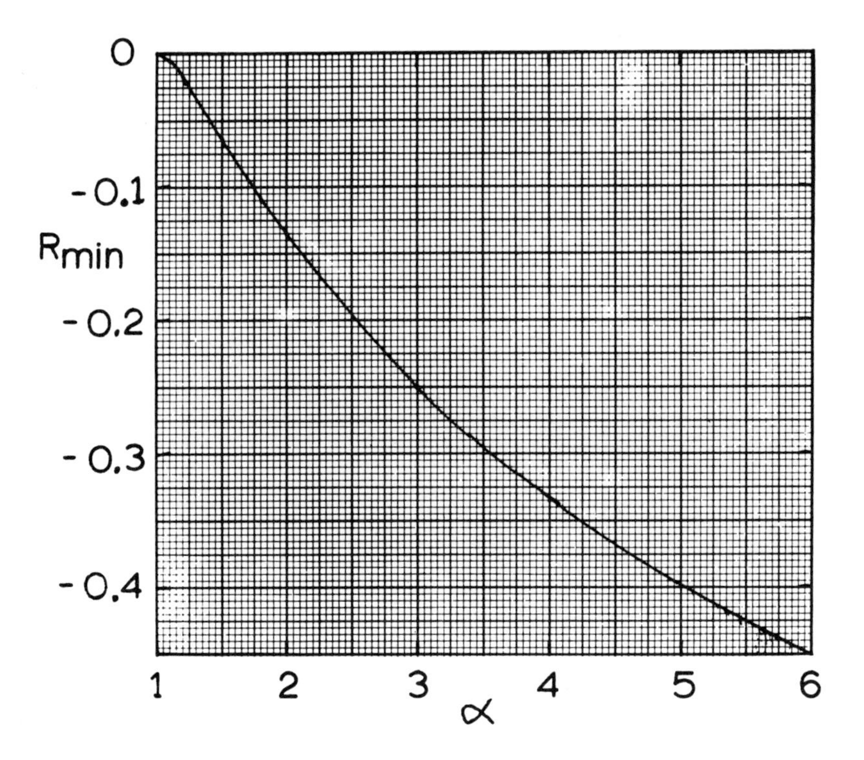

Figure 1. R_{min} as a function of isotope effect (α) according to equations 18 and 20.

LITERATURE CITED

Cardinale, G. J., and Abeles, R. H. 1968. Purification and mechanism of action of proline racemase. Biochemistry 7:3970–3978.

appendix A: A Note on the Use of Fractionation Factors versus Isotope Effects on Rate Constants

The use of fractionation factors is familiar to many chemists but quite foreign to most biochemists, who are used to expressing isotope effects in terms of the effects on the various rate constants for a reaction. Both systems are widely used, and in this book some authors have used one and some the other. In the interest of the reader, the editors present here a comparison of the two systems so that each one can translate what he reads into a system familiar to him. The shorthand notation introduced by Northrop and the β value used by Cleland in this volume are also included.

Consider a representative enzyme mechanism with rate constants assigned to each step:

$$E + A \underset{k_2}{\overset{k_1}{\rightleftharpoons}} EA \underset{k_4}{\overset{k_3}{\rightleftharpoons}} EP \underset{k_6}{\overset{k_5}{\rightleftharpoons}} E + P$$

There will, of course, be a transition state during each of the three steps, which is the same whether approached in a forward or reverse direction. Then, for each intermediate and transition state, we have a fractionation factor that represents the equilibrium constant for reaction of some reference molecule (RD) containing D with this state to give a deuterated state and RH. Fractionation factors greater than 1 represent enrichment of deuterium in this state relative to the reference molecule, while values less than 1 represent depletion. Similar fractionation factors may be defined for other isotopic pairs ($^{13}C/^{12}C$, $^{15}N/^{14}N$, etc.).

Schowen (this volume) uses water as the reference for deuterium fractionation factors, and this is probably the most convenient standard for biochemists. However, Shiner (this volume) uses acetylene (deuterium fractionation factors relative to acetylene are 1.56–1.61 times higher than those relative to water. See footnote *d* to Table 2 of Schowen, this volume). As the following analysis will show, any reference can be used, because all isotope effects are calculated from the ratios of fractionation factors and thus are the same regardless of what reference is used.

For the mechanism given, we have:

$$k_{1H}/k_{1D} = {}^{D}k_1 = \phi_A/\phi_{1,2}$$

where ϕ_A is the fractionation factor for A, and $\phi_{1,2}$ is for the transition state between E + A and EA. Likewise:

$$k_{2H}/k_{2D} = {}^{D}k_2 = \phi_{EA}/\phi_{1,2}$$

Then, if

$$K_{eq\ 1} = k_1/k_2, \quad {}^{D}K_{eq\ 1} = \phi_A/\phi_{EA}$$

Note that for this equilibrium isotope effect, $\phi_{1,2}$ cancels out. Similarly,

$$k_{3H}/k_{3D} = {}^{D}k_3 = \phi_{EA}/\phi_{3,4} \quad {}^{D}k_4 = \phi_{EP}/\phi_{3,4} \quad {}^{D}k_5 = \phi_{EP}/\phi_{5,6} \quad {}^{D}k_6 = \phi_P/\phi_{5,6}$$

If

$$K_{eq\ 2} = k_3/k_4 \quad K_{eq\ 3} = k_5/k_6$$

we have:

$${}^{D}K_{eq\ 2} = \phi_{EA}/\phi_{EP} \quad {}^{D}K_{eq\ 3} = \phi_{EP}/\phi_P$$

Then:

$${}^{D}K_{eq} = {}^{D}K_{eq\ 1}\,{}^{D}K_{eq\ 2}\,{}^{D}K_{eq\ 3} = \phi_A/\phi_P = 1/\beta$$

Thus, equilibrium isotope effects are functions only of fractionation factors for the reactants, and equilibrium effects on individual steps in the mechanism are functions of fractionation factors for intermediates and reactants. Only the isotope effects on individual rate constants (or on collections of rate constants which form an effective rate constant, such as k_1k_3/k_2) are functions of fractionation factors for transition states (as well as fractionation factors for reactants or intermediates).

Fractionation factors can be expressed in terms of isotope effects on rate constants or equilibrium constants as:

$$\phi_{EA} = \phi_A / {}^D K_{eq\,1}$$

$$\phi_{EP} = \phi_P {}^D K_{eq\,3}$$

$$\phi_{1,2} = \phi_A / {}^D k_1$$

$$\phi_{3,4} = \phi_A / {}^D k_3 {}^D K_{eq\,1} = \phi_P {}^D K_{eq\,3} / {}^D k_4$$

$$\phi_{5,6} = \phi_P / {}^D k_6$$

Schowen (this volume) gives fractionation factors for a number of molecules. Values for hydrogen bonded to various types of carbon atoms are given in more detail in Table 1, which is based on several good experimental values plus the calculations of Shiner (this volume).

Table 1. Fractionation factors for D and T between water and various molecules[a]

Molecule	Structure	D	T
Pyruvate-3-[H]	C—CH_2[H]	0.84	0.78[b]
Malate-3-[H]	C—CH[H]—C	0.93[c]	0.90
Dihydroxyacetone-3-P-1-[H]	C—CH[H]—OH	1.00	1.00[d]
Aspartate-2-[H]	C—C[H]NH_3^+—C	1.07	1.10
Glyceraldehyde-3-P-2—[H]	C—C[H]OH—C	1.10	1.15[e]
Glyceraldehyde-3-P-1-[H][f]	C—C[H]$(OH)_2$[g]	1.18	1.27
	C—C[H] = O[h]	0.86	0.80

[a]Except for the first two entries, the values are calculated from the experimental ones (b,c) using the measured fractionation factor for deuterium of 1.18 between C-4 of DPNH and the CHOH carbon of isopropanol, cyclohexanol, and L-malate (Cook and Cleland, unpublished results) and the factors of Hartshorn and Shiner (1972). Thus, replacement of H on the carbon bearing the CH or CD bond of interest with C, N, or O is assumed to increase the fractionation factor by 1.10, 1.15, and 1.18. Tritium effects are assumed to be the 1.442 power of deuterium ones.

[b]From Meloche, Monti, and Cleland, 1977.

[c]From Thomson, 1960.

[d]Experimental value with triose-P isomerase, 1.03 ± 0.04 (Fletcher et al., 1976).

[e]Experimental value with triose-P isomerase, 1.0 ± 0.2 (Fletcher, et al., 1976).

[f]Glyceraldehyde-3-P is 97% hydrated in water at neutral pH (Trentham, McMurray, and Pogson, 1969), so the observed value will be that for the hydrated form. For any specific aldehyde, the observed value thus will depend on the degree of hydration.

[g]The values given here are for use in H_2O; in D_2O, where the OH groups become OD, the deuterium value will be higher by about 15% (the value seen for acetaldehyde by Lewis and Wolfenden, personal communication).

[h]Based on the measured deuterium isotope effect of 1.37 on the equilibrium constant for acetaldehyde hydration (Lewis and Wolfenden, personal communication).

LITERATURE CITED

Fletcher, S. J., Herlihy, J. M., Albery, W. J., and Knowles, J. R. 1976. Energetics of triosephosphate isomerase: The appearance of solvent tritium in substrate glyceraldehyde 3-phosphate and in product. Biochemistry 15:5612–5617.

Hartshorn, S. R., and Shiner, Jr., V. J. 1972. Calculation of H/D, $^{12}C/^{13}C$, and $^{12}C/^{14}C$ fractionation factors from valence force fields derived for a series of simple organic molecules. J. Am. Chem. Soc. 94:9002–9012.

Meloche, H. P., Monti, C. T., and Cleland, W. W. 1977. Magnitude of the equilibrium isotope effect in carbon-tritium bond synthesis. Biochim. Biophys. Acta 480:517–519.

Thomson, J. F. 1960. Fumarase activity in D_2O. Arch. Biochem. Biophys. 90:1–6.

Trentham, D. R., McMurray, C. H., and Pogson, C. I. 1969. The active chemical state of D-glyceraldehyde-3-phosphate in its reactions with D-glyceraldehyde-3-phosphate dehydrogenase, aldolase, and triose phosphate isomerase. Biochem. J. 114:19–24.

appendix B: Computer Programs for Determining Deuterium Isotope Effects

These computer programs assume that one has run at the same time a reciprocal plot for both the deuterated and nondeuterated substrates, and that one wants to compare the slopes to determine the V/K isotope effect and the intercepts to determine the isotope effect on V. The first program assumes that effects are present on both V and V/K and should be tried first. Where one effect seems absent or very poorly determined (i.e., 1.10 ± 0.20, 0.8 ± 0.5, or 3 ± 10), one should then try one of the other two programs which assumes an effect only on V/K or V. The SIGMA values of the two fits are then compared, and the one with the lowest value is the one that should be used.

When there are two (or more) substrates, one may vary either one. If the mechanism is ordered, one normally varies the substrate that adds second, with the one adding first being held at saturating levels, regardless of which substrate contains the deuterium. Thus, with alcohol dehydrogenase, Bush, Shiner, and Mahler (1973) varied acetaldehyde at fixed DPNH or DPND levels and saw an isotope effect on $V/K_{\text{acetaldehyde}}$, but not on V. When the first substrate to add (A) is varied, no V/K_a isotope effect is seen when the second (B) is saturating; as the level of B is dropped, however, the V/K isotope effect increases until, at near-zero levels of B, it equals V/K_b. For a random mechanism with substrates A and B, varying A at saturating B gives the effect on V/K_a and V, while varying B at saturating A gives isotope effects on V/K_b and V.

USING THE PROGRAMS

These programs are written in simple FORTRAN and should work with any computer. All three use the same group of statements for matrix inversion and ending of the program; these are listed separately under "Matrix Solution Subroutine" and should be added to the end of each program.

For each experiment, the data deck consists of a title card and one data card for each experimental velocity. As many sets of such data as desired may be processed, and the program is made to stop by placing a blank card on the end of the data deck. The formats are as follows:

Title card.

Column 1-3. I3 format. Number of data cards to follow.

Column 4-19. Leave blank.

Column 20. Blank if velocities are used. When reciprocal velocities are used as input data, place a 1 in this column.

Column 21–68. Anything placed here is printed out as a title to identify the output.

Data Cards:

Column 1–10. F10.5 format. Velocities or reciprocal velocities (see above). Velocities should be scaled to a convenient size.

Column 11–20. F10.5 format. Concentrations of the substrate that is being varied (not necessarily the one that is labeled). Use M, mM, or μM units as appropriate to give numbers of convenient size.

Column 21–30. F10.5 format. Fraction of labeling with deuterium in the labeled substrate. For the velocities corresponding to the unlabeled substrate, this is normally left blank (i.e., equal to zero). For the velocities corresponding to the labeled substrate, insert 1.0, or whatever value corresponds to the actual degree of labeling (0.95, for example).

Note that even rather low degrees of labeling can be used in such experiments (as long as the isotope effects are large) and still permit calculation of the deuterium isotope effects. Note also that the data from the reciprocal plots for both labeled and unlabeled substrates are all placed together following one title card. The order of the data cards is not important.

The output will consist of:

1. A listing of the equation that is being fitted.
2. A fit to each set of data as follows:
   a. First line: Column 1–3. A number that counts the number of data sets processed. Column 4–20: The number of data points. Column 21–68: Title.

b. A table containing, for each data point, the concentration of the varied substrate, the fraction of deuterium labeling used in the labeled substrate, the experimental and calculated velocities and the difference between them, and reciprocal calculated and experimental velocities and substrate concentrations (the latter three to enable easy plotting of the data). Examination of this table provides a ready check on the fact that a fit has in fact been obtained and that the differences are small and randomly distributed. An error on a data card normally will stand out very clearly as a large difference for that point only, and in addition, will usually cause that point to be discarded and a revised fit to be made (see below).

c. A listing of the calculated isotope effects and their standard errors. These values should be rounded off to a degree consistent with the size of the standard error in each case. Other parameters (*K*, *V*, *V/K* for the deuterated and nondeuterated systems) are not computed, because the two reciprocal plots can easily be fitted individually to the HYPER program of Cleland (1967) to give this information.

d. SIGMA and VARIANCE. When the sum of squares of differences between calculated and experimental velocities is divided by degrees of freedom (number of data points minus number of constants determined in the fit) one gets VARIANCE; it is the average residual least square. SIGMA is the square root of this and is in the units used for velocity. As noted above, the fit with the lowest SIGMA is the better fit.

e. One expects 99 out of 100 of the individual differences between experimental and calculated velocities to be less than 2.6 × SIGMA. Thus, each difference is examined, and those data points which differ more than 2.6 × SIGMA from the calculated value are discarded. If no points are discarded, the program goes on to the next set of data, but if any points are discarded, these points are listed as deviating more than 2.6*Sl and a revised fit is made. As noted above, the most common cause for a bad point is a typing error on a data card, but a bad experimental point will also be discarded.

3. After all data sets have been fitted, the line "Program completed for n lines" will be printed.

NOTE: These programs assume equal variances for the velocities and use the general approach of Wilkinson (1961). Any questions or problems concerning these programs should be addressed to Dr. W. W.

Cleland, Department of Biochemistry, University of Wisconsin, Madison, Wis. 53706.

```
C      CALCULATION OF V/K AND V ISOTOPE EFFECTS
       DIMENSION V(100),A(100),CI(100),S(5,6),Q(5)     ,SM(5),SS(5)
       PRINT 100
   100 FORMAT(44H FIT TO    Y = V*A/(K(1+I*VKI) + A(1+I*VI ))  )
       PRINT 101
   101 FORMAT(40H  VKI = V/K ISOTOPE EFFECT - 1.          )
       PRINT 102
   102 FORMAT(40H  VI  = V   ISOTOPE EFFECT - 1.         /)
    11 FORMAT(I3,I17,48H   ANYTHING HERE WILL BE PRINTED DURING OUTPUT   )
     1 FORMAT (8F 10.5)
       JJ = 0
       II = 0
    14 READ 11, NP, NO
       IF (NP) 99,99,12
    12 M = 1
       N = 4
       N1 = N + 1
       N2 = N + 2
       GO TO 2
    15 READ 1, V(I), A(I), CI(I)
       IF (NO) 19, 19, 32
    32 V(I) = 1./V(I)
    19 Q(1)   = V(I)**2/A(I)
       Q(2)   = V(I)**2*CI(I)/A(I)
       Q(3)   = V(I)**2
       Q(4)   = V(I)**2*CI(I)
       Q(5)   = V(I)
       GO TO 13
    16 CK = S(1,1)/S(3,1)
       VKI = S(2,1)/S(1,1)
       VI = S(4,1)/S(3,1)
       JJ = JJ + 1
       PRINT 11, JJ, NP
       NT = 0
       M = 2
       GO TO 2
    17 D = (1.+CI(I)*VKI )*CK/A(I) + 1. + CI(I)*VI
       Q(1)   = 1./D
       Q(2)   = (1.+CI(I)*VKI )/A(I)/D**2
       Q(3)   = CI(I)/A(I)/D**2
       Q(4)   = CI(I)/D**2
       Q(5)   = V(I)
       GO TO 13
    18 CV = S(1,1)
       CK = CK - S(2,1)/S(1,1)
       VKI = VKI - S(3,1)/S(1,1)/CK
       VI = VI - S(4,1)/S(1,1)
       NT = NT + 1
       IF (NT - 5) 2, 87, 87
    87 S2 = 0
       PRINT 88
    88 FORMAT ( 80H    A CONC      D LABEL      V EXP      V CAL       DIFF   1/V
      1CALC   1/V EXPTL       1/A           /)
       DO 82 I = 1,NP
       X=S(1,1)/((CK/A(I)*(1.+CI(I)*VKI ))+1.+CI(I)*VI   )
       DX=V(I)-X
```

```
      RX=1./X
      X1 = 1./V(I)
      X2 = 1./A(I)
      PRINT 1, A(I),CI(I), V(I), X, DX, RX,X1,X2
   82 S2=S2+DX**2
      P = NP - N
      S2 = S2/P
      S1 = SQRT (S2)
      DO 10 J = 2,N1
      DO 10 K = 1,N
   10 S(K,J) = S(K,J)*SM(K)*SM(J-1)
      VKIO = 1. + VKI
      VIO = 1. + VI
      SEVKI = S1*SQRT(S(3,4))/S(1,1)/CK
      SEVI = S1*SQRT(S(4,5))/S(1,1)
      PRINT 140, VKIO, SEVKI
  140 FORMAT(22H V/K ISOTOPE EFFECT = F10.5,10H    S.E. = F10.5)
      PRINT 141, VIO, SEVI
  141 FORMAT(22H V   ISOTOPE EFFECT = F10.5,10H    S.E. = F10.5)
      PRINT 98, S1
   98 FORMAT(9H SIGMA = F13.9)
      PRINT 44, S2
   44 FORMAT(12H VARIANCE = E14.5//)
      IF(II)30,30,14
   30 S1 = 2.6*S1
      DO 24 I = 1,NP
      IF(ABS (V(I)-CV*A(I)/(CK*(1.+(CI(I)*VKI ))+A(I)*(1.+CI(I)*VI   )))-
     1S1)25,20,20
   20 PRINT 21, V(I), A(I), CI(I)
   21 FORMAT(35H POINT DEVIATES MORE THAN 2.6*S1    3F10.5/)
      GO TO 24
   25 II = II + 1
      A(II) = A(I)
      V(II) = V(I)
      CI(II) = CI(I)
   24 CONTINUE
      IF(NP - II) 14,14,22
   22 NP = II
      NT = 0
      PRINT 23
   23 FORMAT(14H   REVISED FIT//)
      GO TO 2
C     Matrix Solution Subroutine
```

```
C      CALCULATION OF V/K ISOTOPE EFFECT ONLY
       DIMENSION V(100),A(100),CI(100),S(5,6),Q(5)     ,SM(5),SS(5)
       PRINT 100
   100 FORMAT(44H FIT TO    Y = V*A/(K(1+I*VKI) + A)              /)
       PRINT 101
   101 FORMAT(40H  VKI = V/K ISOTOPE EFFECT - 1.          )
    11 FORMAT(I3,I17,48H   ANYTHING HERE WILL BE PRINTED DURING OUTPUT  )
     1 FORMAT (8F 10.5)
       JJ = 0
    14 READ 11, NP, NO
       IF (NP) 99,99,12
    12 M = 1
       N = 3
       N1 = N + 1
       N2 = N + 2
       II = 0
       GO TO 2
    15 READ 1, V(I),A(I),CI(I)
       IF (NO) 19, 19, 32
    32 V(I) = 1./V(I)
    19 Q(1)    = V(I)**2/A(I)
       Q(2)    = V(I)**2*CI(I)/A(I)
       Q(3)    = V(I)**2
       Q(4)    = V(I)
       GO TO 13
    16 CK = S(1,1)/S(3,1)
       VKI = S(2,1)/S(1,1)
       JJ = JJ + 1
       PRINT 11, JJ, NP
       NT = 0
       M = 2
       GO TO 2
    17 D = (1.+CI(I)*VKI  )*CK/A(I) + 1.)
       Q(1)    = 1./D
       Q(2)    = (1.+CI(I)*VKI  )/A(I)/D**2
       Q(3)    = CI(I)/A(I)/D**2
       Q(4)    = V(I)
       GO TO 13
    18 CV = S(1,1)
       CK = CK - S(2,1)/S(1,1)
       VKI = VKI - S(3,1)/S(1,1)/CK
       NT = NT + 1
       IF (NT - 5) 2, 87, 87
    87 S2 = 0
        PRINT 88
    88 FORMAT ( 80H   A CONC     D LABEL      V EXP      V CAL        DIFF  1/V
      1CALC  1/V EXPTL      1/A           /)
       DO 82 I = 1,NP
       X=S(1,1)/(CK/A(I)*(1.+CI(I)*VKI  )+1.)
       DX=V(I)-X
       RX=1./X
       X1 = 1./V(I)
       X2 = 1./A(I)
       PRINT 1, A(I),CI(I), V(I), X, DX, RX,X1,X2
    82 S2=S2+DX**2
```

```
      P = NP - N
      S2 = S2/P
      S1 = SQRT (S2)
      DO 10 J = 2,N1
      DO 10 K = 1,N
   10 S(K,J) = S(K,J)*SM(K)*SM(J-1)
      VKIO = 1. + VKI
      SEVKI = S1*SQRT(S(3,4))/S(1,1)/CK
      PRINT 140, VKIO, SEVKI
  140 FORMAT(22H V/K ISOTOPE EFFECT = F10.5,10H    S.E. = F10.5)
      PRINT 98, S1
   98 FORMAT(9H SIGMA = F13.9)
      PRINT 44, S2
   44 FORMAT(12H VARIANCE = E14.5//)
      IF(II)30,30,14
   30 S1 = 2.6*S1
      DO 24 I = 1,NP
      IF(ABS (V(I)-CV*A(I)/(CK*(1.+CI(I)*VKI )+A(I))) - S1)25,20,20
   20 PRINT 21, V(I), A(I), CI(I)
   21 FORMAT(35H POINT DEVIATES MORE THAN 2.6*S1    3F10.5/)
      GO TO 24
   25 II = II + 1
      V(II) = V(I)
      A(II) = A(I)
      CI(II) = CI(I)
   24 CONTINUE
      IF(NP - II) 14,14,22
   22 NP = II
      NT = 0
      PRINT 23
   23 FORMAT(14H    REVISED FIT//)
      GO TO 2
C     Matrix Solution Subroutine
```

```
C     CALCULATION OF V ISOTOPE EFFECT ONLY
      DIMENSION V(100),A(100),CI(100),S(5,6),Q(5),SM(5),SS(5)
      PRINT 100
  100 FORMAT(44H FIT TO   Y = V*A/(K           + A(1+I*VI )) /)
      PRINT 102
  102 FORMAT(40H  VI  = V   ISOTOPE EFFECT - 1.          /)
   11 FORMAT(I3,I17,48H   ANYTHING HERE WILL BE PRINTED DURING OUTPUT  )
    1 FORMAT (8F 10.5)
      JJ = 0
      II = 0
   14 READ 11, NP, NO
      IF (NP) 99,99,12
   12 M = 1
      N = 3
      P = NP - N
      N1 = N + 1
      N2 = N + 2
      GO TO 2
   15 READ 1, V(I), A(I), CI(I)
      IF (NO) 19, 19, 32
   32 V(I) = 1./V(I)
   19 Q(1)   = V(I)**2/A(I)
      Q(2)   = V(I)**2
      Q(3)   = V(I)**2*CI(I)
      Q(4)   = V(I)
      GO TO 13
   16 CK = S(1,1)/S(2,1)
      VI = S(3,1)/S(2,1)
      JJ = JJ + 1
      PRINT 11, JJ, NP
      NT = 0
      M = 2
      GO TO 2
   17 D = CK/A(I) + 1. + CI(I)*VI
      Q(1)   = 1./D
      Q(2)   = 1./A(I)/D**2
      Q(3)   = CI(I)/D**2
      Q(4)   = V(I)
      GO TO 13
   18 CV = S(1,1)
      CK = CK - S(2,1)/S(1,1)
      VI = VI - S(3,1)/S(1,1)
      NT = NT + 1
      IF (NT - 5) 2, 87, 87
   87 S2 = 0
       PRINT 88
   88 FORMAT ( 80H   A CONC    D LABEL    V EXP      V CAL        DIFF  1/V
     1CALC  1/V EXPTL      1/A            /)
      DO 82 I = 1,NP
      X=CV/(CK/A(I)+1.+CI(I)*VI)
      DX=V(I)-X
      RX=1./X
      X1 = 1./V(I)
      X2 = 1./A(I)
      PRINT 1, A(I),CI(I), V(I), X, DX, RX,X1,X2
   82 S2=S2+DX**2
```

```
      S2 = S2/P
      S1 = SQRT(S2)
      DO 10 J = 2,N1
      DO 10 K = 1,N
   10 S(K,J) = S(K,J)*SM(K)*SM(J-1)
      VIO = 1. + VI
      SEVI = S1*SQRT(S(3,4))/S(1,1)
      PRINT 141, VIO, SEVI
  141 FORMAT(22H V   ISOTOPE EFFECT = F10.5,10H   S.E. = F10.5)
      PRINT 98, S1
   98 FORMAT(9H SIGMA = F13.9)
      PRINT 44, S2
   44 FORMAT(12H VARIANCE = E14.5//)
      IF(II)30,30,14
   30 S1 = 2.6*S1
      DO 24 I = 1,NP
      IF(ABS (V(I)-CV/(CK/A(I)+1.+CI(I)*VI   ))-S1)25,20,20
   20 PRINT 21, V(I), A(I), CI(I)
   21 FORMAT(35H POINT DEVIATES MORE THAN 2.6*S1   3F10.5/)
      GO TO 24
   25 II = II + 1
      V(II) = V(I)
      A(II) = A(I)
      CI(II) = CI(I)
   24 CONTINUE
      IF(NP - II) 14,14,22
   22 NP = II
      NT = 0
      PRINT 23
   23 FORMAT(14H   REVISED FIT//)
      GO TO 2
```

C Matrix solution subroutine

The following statements should be added to the end of each of the three programs.

```
C       MATRIX SOLUTION SUBROUTINE
      2 DO 3 J = 1,N2
        DO 3 K = 1,N1
      3 S(K,J) = 0
        DO 4  I = 1,NP
        GO TO  (15,17), M
     13 DO 4 J = 1,N1
        DO 4 K = 1,N
      4 S(K,J) = S(K,J) + Q(K)*Q(J)
        DO 5 K = 1,N
      5 SM(K) = 1./SQRT(S(K,K))
        SM(N1) = 1.
        DO 6 J = 1,N1
        DO 6 K = 1,N
      6 S(K,J) = S(K,J)*SM(K)*SM(J)
        SS(N1) = -1.
        S(1,N2) = 1.
        DO 8 L = 1,N
        DO 7 K = 1,N
      7 SS(K) = S(K,1)
        DO 8 J = 1,N1
        DO 8 K = 1,N
      8 S(K,J) = S(K+1,J+1) - SS(K+1)*S(1,J+1)/SS(1)
        DO 9 K = 1,N
      9 S(K,1) = S(K,1)*SM(K)
        GO TO (16,18), M
     36 FORMAT(23H PROGRAM COMPLETED FOR I4, 6H LINES   )
     99 PRINT 36, JJ
        STOP
        END
```

LITERATURE CITED

Bush, K., Shiner, V. J., Jr., and Mahler, H. R. 1973. Deuterium isotope effects on initial rates of the liver alcohol dehydrogenase reaction. Biochemistry 12: 4802-4805.

Cleland, W. W. 1967. The statistical analysis of enzyme kinetic data. Adv. Enzymol. 29:1-32.

Wilkinson, G. N. 1961. Statistical estimations in enzyme kinetics. Biochem. J. 80:324-332.

appendix C: FORTRAN Program and Table for Making Equilibrium Perturbation Calculations

FORTRAN PROGRAM

1. The data for an experiment are placed on two cards: a title card and a data card.
2. Title card.

   Column 1–3. I3 format. Number of reactants other than those involved in the perturbation. For malate dehydrogenase, for example, this is 2, while for glutamic dehydrogenase it is 3. Maximum value, 4; minimum value, 1. (If water is the only other reactant, put 1 here, and 55 M on the data card.)

   Column 4–20. F17.5 format. The fraction of isotope used. For example, for 90% C-13, use 0.90; for 99% D, use 0.99. If nothing is put here, the fraction is assumed to be 1.00.

   Column 21–67. Anything put here will be printed out verbatim as a title.
3. Data card.

   Column 1-10. F10.5 format. The size of the perturbation in micromolar units. If the perturbation is away from rather than towards the reactant with the heavier isotope (i.e., an inverse isotope effect is observed), a minus sign must precede the value here.

   Column 11–20. F10.5 format. The value of β (K_{eq_D}/K_{eq_H}). If left blank, a value of 1.00 is assumed.

Column 21–30. E10.5 format. Concentration of A in molar units (A is the molecule containing the heavy isotope).

Column 31–40. E10.5 format. Concentration of P in molar units (P is the molecule with the normal isotope).

Columns 41–50, 51–60, 61–70, 71–80. E10.5 formats. Concentrations of other reactants in molar units. Put as many values here as are called for in column 1–3 of the title card, and leave the other fields blank.

CAUTION: E format requires that the exponent be placed in the last 4 columns of the field. Example: 1.52E-03 for 1.52 mM.

4. When more than one experiment is run, the data deck consists of as many pairs of title and data cards as there are experiments. In each case, however, a blank card is placed on the end of the total data deck to stop the program.
5. The first line of output will contain the title preceded by: a) a number, which simply counts the experiments, and b) the value of β.
6. The second line gives $1/\Sigma(1/C_0)$, P_0', P_0'', and y under the labels AP, RP, RPP, and Y.
7. The third line will give the value of Δ/P_0'' to use in the accompanying table to determine apparent α. Multiply apparent α by z to get true α, and then by β to get $\alpha*\beta$.
8. When apparent α is less than 1.1, the fourth line will give a reasonably close estimate of α and $\alpha*\beta$, although for full precision, the table should be used.

NOTE: This program was written to be distributed with reprints of the original article by Schimerlik, Rife, and Cleland (1975). In that article, B and R were molecules involved in the perturbation, and they correspond to A and P here. R_0' and R_0'' were used in place of P_0' and P_0'', and A_0' was used for $1/\Sigma(1/C_0)$. Thus, in this program the variables B and R are concentrations of A and P; A_i is used for concentrations of the other reactants, RP is P_0', RPP is P_0'', and AP is $1/\Sigma(1/C_0)$. FI is fractional labeling with the isotope, AONE is the value with which to enter the table, and PERT is the size of the perturbation. BETA, Y, and Z are those parameters.

EQUILIBRIUM PERTURBATION TABLE

The following table tabulates solutions to equation 26 of Cleland's chapter (this volume) as values of $(P_{max} - P_0)/P_0''$ (listed in the tables as Δ/P_0'') versus apparent α. When the isotope effect is inverse (that is, during the perturbation the reaction shifts away from the compound containing the heavier isotope, rather than toward it), consider Δ/P_0'' to be negative; for normal isotope effects, consider it positive.

```
      DIMENSION A(4)
      PRINT 100
  100 FORMAT(40H CALCULATION OF EQUILIBRIUM PERTURBATION //)
  101 FORMAT( I3,F17.5,46H                TITLE
    1 FORMAT(2F10.5,6E10.5)
      JJ = 0
  114 READ 101, NA, FI
      IF (NA) 99, 99, 112
  112 READ 1, PERT, BETA, B, R, A
      IF (FI) 80, 80, 81
   80 FI = 1.
   81 SA = 0
      DO 2 I = 1, NA
    2 SA = SA + 1./A(I)
      AP = 1./SA
      BI = 1./B
      RI = 1./R
      IF(BETA)20,20,21
   20 BETA = 1.
   21 IF (BETA - 1.) 3, 4, 3
    4 RP = 1./(BI + RI)
      Z = 1.
      GO TO 5
    3 RP = 2./(RI + BI + SQRT((RI+BI)**2+4.*(1./BETA-1.)*RI*BI))
      BR = B/R
      Z = (1.-BR+SQRT((1.-BR)**2+4.*BR/BETA))/2.
    5 IF (AP/RP - .3) 6, 6, 8
    8 Y = 2.2
      GO TO 15
    6 IF (AP/RP - .16) 9, 9, 11
   11 Y = 2.3
      GO TO 15
    9 IF (AP/RP - .08) 12, 12, 14
   14 Y = 2.4
      GO TO 15
   12 Y = 2.5
   15 RPP = 1./(1./RP + 1./Y/AP)
      JJ = JJ + 1
      PRINT 101, JJ, BETA
      PRINT 102,AP,RP,RPP,Y
  102 FORMAT(6H AP = E14.5,8H    RP = E14.5,9H    RPP = E14.5,
     1 7H    Y = F5.2)
      AONE = 1.00E-06*PERT/RPP/FI
      PRINT 18, AONE, Z
   18 FORMAT(25H ENTER TABLE WITH VALUE   F10.5, 9H      Z = F10.5//)
      AONE = AONE*2.72
      IF (AONE - .1) 16, 16, 114
   16 ALPHA = (AONE + 1.)*Z
      AB = ALPHA*BETA
      PRINT 30, ALPHA, AB
   30 FORMAT(9H ALPHA = F10.5,18H      ALPHA*BETA = F10.5//)
      GO TO 114
   99 PRINT 50
   50 FORMAT(18H PROGRAM COMPLETED /)
      STOP
      END
```

Equilibrium Perturbation Table

Δ/Po"	app α	Δ/Po"	app α	Δ/Po"	app α	Δ/Po"	app α
-.1573	.650	-.0997	.762	-.0495	.874	-.0208	.945
-.1562	.652	-.0987	.764	-.0487	.876	-.0204	.946
-.1551	.654	-.0978	.766	-.0478	.878	-.0200	.947
-.1540	.656	-.0968	.768	-.0470	.880	-.0196	.948
-.1529	.658	-.0959	.770	-.0462	.882	-.0193	.949
-.1518	.660	-.0949	.772	-.0453	.884	-.0189	.950
-.1507	.662	-.0940	.774	-.0445	.886	-.0185	.951
-.1496	.664	-.0930	.776	-.0437	.888	-.0181	.952
-.1485	.666	-.0921	.778	-.0428	.890	-.0177	.953
-.1474	.668	-.0912	.780	-.0420	.892	-.0173	.954
-.1464	.670	-.0902	.782	-.0412	.894	-.0169	.955
-.1453	.672	-.0893	.784	-.0404	.896	-.0166	.956
-.1442	.674	-.0884	.786	-.0396	.898	-.0162	.957
-.1431	.676	-.0874	.788	-.0387	.900	-.0158	.958
-.1421	.678	-.0865	.790	-.0379	.902	-.0154	.959
-.1410	.680	-.0856	.792	-.0371	.904	-.0150	.960
-.1399	.682	-.0847	.794	-.0367	.905	-.0146	.961
-.1389	.684	-.0838	.796	-.0363	.906	-.0143	.962
-.1378	.686	-.0828	.798	-.0359	.907	-.0139	.963
-.1368	.688	-.0819	.800	-.0355	.908	-.0135	.964
-.1357	.690	-.0810	.802	-.0351	.909	-.0131	.965
-.1347	.692	-.0801	.804	-.0347	.910	-.0127	.966
-.1336	.694	-.0792	.806	-.0343	.911	-.0123	.967
-.1326	.696	-.0783	.808	-.0339	.912	-.0120	.968
-.1316	.698	-.0774	.810	-.0335	.913	-.0116	.969
-.1305	.700	-.0765	.812	-.0331	.914	-.0112	.970
-.1295	.702	-.0756	.814	-.0327	.915	-.0108	.971
-.1285	.704	-.0747	.816	-.0323	.916	-.0104	.972
-.1274	.706	-.0738	.818	-.0319	.917	-.0101	.973
-.1264	.708	-.0729	.820	-.0315	.918	-.0097	.974
-.1254	.710	-.0720	.822	-.0311	.919	-.0093	.975
-.1244	.712	-.0711	.824	-.0307	.920	-.0089	.976
-.1233	.714	-.0702	.826	-.0303	.921	-.0086	.977
-.1223	.716	-.0693	.828	-.0299	.922	-.0082	.978
-.1213	.718	-.0684	.830	-.0295	.923	-.0078	.979
-.1203	.720	-.0676	.832	-.0291	.924	-.0074	.980
-.1193	.722	-.0667	.834	-.0287	.925	-.0071	.981
-.1183	.724	-.0658	.836	-.0283	.926	-.0067	.982
-.1173	.726	-.0649	.838	-.0279	.927	-.0063	.983
-.1163	.728	-.0641	.840	-.0275	.928	-.0059	.984
-.1153	.730	-.0632	.842	-.0271	.929	-.0056	.985
-.1143	.732	-.0623	.844	-.0267	.930	-.0052	.986
-.1133	.734	-.0615	.846	-.0263	.931	-.0048	.987
-.1123	.736	-.0606	.848	-.0259	.932	-.0044	.988
-.1113	.738	-.0597	.850	-.0255	.933	-.0041	.989
-.1104	.740	-.0589	.852	-.0251	.934	-.0037	.990
-.1094	.742	-.0580	.854	-.0247	.935	-.0033	.991
-.1084	.744	-.0571	.856	-.0243	.936	-.0030	.992
-.1074	.746	-.0563	.858	-.0239	.937	-.0026	.993
-.1064	.748	-.0554	.860	-.0235	.938	-.0022	.994
-.1055	.750	-.0546	.862	-.0232	.939	-.0018	.995
-.1045	.752	-.0537	.864	-.0228	.940	-.0015	.996
-.1035	.754	-.0529	.866	-.0224	.941	-.0011	.997
-.1026	.756	-.0520	.868	-.0220	.942	-.0007	.998
-.1016	.758	-.0512	.870	-.0216	.943	-.0004	.999
-.1006	.760	-.0503	.872	-.0212	.944	.0000	1.000

Δ/Po"	app α	Δ/Po"	app α	Δ/Po"	app α	Δ/Po"	app α
.0004	1.001	.0204	1.057	.0394	1.113	.0574	1.169
.0007	1.002	.0207	1.058	.0397	1.114	.0577	1.170
.0011	1.003	.0211	1.059	.0400	1.115	.0580	1.171
.0015	1.004	.0214	1.060	.0404	1.116	.0583	1.172
.0018	1.005	.0218	1.061	.0407	1.117	.0586	1.173
.0022	1.006	.0221	1.062	.0410	1.118	.0590	1.174
.0026	1.007	.0225	1.063	.0413	1.119	.0593	1.175
.0029	1.008	.0228	1.064	.0417	1.120	.0596	1.176
.0033	1.009	.0232	1.065	.0420	1.121	.0599	1.177
.0037	1.010	.0235	1.066	.0423	1.122	.0602	1.178
.0040	1.011	.0239	1.067	.0427	1.123	.0605	1.179
.0044	1.012	.0242	1.068	.0430	1.124	.0608	1.180
.0048	1.013	.0245	1.069	.0433	1.125	.0611	1.181
.0051	1.014	.0249	1.070	.0436	1.126	.0614	1.182
.0055	1.015	.0252	1.071	.0440	1.127	.0618	1.183
.0058	1.016	.0256	1.072	.0443	1.128	.0621	1.184
.0062	1.017	.0259	1.073	.0446	1.129	.0624	1.185
.0066	1.018	.0263	1.074	.0449	1.130	.0627	1.186
.0069	1.019	.0266	1.075	.0453	1.131	.0630	1.187
.0073	1.020	.0269	1.076	.0456	1.132	.0633	1.188
.0076	1.021	.0273	1.077	.0459	1.133	.0636	1.189
.0080	1.022	.0276	1.078	.0462	1.134	.0639	1.190
.0084	1.023	.0280	1.079	.0466	1.135	.0642	1.191
.0087	1.024	.0283	1.080	.0469	1.136	.0645	1.192
.0091	1.025	.0286	1.081	.0472	1.137	.0648	1.193
.0094	1.026	.0290	1.082	.0475	1.138	.0651	1.194
.0098	1.027	.0293	1.083	.0478	1.139	.0654	1.195
.0102	1.028	.0297	1.084	.0482	1.140	.0658	1.196
.0105	1.029	.0300	1.085	.0485	1.141	.0661	1.197
.0109	1.030	.0303	1.086	.0488	1.142	.0664	1.198
.0112	1.031	.0307	1.087	.0491	1.143	.0667	1.199
.0116	1.032	.0310	1.088	.0495	1.144	.0670	1.200
.0119	1.033	.0314	1.089	.0498	1.145	.0676	1.202
.0123	1.034	.0317	1.090	.0501	1.146	.0682	1.204
.0127	1.035	.0320	1.091	.0504	1.147	.0688	1.206
.0130	1.036	.0324	1.092	.0507	1.148	.0694	1.208
.0134	1.037	.0327	1.093	.0511	1.149	.0700	1.210
.0137	1.038	.0330	1.094	.0514	1.150	.0706	1.212
.0141	1.039	.0334	1.095	.0517	1.151	.0712	1.214
.0144	1.040	.0337	1.096	.0520	1.152	.0718	1.216
.0148	1.041	.0340	1.097	.0523	1.153	.0724	1.218
.0151	1.042	.0344	1.098	.0526	1.154	.0730	1.220
.0155	1.043	.0347	1.099	.0530	1.155	.0736	1.222
.0158	1.044	.0350	1.100	.0533	1.156	.0742	1.224
.0162	1.045	.0354	1.101	.0536	1.157	.0748	1.226
.0165	1.046	.0357	1.102	.0539	1.158	.0754	1.228
.0169	1.047	.0360	1.103	.0542	1.159	.0760	1.230
.0172	1.048	.0364	1.104	.0545	1.160	.0766	1.232
.0176	1.049	.0367	1.105	.0549	1.161	.0772	1.234
.0179	1.050	.0370	1.106	.0552	1.162	.0778	1.236
.0183	1.051	.0374	1.107	.0555	1.163	.0784	1.238
.0186	1.052	.0377	1.108	.0558	1.164	.0790	1.240
.0190	1.053	.0380	1.109	.0561	1.165	.0796	1.242
.0193	1.054	.0384	1.110	.0564	1.166	.0802	1.244
.0197	1.055	.0387	1.111	.0568	1.167	.0807	1.246
.0200	1.056	.0390	1.112	.0571	1.168	.0813	1.248

Δ/Po"	app α	Δ/Po"	app α	Δ/Po"	app α	Δ/Po"	app α
.0819	1.250	.1132	1.362	.1418	1.474	.1961	1.715
.0825	1.252	.1137	1.364	.1423	1.476	.1971	1.720
.0831	1.254	.1143	1.366	.1428	1.478	.1981	1.725
.0837	1.256	.1148	1.368	.1433	1.480	.1992	1.730
.0843	1.258	.1153	1.370	.1438	1.482	.2002	1.735
.0848	1.260	.1159	1.372	.1443	1.484	.2012	1.740
.0854	1.262	.1164	1.374	.1448	1.486	.2022	1.745
.0860	1.264	.1169	1.376	.1453	1.488	.2032	1.750
.0866	1.266	.1175	1.378	.1457	1.490	.2042	1.755
.0871	1.268	.1180	1.380	.1462	1.492	.2052	1.760
.0877	1.270	.1185	1.382	.1467	1.494	.2062	1.765
.0883	1.272	.1190	1.384	.1472	1.496	.2072	1.770
.0889	1.274	.1196	1.386	.1477	1.498	.2082	1.775
.0894	1.276	.1201	1.388	.1481	1.500	.2092	1.780
.0900	1.278	.1206	1.390	.1493	1.505	.2102	1.785
.0906	1.280	.1211	1.392	.1505	1.510	.2112	1.790
.0912	1.282	.1216	1.394	.1517	1.515	.2122	1.795
.0917	1.284	.1222	1.396	.1529	1.520	.2132	1.800
.0923	1.286	.1227	1.398	.1541	1.525	.2141	1.805
.0929	1.288	.1232	1.400	.1553	1.530	.2151	1.810
.0934	1.290	.1237	1.402	.1565	1.535	.2161	1.815
.0940	1.292	.1242	1.404	.1576	1.540	.2171	1.820
.0946	1.294	.1248	1.406	.1588	1.545	.2180	1.825
.0951	1.296	.1253	1.408	.1599	1.550	.2190	1.830
.0957	1.298	.1258	1.410	.1611	1.555	.2199	1.835
.0962	1.300	.1263	1.412	.1623	1.560	.2209	1.840
.0968	1.302	.1268	1.414	.1634	1.565	.2219	1.845
.0974	1.304	.1273	1.416	.1645	1.570	.2228	1.850
.0979	1.306	.1278	1.418	.1657	1.575	.2238	1.855
.0985	1.308	.1283	1.420	.1668	1.580	.2247	1.860
.0990	1.310	.1289	1.422	.1680	1.585	.2256	1.865
.0996	1.312	.1294	1.424	.1691	1.590	.2266	1.870
.1001	1.314	.1299	1.426	.1702	1.595	.2275	1.875
.1007	1.316	.1304	1.428	.1713	1.600	.2284	1.880
.1013	1.318	.1309	1.430	.1724	1.605	.2294	1.885
.1018	1.320	.1314	1.432	.1736	1.610	.2303	1.890
.1024	1.322	.1319	1.434	.1747	1.615	.2312	1.895
.1029	1.324	.1324	1.436	.1758	1.620	.2321	1.900
.1035	1.326	.1329	1.438	.1769	1.625	.2331	1.905
.1040	1.328	.1334	1.440	.1780	1.630	.2340	1.910
.1046	1.330	.1339	1.442	.1791	1.635	.2349	1.915
.1051	1.332	.1344	1.444	.1802	1.640	.2358	1.920
.1057	1.334	.1349	1.446	.1812	1.645	.2367	1.925
.1062	1.336	.1354	1.448	.1823	1.650	.2376	1.930
.1067	1.338	.1359	1.450	.1834	1.655	.2385	1.935
.1073	1.340	.1364	1.452	.1845	1.660	.2394	1.940
.1078	1.342	.1369	1.454	.1855	1.665	.2403	1.945
.1084	1.344	.1374	1.456	.1866	1.670	.2412	1.950
.1089	1.346	.1379	1.458	.1877	1.675	.2421	1.955
.1095	1.348	.1384	1.460	.1887	1.680	.2430	1.960
.1100	1.350	.1389	1.462	.1898	1.685	.2439	1.965
.1105	1.352	.1394	1.464	.1908	1.690	.2448	1.970
.1111	1.354	.1399	1.466	.1919	1.695	.2456	1.975
.1116	1.356	.1404	1.468	.1929	1.700	.2465	1.980
.1121	1.358	.1409	1.470	.1940	1.705	.2474	1.985
.1127	1.360	.1414	1.472	.1950	1.710	.2483	1.990

Δ/Po"	app α	Δ/Po"	app α	Δ/Po"	app α	Δ/Po"	app α
.2491	1.995	.3323	2.55	.4071	3.22	.4959	4.34
.2500	2.00	.3336	2.56	.4091	3.24	.4972	4.36
.2517	2.01	.3349	2.57	.4110	3.26	.4985	4.38
.2534	2.02	.3361	2.58	.4129	3.28	.4998	4.40
.2552	2.03	.3374	2.59	.4147	3.30	.5011	4.42
.2569	2.04	.3387	2.60	.4166	3.32	.5023	4.44
.2585	2.05	.3399	2.61	.4185	3.34	.5036	4.46
.2602	2.06	.3412	2.62	.4203	3.36	.5048	4.48
.2619	2.07	.3424	2.63	.4221	3.38	.5061	4.50
.2635	2.08	.3437	2.64	.4239	3.40	.5073	4.52
.2652	2.09	.3449	2.65	.4257	3.42	.5086	4.54
.2668	2.10	.3462	2.66	.4275	3.44	.5098	4.56
.2685	2.11	.3474	2.67	.4293	3.46	.5110	4.58
.2701	2.12	.3486	2.68	.4310	3.48	.5122	4.60
.2717	2.13	.3498	2.69	.4328	3.50	.5134	4.62
.2733	2.14	.3510	2.70	.4345	3.52	.5146	4.64
.2749	2.15	.3522	2.71	.4362	3.54	.5158	4.66
.2765	2.16	.3534	2.72	.4379	3.56	.5170	4.68
.2781	2.17	.3546	2.73	.4396	3.58	.5182	4.70
.2796	2.18	.3558	2.74	.4413	3.60	.5193	4.72
.2812	2.19	.3570	2.75	.4429	3.62	.5205	4.74
.2828	2.20	.3582	2.76	.4446	3.64	.5216	4.76
.2843	2.21	.3593	2.77	.4462	3.66	.5228	4.78
.2858	2.22	.3605	2.78	.4479	3.68	.5239	4.80
.2874	2.23	.3617	2.79	.4495	3.70	.5251	4.82
.2889	2.24	.3628	2.80	.4511	3.72	.5262	4.84
.2904	2.25	.3640	2.81	.4527	3.74	.5273	4.86
.2919	2.26	.3651	2.82	.4543	3.76	.5284	4.88
.2934	2.27	.3663	2.83	.4559	3.78	.5295	4.90
.2949	2.28	.3674	2.84	.4574	3.80	.5306	4.92
.2964	2.29	.3685	2.85	.4590	3.82	.5317	4.94
.2978	2.30	.3696	2.86	.4605	3.84	.5328	4.96
.2993	2.31	.3708	2.87	.4620	3.86	.5339	4.98
.3007	2.32	.3719	2.88	.4636	3.88	.5350	5.00
.3022	2.33	.3730	2.89	.4651	3.90	.5377	5.05
.3036	2.34	.3741	2.90	.4666	3.92	.5403	5.10
.3051	2.35	.3752	2.91	.4681	3.94	.5429	5.15
.3065	2.36	.3763	2.92	.4695	3.96	.5455	5.20
.3079	2.37	.3774	2.93	.4710	3.98	.5480	5.25
.3093	2.38	.3785	2.94	.4725	4.00	.5505	5.30
.3107	2.39	.3796	2.95	.4739	4.02	.5530	5.35
.3121	2.40	.3806	2.96	.4754	4.04	.5554	5.40
.3135	2.41	.3817	2.97	.4768	4.06	.5578	5.45
.3149	2.42	.3828	2.98	.4782	4.08	.5602	5.50
.3163	2.43	.3838	2.99	.4796	4.10	.5625	5.55
.3177	2.44	.3849	3.00	.4810	4.12	.5648	5.60
.3190	2.45	.3870	3.02	.4824	4.14	.5671	5.65
.3204	2.46	.3891	3.04	.4838	4.16	.5694	5.70
.3217	2.47	.3912	3.06	.4852	4.18	.5716	5.75
.3231	2.48	.3932	3.08	.4866	4.20	.5738	5.80
.3244	2.49	.3953	3.10	.4879	4.22	.5760	5.85
.3257	2.50	.3973	3.12	.4893	4.24	.5781	5.90
.3271	2.51	.3993	3.14	.4906	4.26	.5803	5.95
.3284	2.52	.4013	3.16	.4919	4.28	.5824	6.00
.3297	2.53	.4032	3.18	.4933	4.30	.5844	6.05
.3310	2.54	.4052	3.20	.4946	4.32	.5865	6.10

Δ/Po"	app α
.5885	6.15
.5905	6.20
.5925	6.25
.5945	6.30
.5964	6.35
.5983	6.40
.6002	6.45
.6021	6.50
.6039	6.55
.6058	6.60
.6076	6.65
.6094	6.70
.6111	6.75
.6129	6.80
.6146	6.85
.6163	6.90
.6180	6.95
.6197	7.00
.6214	7.05
.6230	7.10
.6247	7.15
.6263	7.20
.6279	7.25
.6295	7.30
.6310	7.35
.6326	7.40
.6341	7.45
.6357	7.50
.6372	7.55
.6387	7.60
.6401	7.65
.6416	7.70
.6431	7.75
.6445	7.80
.6459	7.85
.6473	7.90
.6487	7.95
.6501	8.00
.6515	8.05
.6529	8.10
.6542	8.15
.6555	8.20
.6569	8.25
.6582	8.30
.6595	8.35
.6608	8.40
.6620	8.45
.6633	8.50
.6646	8.55
.6658	8.60
.6671	8.65
.6683	8.70
.6695	8.75
.6707	8.80
.6719	8.85
.6731	8.90

Δ/Po"	app α
.6742	8.95
.6754	9.00
.6766	9.05
.6777	9.10
.6788	9.15
.6800	9.20
.6811	9.25
.6822	9.30
.6833	9.35
.6844	9.40
.6855	9.45
.6865	9.50
.6876	9.55
.6887	9.60
.6897	9.65
.6908	9.70
.6918	9.75
.6928	9.80
.6938	9.85
.6948	9.90
.6958	9.95
.6968	10.00
.6978	10.05
.6988	10.10
.6998	10.15
.7007	10.20
.7017	10.25
.7026	10.30
.7036	10.35
.7045	10.40
.7055	10.45
.7064	10.50
.7073	10.55
.7082	10.60
.7091	10.65
.7100	10.70
.7109	10.75
.7118	10.80
.7127	10.85
.7135	10.90
.7144	10.95
.7153	11.00
.7161	11.05
.7170	11.10
.7178	11.15
.7187	11.20
.7195	11.25
.7203	11.30
.7211	11.35
.7219	11.40
.7228	11.45
.7236	11.50
.7244	11.55
.7251	11.60
.7259	11.65
.7267	11.70

Δ/Po"	app α
.7275	11.75
.7283	11.80
.7290	11.85
.7298	11.90
.7306	11.95
.7313	12.00
.7321	12.05
.7328	12.10
.7335	12.15
.7343	12.20
.7350	12.25
.7357	12.30
.7365	12.35
.7372	12.40
.7379	12.45
.7386	12.50
.7393	12.55
.7400	12.60
.7407	12.65
.7414	12.70
.7421	12.75
.7427	12.80
.7434	12.85
.7441	12.90
.7448	12.95
.7454	13.00
.7461	13.05
.7468	13.10
.7474	13.15
.7481	13.20
.7487	13.25
.7493	13.30
.7500	13.35
.7506	13.40
.7513	13.45
.7519	13.50
.7525	13.55
.7531	13.60
.7537	13.65
.7544	13.70
.7550	13.75
.7556	13.80
.7562	13.85
.7568	13.90
.7574	13.95
.7580	14.00
.7586	14.05
.7591	14.10
.7597	14.15
.7603	14.20
.7609	14.25
.7615	14.30
.7620	14.35
.7626	14.40
.7632	14.45
.7637	14.50

Δ/Po"	app α
.7643	14.55
.7648	14.60
.7654	14.65
.7659	14.70
.7665	14.75
.7670	14.80
.7676	14.85
.7681	14.90
.7687	14.95
.7692	15.00
.7697	15.05

LITERATURE CITED

Schimerlik, M. I., Rife, J. E., and Cleland, W. W. 1975. Equilibrium perturbation by isotope substitution. Biochemistry 14:5347-5354.

appendix D: Table for Use with Northrop's Method

This table tabulates solutions of the equation:

$$y = \frac{(V/K)_H/(V/K)_D - 1}{(V/K)_H/(V/K)_T - 1} = \frac{k_H/k_D - 1}{(k_H/k_D)^{1.442} - 1}$$

where $k_H/k_T = (k_H/k_D)^{1.442}$

Entry of the table with the experimental value of y gives the values of k_H/k_D and k_H/k_T for the bond-breaking step.

NOTE: An exact solution is obtained only if $\beta = K_{eq_D}/K_{eq_H} = 1$. If $\beta \neq 1$, k_H/k_D is the isotope effect in the forward direction, and $\beta\, k_H/k_D$ is the value in the reverse direction. Likewise, $\beta^{1.442} k_H/k_T$ is the reverse tritium isotope effect. In this case, the observed V/K effects are used to calculate y and an apparent value of k_H/k_D is obtained from the table. $(V/K)_H/(V/K)_D$ is then multiplied by β, and $(V/K)_H/(V/K)_T$ by $\beta^{1.442}$, and an apparent value of $\beta\, k_H/k_D$ is then obtained by entering the table with y′:

$$y' = \frac{\beta\,(V/K)_H/(V/K)_D - 1}{\beta^{1.442}\,(V/K)_H/(V/K)_T - 1}$$

The following limits can then be determined:

$$\frac{(\text{app } \beta\, k_H/k_D)}{\beta} \geqslant k_H/k_D \geqslant (\text{app } k_H/k_D)$$

$$(\text{app } \beta\, k_H/k_D) \geqslant \beta\, k_H/k_D \geqslant \beta\,(\text{app } k_H/k_D)$$

These equations apply when $\beta < 1$; when $\beta > 1$, the inequalities are reversed.

This table may also be used to calculate deuterium effects corresponding to tritium ones, or vice versa, by comparing the last two columns. The comparison is valid, of course, only for true kinetic isotope effects on the bond-breaking step or for equilibrium effects.

Table for Northrop's Method. $y = \dfrac{(V/K)_H\ /(V/K)_D - 1}{(V/K)_H\ /(V/K)_T - 1}$

y	k_H/k_D	k_H/k_T	y	k_H/k_D	k_H/k_T	y	k_H/k_D	k_H/k_T
.6920	1.01	1.014	.6220	1.57	1.916	.5624	2.26	3.241
.6904	1.02	1.029	.6210	1.58	1.934	.5609	2.28	3.282
.6889	1.03	1.044	.6199	1.59	1.952	.5595	2.30	3.324
.6874	1.04	1.058	.6189	1.60	1.969	.5581	2.32	3.365
.6860	1.05	1.073	.6179	1.61	1.987	.5566	2.34	3.407
.6845	1.06	1.088	.6169	1.62	2.005	.5552	2.36	3.449
.6830	1.07	1.102	.6159	1.63	2.023	.5539	2.38	3.492
.6816	1.08	1.117	.6149	1.64	2.041	.5525	2.40	3.534
.6802	1.09	1.132	.6139	1.65	2.059	.5511	2.42	3.577
.6787	1.10	1.147	.6129	1.66	2.077	.5498	2.44	3.619
.6773	1.11	1.162	.6119	1.67	2.095	.5484	2.46	3.662
.6759	1.12	1.178	.6110	1.68	2.113	.5471	2.48	3.705
.6745	1.13	1.193	.6100	1.69	2.131	.5458	2.50	3.748
.6732	1.14	1.208	.6090	1.70	2.149	.5445	2.52	3.792
.6718	1.15	1.223	.6081	1.71	2.168	.5432	2.54	3.835
.6704	1.16	1.239	.6071	1.72	2.186	.5419	2.56	3.879
.6691	1.17	1.254	.6062	1.73	2.204	.5406	2.58	3.922
.6678	1.18	1.270	.6052	1.74	2.223	.5394	2.60	3.966
.6664	1.19	1.285	.6043	1.75	2.241	.5381	2.62	4.010
.6651	1.20	1.301	.6034	1.76	2.260	.5369	2.64	4.055
.6638	1.21	1.316	.6024	1.77	2.278	.5357	2.66	4.099
.6625	1.22	1.332	.6015	1.78	2.297	.5344	2.68	4.144
.6612	1.23	1.348	.6006	1.79	2.315	.5332	2.70	4.188
.6599	1.24	1.364	.5997	1.80	2.334	.5320	2.72	4.233
.6586	1.25	1.380	.5988	1.81	2.353	.5308	2.74	4.278
.6574	1.26	1.396	.5979	1.82	2.371	.5296	2.76	4.323
.6561	1.27	1.412	.5970	1.83	2.390	.5285	2.78	4.368
.6549	1.28	1.428	.5961	1.84	2.409	.5273	2.80	4.414
.6536	1.29	1.444	.5952	1.85	2.428	.5261	2.82	4.459
.6524	1.30	1.460	.5943	1.86	2.447	.5250	2.84	4.505
.6512	1.31	1.476	.5934	1.87	2.466	.5238	2.86	4.551
.6500	1.32	1.492	.5926	1.88	2.485	.5227	2.88	4.597
.6488	1.33	1.509	.5917	1.89	2.504	.5216	2.90	4.643
.6476	1.34	1.525	.5908	1.90	2.523	.5205	2.92	4.689
.6464	1.35	1.541	.5900	1.91	2.542	.5194	2.94	4.735
.6452	1.36	1.558	.5891	1.92	2.562	.5183	2.96	4.782
.6440	1.37	1.575	.5883	1.93	2.581	.5172	2.98	4.829
.6428	1.38	1.591	.5874	1.94	2.600	.5161	3.00	4.875
.6417	1.39	1.608	.5866	1.95	2.620	.5150	3.02	4.922
.6405	1.40	1.624	.5857	1.96	2.639	.5139	3.04	4.969
.6394	1.41	1.641	.5849	1.97	2.658	.5129	3.06	5.017
.6382	1.42	1.658	.5841	1.98	2.678	.5118	3.08	5.064
.6371	1.43	1.675	.5832	1.99	2.697	.5108	3.10	5.111
.6360	1.44	1.692	.5824	2.00	2.717	.5097	3.12	5.159
.6349	1.45	1.709	.5808	2.02	2.756	.5087	3.14	5.207
.6338	1.46	1.726	.5792	2.04	2.796	.5077	3.16	5.255
.6327	1.47	1.743	.5776	2.06	2.835	.5067	3.18	5.303
.6316	1.48	1.760	.5760	2.08	2.875	.5056	3.20	5.351
.6305	1.49	1.777	.5744	2.10	2.915	.5046	3.22	5.399
.6294	1.50	1.794	.5729	2.12	2.955	.5036	3.24	5.448
.6283	1.51	1.812	.5713	2.14	2.995	.5027	3.26	5.496
.6272	1.52	1.829	.5698	2.16	3.036	.5017	3.28	5.545
.6262	1.53	1.846	.5683	2.18	3.076	.5007	3.30	5.594
.6251	1.54	1.864	.5668	2.20	3.117	.4997	3.32	5.643
.6241	1.55	1.881	.5653	2.22	3.158	.4988	3.34	5.692
.6230	1.56	1.899	.5638	2.24	3.199	.4978	3.36	5.741

y	k_H/k_D	k_H/k_T	y	k_H/k_D	k_H/k_T	y	k_H/k_D	k_H/k_T
.4968	3.38	5.790	.4517	4.50	8.748	.4180	5.62	12.054
.4959	3.40	5.840	.4510	4.52	8.805	.4174	5.64	12.116
.4950	3.42	5.889	.4503	4.54	8.861	.4169	5.66	12.178
.4940	3.44	5.939	.4497	4.56	8.917	.4164	5.68	12.240
.4931	3.46	5.989	.4490	4.58	8.974	.4159	5.70	12.302
.4922	3.48	6.039	.4483	4.60	9.030	.4153	5.72	12.364
.4913	3.50	6.089	.4476	4.62	9.087	.4148	5.74	12.427
.4903	3.52	6.139	.4470	4.64	9.144	.4143	5.76	12.489
.4894	3.54	6.190	.4463	4.66	9.201	.4138	5.78	12.552
.4885	3.56	6.240	.4457	4.68	9.258	.4133	5.80	12.614
.4876	3.58	6.291	.4450	4.70	9.315	.4128	5.82	12.677
.4868	3.60	6.341	.4443	4.72	9.372	.4123	5.84	12.740
.4859	3.62	6.392	.4437	4.74	9.429	.4118	5.86	12.803
.4850	3.64	6.443	.4431	4.76	9.487	.4113	5.88	12.866
.4841	3.66	6.494	.4424	4.78	9.544	.4108	5.90	12.929
.4833	3.68	6.546	.4418	4.80	9.602	.4103	5.92	12.992
.4824	3.70	6.597	.4411	4.82	9.659	.4098	5.94	13.056
.4815	3.72	6.648	.4405	4.84	9.717	.4093	5.96	13.119
.4807	3.74	6.700	.4399	4.86	9.775	.4088	5.98	13.183
.4798	3.76	6.752	.4392	4.88	9.833	.4083	6.00	13.246
.4790	3.78	6.804	.4386	4.90	9.892	.4071	6.05	13.406
.4782	3.80	6.856	.4380	4.92	9.950	.4059	6.10	13.566
.4773	3.82	6.908	.4374	4.94	10.008	.4047	6.15	13.726
.4765	3.84	6.960	.4368	4.96	10.067	.4035	6.20	13.888
.4757	3.86	7.012	.4362	4.98	10.125	.4023	6.25	14.049
.4749	3.88	7.065	.4355	5.00	10.184	.4012	6.30	14.212
.4741	3.90	7.117	.4349	5.02	10.243	.4000	6.35	14.375
.4733	3.92	7.170	.4343	5.04	10.302	.3989	6.40	14.538
.4725	3.94	7.223	.4337	5.06	10.361	.3977	6.45	14.702
.4717	3.96	7.276	.4331	5.08	10.420	.3966	6.50	14.867
.4709	3.98	7.329	.4325	5.10	10.479	.3955	6.55	15.032
.4701	4.00	7.382	.4319	5.12	10.538	.3944	6.60	15.198
.4693	4.02	7.435	.4314	5.14	10.598	.3933	6.65	15.364
.4685	4.04	7.489	.4308	5.16	10.657	.3923	6.70	15.531
.4677	4.06	7.542	.4302	5.18	10.717	.3912	6.75	15.698
.4670	4.08	7.596	.4296	5.20	10.776	.3901	6.80	15.866
.4662	4.10	7.650	.4290	5.22	10.836	.3891	6.85	16.035
.4654	4.12	7.703	.4284	5.24	10.896	.3881	6.90	16.204
.4647	4.14	7.757	.4279	5.26	10.956	.3870	6.95	16.374
.4639	4.16	7.811	.4273	5.28	11.016	.3860	7.00	16.544
.4632	4.18	7.866	.4267	5.30	11.077	.3850	7.05	16.714
.4624	4.20	7.920	.4262	5.32	11.137	.3840	7.10	16.886
.4617	4.22	7.974	.4256	5.34	11.197	.3830	7.15	17.057
.4609	4.24	8.029	.4250	5.36	11.258	.3820	7.20	17.230
.4602	4.26	8.084	.4245	5.38	11.318	.3810	7.25	17.402
.4595	4.28	8.138	.4239	5.40	11.379	.3801	7.30	17.576
.4588	4.30	8.193	.4234	5.42	11.440	.3791	7.35	17.750
.4580	4.32	8.248	.4228	5.44	11.501	.3782	7.40	17.924
.4573	4.34	8.303	.4223	5.46	11.562	.3772	7.45	18.099
.4566	4.36	8.359	.4217	5.48	11.623	.3763	7.50	18.274
.4559	4.38	8.414	.4212	5.50	11.684	.3754	7.55	18.450
.4552	4.40	8.470	.4206	5.52	11.746	.3744	7.60	18.627
.4545	4.42	8.525	.4201	5.54	11.807	.3735	7.65	18.804
.4538	4.44	8.581	.4196	5.56	11.869	.3726	7.70	18.981
.4531	4.46	8.637	.4190	5.58	11.930	.3717	7.75	19.159
.4524	4.48	8.692	.4185	5.60	11.992	.3708	7.80	19.337

y	k_H/k_D	k_H/k_T
.3699	7.85	19.516
.3691	7.90	19.696
.3682	7.95	19.876
.3673	8.00	20.057
.3665	8.05	20.238
.3656	8.10	20.419
.3648	8.15	20.601
.3639	8.20	20.784
.3631	8.25	20.967
.3623	8.30	21.150
.3615	8.35	21.334
.3607	8.40	21.518
.3598	8.45	21.703
.3590	8.50	21.889
.3582	8.55	22.075
.3575	8.60	22.261
.3567	8.65	22.448
.3559	8.70	22.635
.3551	8.75	22.823
.3544	8.80	23.011
.3536	8.85	23.200
.3528	8.90	23.389
.3521	8.95	23.579
.3513	9.00	23.769
.3506	9.05	23.960
.3499	9.10	24.151
.3491	9.15	24.343
.3484	9.20	24.535
.3477	9.25	24.727
.3470	9.30	24.920
.3463	9.35	25.114
.3456	9.40	25.308
.3449	9.45	25.502
.3442	9.50	25.697
.3435	9.55	25.892
.3428	9.60	26.088
.3421	9.65	26.284
.3414	9.70	26.480
.3408	9.75	26.677
.3401	9.80	26.875
.3394	9.85	27.073
.3388	9.90	27.271
.3381	9.95	27.470
.3375	10.00	27.669
.3368	10.05	27.869
.3362	10.10	28.069
.3355	10.15	28.270
.3349	10.20	28.471
.3343	10.25	28.672
.3336	10.30	28.874
.3330	10.35	29.077
.3324	10.40	29.279
.3318	10.45	29.483
.3312	10.50	29.686
.3306	10.55	29.890
.3300	10.60	30.095
.3294	10.65	30.300
.3288	10.70	30.505
.3282	10.75	30.711
.3276	10.80	30.917
.3270	10.85	31.124
.3264	10.90	31.331
.3258	10.95	31.538
.3252	11.00	31.746
.3241	11.10	32.163
.3230	11.20	32.582
.3219	11.30	33.002
.3208	11.40	33.424
.3197	11.50	33.847
.3186	11.60	34.273
.3175	11.70	34.700
.3165	11.80	35.128
.3154	11.90	35.558
.3144	12.00	35.990
.3134	12.10	36.423
.3123	12.20	36.858
.3113	12.30	37.294
.3104	12.40	37.732
.3094	12.50	38.172
.3084	12.60	38.613
.3074	12.70	39.056
.3065	12.80	39.500
.3056	12.90	39.946
.3046	13.00	40.393
.3037	13.10	40.842
.3028	13.20	41.292
.3019	13.30	41.744
.3010	13.40	42.197
.3001	13.50	42.652
.2992	13.60	43.109
.2984	13.70	43.566
.2975	13.80	44.026
.2966	13.90	44.486
.2958	14.00	44.949
.2950	14.10	45.412
.2941	14.20	45.877
.2933	14.30	46.344
.2925	14.40	46.812
.2917	14.50	47.282
.2909	14.60	47.753
.2901	14.70	48.225
.2893	14.80	48.699
.2885	14.90	49.174
.2878	15.00	49.650
.2870	15.10	50.128
.2862	15.20	50.608
.2855	15.30	51.089
.2847	15.40	51.571
.2840	15.50	52.054
.2833	15.60	52.539
.2826	15.70	53.026
.2818	15.80	53.513
.2811	15.90	54.003
.2804	16.00	54.493
.2797	16.10	54.985
.2790	16.20	55.478
.2783	16.30	55.972
.2776	16.40	56.468
.2770	16.50	56.965
.2763	16.60	57.464
.2756	16.70	57.964
.2749	16.80	58.465
.2743	16.90	58.967
.2736	17.00	59.471
.2730	17.10	59.976
.2723	17.20	60.483
.2717	17.30	60.991
.2711	17.40	61.500
.2704	17.50	62.010
.2698	17.60	62.521
.2692	17.70	63.034
.2686	17.80	63.549
.2680	17.90	64.064
.2674	18.00	64.581
.2668	18.10	65.099
.2662	18.20	65.618
.2656	18.30	66.138
.2650	18.40	66.660
.2644	18.50	67.183
.2638	18.60	67.708
.2633	18.70	68.233
.2627	18.80	68.760
.2621	18.90	69.288
.2616	19.00	69.817
.2610	19.10	70.348
.2604	19.20	70.879
.2599	19.30	71.412
.2594	19.40	71.947
.2588	19.50	72.482
.2583	19.60	73.019
.2577	19.70	73.556
.2572	19.80	74.095
.2567	19.90	74.636
.2561	20.00	75.177
.2556	20.10	75.720
.2551	20.20	76.263
.2546	20.30	76.808
.2541	20.40	77.355
.2536	20.50	77.902
.2531	20.60	78.451
.2526	20.70	79.000
.2521	20.80	79.551
.2516	20.90	80.103
.2511	21.00	80.657

Bibliographies and Reviews of Isotope Effects

The following bibliography lists the published bibliographies concerning isotopes and isotope effects. The second part is a chronological listing of reviews of isotope effects in chemical reactions and in physical processes. Review articles more than 15 years old have been included only when they are still of particular interest, because most such reviews have been supplanted by more recent ones.

BIBLIOGRAPHIES

Samuel, D., and Steckel, F. 1959. Bibliography on the Stable Isotopes of Oxygen (O^{17} and O^{18}). Pergamon Press, New York. A complete bibliography through 1957.

Samuel, D., and Steckel, F. 1961. Research with the stable isotopes of oxygen (O^{17} and O^{18}) during 1958-1960. Int. J. Appl. Radiat. Isot. 11:190-229. A complete bibliography with indices.

Borowitz, J., Samuel, D., and Steckel, F. 1965. Research with the isotopes of oxygen (O^{15}, O^{17}, and O^{18}) during 1961-1963. Int. J. Appl. Radiat. Isot. 16:97-142. A complete bibliography with indices.

Samuel, D., and Steckel, F. 1968. Research with the isotopes of oxygen (O^{15}, O^{17}, and O^{18}) during 1963-1966. Int. J. Appl. Radiat. Isot. 19:175-217. A complete bibliography with indices.

Stern, M. J., and Wolfsberg, M. 1972. Heavy-atom kinetic isotope effects, an indexed bibliography. NBS Special Publication No. 349, U.S. Department of Commerce, Washington. A complete bibliography of heavy-atom kinetic isotope effects through 1968.

REVIEWS

Wiberg, K. B. 1955. The deuterium isotope effect. Chem. Rev. 55:713-743. An extensive survey of a variety of reactions.

Roginskii, S. Z. 1957. Theoretical Principles of Isotope Methods for Investigating Chemical Reactions. Consultants Bureau, New York. 456 p. English translation of a 1956 Russian book. Although rather out of date, it is important because of its discussion of a considerable amount of early Soviet work.

Bigeleisen, J., and Wolfsberg, M. 1958. Theoretical and experimental aspects of isotope effects in chemical kinetics. Adv. Chem. Phys. 1:15–76. A classic review article, with particular emphasis on theory and experimental methods.

Murray, A., III, and Williams, D. L. 1958. Organic Syntheses with Isotopes. Interscience, New York. 2 vols., 2096 p. An extensive compendium of methods for the synthesis of compounds labeled both with the hydrogen isotopes and with isotopes of heavy atoms.

Melander, L. 1960. Isotope Effects on Reaction Rates. Ronald Press, New York. 181 p. The classic introduction to the field. Covers both calculation and interpretation of isotope effects.

Westheimer, F. H. 1961. The magnitude of the primary kinetic isotope effect for compounds of hydrogen and deuterium. Chem. Rev. 61:265–273. The classic paper presenting the theoretical explanation for small hydrogen isotope effects.

Weston, R. E. 1961. Isotope effects in chemical reactions. Annu. Rev. Nucl. Sci. 11:439–460. A general introduction and survey with particular emphasis on solvent isotope effects.

International Atomic Energy Agency (Vienna). 1962. Tritium in the Physical and Biological Sciences. International Atomic Energy Agency, Vienna. 2 vols., 807 p. Proceedings of a symposium held in Vienna in 1961. Session topics include distribution of tritium in nature, tritium in chemistry and physics, detection and counting of tritium, preparation of tritiated compounds, general aspects of tritium in biological studies, synthesis of tritiated biological compounds, radiation effects of tritium, distribution and metabolism of tritiated thymidine and related compounds for studying cell metabolism, use of tritiated thymidine and related compounds in radiobiology, and use of other tritiated compounds for metabolic studies.

Halevi, E. A. 1963. Secondary isotope effects. Progr. Phys. Org. Chem. 1:109–221. An extensive discussion, including isotope effects on molecular properties, theoretical considerations, and secondary isotope effects on reaction rates.

Collins, C. J. 1964. Isotopes and organic reaction mechanisms. Adv. Phys. Org. Chem. 2:3–91. Particularly good discussion of methods of measurement.

Craig, H., Miller, S. L., and Wasserburg, G. J. 1964. Isotopic and Cosmic Chemistry. North-Holland Publishing Co., Amsterdam. 553 p. A collection of papers dedicated to H. C. Urey by his former students and associates. Included are articles on isotopes in phase equilibria, isotopes in biology, and isotopes in geochemistry.

Proceedings of a symposium on isotope effects held in Vienna in 1964. Pure Appl. Chem. 8:217–552. Sections include theory and interpretation of isotope effects, investigations of isotope effects in chemical systems, investigations of isotope effects in biochemical systems, and investigations of isotope effects in biological systems.

Roth, L. J. 1964. Isotopes in Experimental Pharmacology. University of Chicago Press, Chicago. 488 p. Proceedings of a symposium held in Chicago in 1964. Includes several articles on isotope effects.

Zollinger, H. 1964. Hydrogen isotope effects in aromatic substitution reactions. Adv. Phys. Org. Chem. 2:163–200.

Evans, E. A. 1966. Tritium and its compounds. Van Nostrand-Reinhold, New

York. 441 p. Discusses preparation and handling of tritiated compounds. Includes a short chapter on isotope effects.

Laszlo, P., and Welvart, Z. 1966. Secondary deuterium isotope effects. Bull Soc. Chim. France 2412–2438. Extensive review with many references. In French.

Saunders, W. H., Jr. 1966. Kinetic isotope effects. Survey Progr. Chem. 3: 109–146. A general introduction to the subject.

Simon, H., and Palm, D. 1966. Isotope effects in organic chemistry and biochemistry. Angew. Chem. Int. Edn. 5:920–933. Surveys results on a number of enzyme-catalyzed reactions.

Thornton, E. R. 1966. Physical organic chemistry. Annu. Rev. Phys. Chem. 17:349–372. Primarily concerned with secondary deuterium isotope effects and solvent isotope effects in organic reactions.

Albery, W. J. 1967. The role of the solvent in aqueous proton transfers. Progr. React. Kinet. 4:353–398. Includes a useful section on isotope effects.

Conway, B. E. 1967. Some chemical factors in the kinetics of processes at electrodes. Progr. React. Kinet. 4:399–483. Includes a section on hydrogen isotope effects in electrochemical reactions.

Simon, H., and Floss, H. G. 1967. Application of isotopes in organic chemistry and biochemistry. Springer-Verlag, New York. An ongoing series. Vol. 1 is concerned with determination of isotopes in labeled compounds, Vol. 2 with measurement of radioactive and stable isotopes. In German.

Weston, R. E. 1967. Transition-state models and hydrogen isotope effects. Science 158:332–342. Prediction of isotope effects.

Duncan, J. F., and Cook, G. B. 1968. Isotopes in Chemistry. Clarendon Press, Oxford. 258 p. A general introduction to isotope studies, including isotope analysis, isotope exchange, and isotope effects.

Raaen, V. F., Ropp, G. A., and Raaen, H. P. 1968. Carbon-14. McGraw-Hill, New York. Ch. 5. An excellent discussion of carbon isotope effects, including a complete coverage of the literature through 1962.

Arnett, E. M., and McKelvey, D. R. 1969. Solvent isotope effect on thermodynamics of nonreacting solutes. In: J. F. Coetzee and C. D. Ritchie (eds.), Solute-solvent interactions, pp. 344–398. Dekker, New York. Useful discussion of thermodynamic properties of H_2O and D_2O.

Caldin, E. F. 1969. Tunneling in proton-transfer reactions in solution. Chem. Rev. 69:135–156. Includes a useful discussion of the application of hydrogen isotope effects to the study of this phenomenon.

Gold, V. 1969. Protolytic processes in H_2O—D_2O mixtures. Adv. Phys. Org. Chem. 7:259–331. Development of the theory of solvent isotope effects for mixed solvents.

Höpfner, A. 1969. Vapor pressure isotope effects. Angew. Chem. Int. Edn. 8:689–699. Extensive discussion of theory, with experimental examples.

Laughton, P. M., and Robertson, R. E. 1969. Solvent isotope effects for equilibria and reactions. In: J. F. Coetzee and C. D. Ritchie (eds.), Solute-solvent interactions, pp. 399–538. Dekker, New York.

Spindel, W. 1969. Isotope Effects in Chemical Processes. American Chemical Society, Washington, D. C. 278 p. Proceedings of a symposium on theory and practice of isotope separation processes.

Wolfsberg, M. 1969. Isotope effects. Annu. Rev. Phys. Chem. 20:449–478. A survey of recent advances in a number of areas.

Collins, C. J., and Bowman, N. S. 1970. Isotope effects in chemical reactions. Van Nostrand-Reinhold, New York. 435 p. Excellent and extensive discussions of several topics. Contents: W. A. Van Hook, Kinetic isotope effects: introduction and discussion of the theory; V. J. Shiner, Jr., Deuterium isotope effects in solvolytic substitution at saturated carbon; D. E. Sunko and S. Borcic, Secondary deuterium isotope effects and neighboring group participation; E. K. Thornton and E. R. Thornton, Origin and interpretation of isotope effects; J. J. Katz and H. L. Crespi, Isotope effects in biological systems; A. Fry, Heavy atom isotope effects in organic reaction mechanism studies.

Richards, J. H. 1970. Kinetic isotope effects in enzymic reactions. In: P. D. Boyer (ed.), The Enzymes, 3rd ed. Vol. 2. 321–333. Academic Press, New York. A brief, selective discussion of primary and secondary hydrogen isotope effects in some enzyme-catalyzed reactions.

Lacroix, M. 1971. Isotope effects in biology. Part I: Theory of isotope effects and observed variations in natural products. Bull. Soc. Roy. Sci. Liege 40:68–89. In French. Particularly good for discussion of natural isotope variations. Extensively annotated.

Lacroix, M. 1971. Isotope effects in biology. Part II: Biological observations and applications. Bull. Soc. Roy. Sci. Liege 40:179–198. In French. Particularly good discussion of isotopic fractionations which occur on incorporation of isotopes into biological materials.

Thomas, A. F. 1971. Deuterium labeling in organic chemistry. Appleton-Century-Crofts, New York. 518 p. Primarily concerned with synthesis and analysis of deuterium-labeled compounds, but also contains a section on isotope effects.

Fry, A. 1972. Isotope effect studies of elimination reactions. Chem. Soc. Rev. 1:163–210. Summarizes the extensive work with a variety of isotopes on eliminations.

Scheppele, S. E. 1972. Kinetic isotope effects as a valid measure of structure-reactivity relationships. Isotope effects and nonclassical theory. Chem. Rev. 72:511–532. Isotope effects in solvolysis reactions, particularly those involving anchimeric assistance.

Schowen, R. L. 1972. Mechanistic deductions from solvent isotope effects. Progr. Phys. Org. Chem. 9:275–332. Good, lucid discussion of all aspects of this complex subject.

Wolfsberg, M. 1972. Theoretical evaluation of experimentally observed isotope effects. Accts. Chem. Res. 5:225–233. Survey of recent progress in theory of isotope effects.

Bell, R. P. 1973. The proton in chemistry. 2nd Ed. Cornell University Press, Ithaca, N.Y. 310 p. Includes an extensive discussion of hydrogen isotope effects in proton-transfer reactions.

Bigeleisen, J., Lee, M. W., and Mandel, F. 1973. Equilibrium isotope effects. Annu. Rev. Phys. Chem. 24:407–440. Particular emphasis on isotope effects on physical properties.

Carter, R. E., and Melander, L. 1973. Experiments on the nature of steric isotope effects. Adv. Phys. Org. Chem. 10:1–27.

Kirsch, J. F. 1973. Mechanism of enzyme action. Annu. Rev. Biochem. 42: 205–234. Includes a discussion of applications of isotope effects in studies of enzyme reaction mechanisms.

Bell, R. P. 1974. Recent advances in the study of kinetic hydrogen isotope effects. Chem. Soc. Rev. 3:513–544. Particular emphasis on the role of tunneling in proton transfer reactions.

Jansco, G., and Van Hook, A. 1974. Condensed phase isotope effects (especially vapor pressure isotope effects). Chem. Rev. 74:689–750. Very thorough discussion of isotope effects on physical processes.

MacColl, A. 1974. Heavy-atom kinetic isotope effects. Annu. Repts. Chem. Soc. 71B:77–101. An excellent recent review covering theory, experimental methods, and applications to organic reaction mechanisms.

Saunders, W. H., Jr. 1974. Kinetic isotope effects. In: E. S. Lewis (ed.), Investigation of Rates and Mechanisms of Reactions. (A. Weissberger, (ed.), Technique of Chemistry, Vol. 6) Wiley, New York, Part I, ch. 5. A useful and concise introduction to the study of isotope effects, with examples drawn from organic chemistry.

Bigeleisen, J., Lee, M. W., and Mandel, F. 1975. Mean square force in simple liquids and solids from isotope effect studies. Accts. Chem. Res. 8:179–184. Summarizes recent work of Bigeleisen and students on isotope effects on molecular processes.

Buckingham, A. D., and Urland, W. 1975. Isotope effects on molecular properties. Chem. Rev. 75:113–117.

Buncel, E., and Lee, C. C. 1975. Isotopes in molecular rearrangements. Elsevier, Amsterdam. The first volume of this interesting series has recently appeared. Ongoing members of the series promise to provide a variety of articles on isotope effects.

Caldin, E., and Gold, V. 1975. Proton-transfer reactions. Wiley, New York. 448 p. A collection of articles on various aspects of this subject by friends and former collaborators of R. P. Bell. Articles include: R. A. More O'Ferrall, Substrate isotope effects; W. J. Albery, Solvent isotope effects; and other articles not directly related to isotope effects.

Dolbier, W. R., Jr. 1975. Isotope effects in pericyclic reactions. In: E. Buncel and C. C. Lee (eds.), Isotopes in Molecular Rearrangements. Elsevier, Amsterdam, pp. 27–59. Vol. 1.

Faraday Division of the Chemical Society (London). 1975. Proton transfer. Faraday Symp. Chem. Soc. 10:1–169. Proceedings of a symposium on proton transfer, including many aspects of isotope effects.

Klein, F. S. 1975. Isotope effects in chemical kinetics. Annu. Rev. Phys. Chem. 26:191–210. Covers the literature from 1969–1974. Primarily covers hydrogen isotope effects in simple reactions.

Rock, P. A. 1975. Isotopes and chemical principles. American Chemical Society, Washington. 215 p. Proceedings of a symposium. Contents: J. Bigeleisen, Quantum mechanical foundations of isotope chemistry; C. P. Nash, Isotope effects and spectroscopy; R. E. Weston, Jr., Isotope effects and quantum-mechanical tunneling; M. Wolfsberg and L. I. Kleinman, Corrections to the Born-Oppenheimer approximation in the calculation of isotope effects on equilibrium constants; W. Spindel, Isotope separation processes; W. A. Van Hook,

Condensed phase isotope effects, especially vapor pressure isotope effects: aqueous solutions; P. A. Rock, The electrochemical determination of equilibrium constants for isotope-exchange reactions; V. J. Shiner, Jr., Isotope effects and reaction mechanism; J. J. Katz, R. A. Uphaus, H. L. Crespi, and M. I. Blake, Isotope chemistry and biology.

Spicer, L. D., and Poulter, C. D. 1975. Isotopes as probes in determining reaction mechanisms. In: H. Eyring (ed.), Physical Chemistry, Vol. 7., ch. 11. Academic Press, New York. A general introduction to the theory and practice of isotope effects.

Kreevoy, M. M. 1976. The effect of structure on proton-transfer isotope effects. In: E. Buncel and C. C. Lee (eds.), Isotopes in Molecular Rearrangements, pp. 1–32, Vol. 2. Elsevier, Amsterdam.

Mills, R., and Harris, K. R. 1976. The effect of isotopic substitution on diffusion in liquids. Chem. Soc. Rev. 5:215–231.

Simon, H., and Kraus, A. 1976. Hydrogen isotope transfer in biochemical processes. In: E. Buncel and C. C. Lee (eds.), Isotopes in Molecular Rearrangements, pp. 153–230, Vol. 2. Elsevier, Amsterdam.

Smith, P. J. 1976. Isotope effects and transition states in elimination reactions. In: E. Buncel and C. C. Lee (eds.), Isotopes in Molecular Rearrangements, pp. 231–270, Vol. 2. Elsevier, Amsterdam.

Dunn, G. E. ^{13}C isotope effects in decarboxylation reactions. In: E. Buncel and C. C. Lee (eds.), Isotopes in Molecular Rearrangements. Vol. 3. Elsevier, Amsterdam. In preparation.

O'Leary, M. H. 1977. Enzymic catalysis of decarboxylation. Bioorg. Chem. Application of carbon isotope effects to studies of enzyme-catalyzed decarboxylation.

Schowen, R. L., and Gandour, R. Transition States of Biochemical Processes. Plenum Publishing Corp., New York. In press. Contents: E. K. Thornton and E. R. Thornton, The scope and limitations of the concept of the transition state; R. L. Schowen, Catalytic power and transition-state structure; R. E. Christoffersen and G. M. Maggiora, Quantum-mechanical calculations of transition-state structures and reaction pathways; J. P. Klinman, The use of primary hydrogen isotope effects; J. L. Hogg, The use of secondary hydrogen isotope effects; K. B. J. Schowen, The use of solvent hydrogen isotope effects; M. H. O'Leary, The use of heavy-atom isotope effects; A. S. Mildvan, Magnetic-resonance approaches to transition-state structures; E. Shefter, Transition-state structures from crystallography; M. F. Hegazi, D. M. Quinn, and R. L. Schowen, Acyl and methyl transfer reactions; H. Bull and E. H. Cordes, Acetal and ketal hydrolysis: R. M. Pollack, Decarboxylation and related reactions; S. J. Benkovic, Phosphoryl transfer; R. D. Gandour, Intramolecular reactions and the relevance of models; R. V. Wolfenden, The concept of transition-state analog enzyme inhibitors; J. K. Coward, Transition-state structures and reaction mechanism in the design of drugs.

Willi, A. V. Kinetic carbon-13 and other isotope effects in cleavage and formation of bonds to carbon. In: E. Buncel and C. C. Lee (eds.), Isotopes in Molecular Rearrangements. Vol. 3. Elsevier, Amsterdam. In preparation.

Index